LIGHT AND PLANT DEVELOPMENT

*These titles are now out of print

Light and Plant Development

H. SMITH
Department of Physiology and Environmental Studies
University of Nottingham School of Agriculture

BUTTERWORTHS
LONDON - BOSTON
Sydney - Wellington - Durban - Toronto

THE BUTTERWORTH GROUP

UNITED KINGDOM

Butterworth & Co (Publishers) Ltd
London: 88 Kingsway, WC2B 6AB

AUSTRALIA

Butterworths Pty Ltd
Sydney: 586 Pacific Highway, Chatswood, NSW 2067
Also at Melbourne, Brisbane, Adelaide and Perth

SOUTH AFRICA

Butterworth & Co (South Africa) (Pty) Ltd
Durban: 152–154 Gale Street

NEW ZEALAND

Butterworths of New Zealand Ltd
Wellington: 26–28 Waring Taylor Street, 1

CANADA

Butterworth & Co (Canada) Ltd
Toronto: 2265 Midland Avenue, Scarborough, Ontario, MIP 4SI

USA

Butterworths (Publishers) Inc
Boston: 19 Cummings Park, Woburn, Mass. 01801

First published 1976

ISBN 0 408 70719 4

© The several contributors named in the list of contents, 1976

LIBRARY OF CONGRESS CATALOGING IN PUBLICATION DATA

Easter School in Agricultural Science, 22d, University
    of Nottingham, Eng., 1975.
    Light and plant development.

    Bibliography: p.
    Includes index.
    1. Plants, Effect of light on--Congresses. I. Smith,
Harry. II. Title.
QK757.E26 1975   581.1'9'153   76-10178
ISBN 0-408-70719-4

Printed in England by Cox & Wyman Ltd, London, Fakenham and Reading

PREFACE

This volume presents the Proceedings of the 22nd University of Nottingham Easter School in Agricultural Science which was held at Sutton Bonington from April 7th to 10th, 1975. Each Easter School is intended to be a forum in which a specific problem of particular importance to agriculture can be discussed by specialists from science, commerce, industry and practical agriculture. In most cases, the immediacy of the chosen topics for practical agriculture is clearly evident; in others, the subject areas are as yet of potential importance only to agricultural productivity. 'Light and Plant Development' comes into the latter category. In planning and organising this Easter School, I have endeavoured to produce a volume which will provide a fully comprehensive and up-to-date summary of the control of growth and development of higher plants by light. Certain aspects, however, have been omitted, notably photosynthesis, which obviously could take up several volumes in its own right. Similarly, phototropism receives scant attention, although Chapter 2 contains the first report of really good evidence on the chemical nature of the photoreceptor. The concentration of virtually the whole of the volume on phytochrome may seem, at first glance, unbalanced, but it does reflect realistically our relatively high level of understanding of phytochrome compared with our almost complete ignorance of other photoreceptors. It is hoped that this volume will assist crop scientists to come to grips with a topic which is still almost totally studied at the fundamental level, but which nevertheless has great potential importance for crop production.

I am gratified to be able to say that the Easter School, and the Annual European Photomorphogenesis Symposium which followed (April 10th–12th), were highly successful. This was due principally to patient and conscientious organisation by the Conference Secretary, Miss Edna Lord, who carried an extremely heavy burden for several weeks without complaint – I am deeply grateful to her. I also wish to record my appreciation for the assistance given by Mrs. Laurel Dee and Mrs. Carol Stanton, who typed the abstracts, Mrs. Clarice Ingram, Mrs. Lena Parker, Mrs. Valerie Blunt and Mr John Blunt, who organised the visual aids, and Miss Audrey Evans, Mr Geoffrey Holmes, Mr Sandy Giles and Dr Chris Johnson, who looked after various aspects of participant welfare. In planning the two meetings I was advised by Professor J.L. Monteith, Dr W.R. Briggs and Dr Daphne Vince-Prue, to all of whom I am grateful.

Special thanks should also go to the Chairmen of the Sessions, all of whom accepted my invitation to take on this task at very short

Preface

notice; they were Professor J.L. Monteith, Dr W.R. Briggs, Professor
H. Mohr, Professor W. Haupt, Professor A.W. Galston and Dr W.S.
Hillman. I should also like to express my gratitude to Plant Protection
Ltd., Bracknell, for a very welcome financial contribution, and to the
Royal Society, the Agricultural Research Council and the British Council,
who separately assisted three of the overseas contributors to travel to
Britain.

Finally, I should like to thank the 128 participants from 20 countries
who, by their friendliness, created a superb atmosphere for the communication of ideas.

H. SMITH

CONTENTS

Contents

II The Site of Phytochrome Action

III Cellular Aspects of Phytochrome Action

IV Physiological Aspects of Phytochrome Action

Contents

V Photoperiodism, Endogenous Rhythms and Phytochrome

To

H.A. BORTHWICK and S.B. HENDRICKS

By the unanimous approval of the participants at the 22nd University of Nottingham Easter School in Agricultural Science, this volume is dedicated to the late Dr Harry A. Borthwick and to Dr Sterling B. Hendricks, whose joint inspiration and perseverance over more than 35 years have laid the basis of our present knowledge and understanding of photomorphogenesis.

H.A. BORTHWICK AND S.B. HENDRICKS – PIONEERS OF PHOTOMORPHOGENESIS[1]

WINSLOW R. BRIGGS
Department of Plant Biology, Carnegie Institution of Washington, Stanford, California 94305, U.S.A.

The participants in this 22nd Easter School take great pleasure in dedicating this volume to Dr Harry A. Borthwick and Dr Sterling B. Hendricks in honour of their monumental contributions to the field to which this School is devoted. Their early interest in flowering led them into a remarkable series of studies, the results of which are fundamental to most of what is discussed in this volume. Their research was carried out at the Pioneering Laboratory of Plant Physiology at the Plant Industry Station of the United States Department of Agriculture; they and their many colleagues made it one of the most distinguished laboratories in the world, and unique in the Department of Agriculture.

While it is not the purpose of these short paragraphs to review all of their work, it is appropriate to list some of their major accomplishments in photomorphogenesis. In 1946, in one of a series of studies with Parker, they carefully examined the spectral sensitivity of inhibition of flowering by a light break in the middle of the night in the two short-day plants Biloxi soybean and cocklebur (Parker *et al.*, 1946). The results were the first action spectrum for a phytochrome effect, and showed an action maximum near 665 nm. Two years later came an action spectrum for night interruption in Wintex barley, showing for the first time flowering promotion in a long-day plant under non-inductive photoperiod (Borthwick, Hendricks and Parker, 1948). The action spectrum was essentially identical with that for the short-day plants, and constituted the first evidence that the same photoreceptor might regulate flowering in both types of photoperiodically sensitive plants. The next year (Parker *et al.*, 1949) they were joined by F.W. Went to measure the action spectrum for leaf enlargement and stem growth inhibition in dark-grown pea seedlings. They thus obtained the first of many by now familiar action spectra for vegetative responses to phytochrome phototransformation. A year later (Parker, Hendricks and Borthwick, 1950) appeared another study on a long-day plant, *Hyoscyamus*. The work was really a condensed version

[1] CIW–DPB Publication No.553

of the Wintex barley paper, but it is significant in that the authors suggest for the first time a phycocyanin-like pigment as the photo-receptor – a full 15 years before the suggestion was verified experimentally.

They then turned their attention to light-promoted lettuce seed germination (Borthwick *et al.,* 1952), a system first discovered by Flint and McAlister in the mid 1930s at the Smithsonian Institution. An action spectrum for promotion of germination showed a maximum at 665 nm, suggesting the same photoreceptor to be involved. However, since the lettuce seed system was known to be reversible by far-red light, they also obtained the first action spectrum for far-red reversal. The same paper reported repeated photoreversibility following several sequential alternating red and far-red treatments, and the authors proposed for the first time a single photoreversible pigment. A second paper on lettuce seed germination (Borthwick *et al.,* 1954) explored the system in greater detail, demonstrating that the light reactions in both directions were independent of temperature, and showing for the first time escape from photoreversibility with time after far-red but not after red treatment.

Meanwhile, Borthwick, Hendricks and Parker (1952) asked the suddenly obvious question: since the action spectrum for promotion of lettuce seed germination was identical with those measured for the other effects, could these other effects be reversed by far-red light as well? Using cocklebur seedlings grown under day lengths that were sufficiently short for flowering, they showed that far-red light completely cancelled the effect of red light in inhibiting flower formation. The action spectrum for this far-red reversibility was roughly the same as that for lettuce seed germination. Photoreversibility was thus not just a peculiar property of certain varieties of lettuce seeds.

Other workers at Beltsville then demonstrated similar photoreversible control of flavone formation in tomato skins (Piringer and Heinze, 1954) and promotion of leaf enlargement and epicotyl elongation and inhibition of hypocotyl elongation in bean seedlings (Downs, 1955). Then Hendricks, Borthwick and Downs (1956) showed both with Pinto bean internodal elongation and *Lepidium* and lettuce seed germination that the photochemistry followed first-order kinetics for both red and far-red photoresponses. By this time, the Beltsville group was convinced that a single photoreversible pigment was involved, and they set out to find it.

Their success in detecting phytochrome spectrophotometrically in plant tissue and achieving preliminary isolation in buffer is documented in their often cited 1959 paper (Butler *et al.,* 1959), a paper that marked the beginning of *in vitro* work with phytochrome. It also presented instrumentation techniques which are now widely used, not only for phytochrome but also for a wide variety of other types of spectral studies. There followed from Beltsville the first action spectrum for phytochrome phototransformation *in vitro* (Butler, Siegelman and Hendricks, 1964), the first partial purification technique for phytochrome (Siegelman and Firer, 1964), establishment of the chromophore as a bilitriene similar to allophycocyanin (Siegelman and Hendricks, 1965; Siegelman, Turner and Hendricks, 1966), and numerous other papers on phytochrome reactions both *in vivo* and *in vitro*. The demonstration (Fondeville, Borthwick and Hendricks, 1966) that leaflet closing in *Mimosa* was under rapid phytochrome control gave strong impetus to the already existing suspicion that phytochrome was somehow acting in association with a membrane system. This evidence, summarised by Hendricks and Borthwick (1967), was integrated into a useful model for phytochrome action (Borthwick *et al.*, 1969).

Borthwick retained an active interest in phytochrome (Borthwick, 1972a,b) until his death on May 21st, 1974. Hendricks continues a vigorous pursuit of the problem of seed germination (Hendricks and Taylorson, 1975), where new ground is again being broken.

Among other things, what emerges from this listing of contributions is the extraordinary breadth of inquiry that Borthwick and Hendricks applied to plant photomorphogenesis. Their approaches ranged from whole plant physiology through organ physiology, biochemistry, photochemistry and biophysics. We commend them for this breadth, and for the durability, magnitude and insight of their research.

References

BORTHWICK, H.A. (1972a). In *Phytochrome*, p.3. Ed. Mitrakos, K. and Shropshire, W. Jr. Academic Press, New York

BORTHWICK, H.A. (1972b). In *Phytochrome*, p.27. Ed. Mitrakos, K. and Shropshire, W. Jr. Academic Press, New York

BORTHWICK, H.A., HENDRICKS, S.B. and PARKER, M.W. (1948). *Bot. Gaz.*, **110**, 103

BORTHWICK, H.A., HENDRICKS, S.B. and PARKER, M.W. (1952). *Proc. Nat. Acad. Sci. U.S.A.*, **38**, 929

BORTHWICK, H.A., HENDRICKS, S.B., PARKER, M.W., TOOLE, E.H. and TOOLE, V.K. (1952). *Proc. Nat. Acad. Sci. U.S.A.*, **38**, 662

BORTHWICK, H.A., HENDRICKS, S.B., SCHNEIDER, M.J., TAYLORSON, R.B. and TOOLE, V.K. (1969). *Proc. Nat. Acad. Sci. U.S.A.*, **64**, 479

BORTHWICK, H.A., HENDRICKS, S.B., TOOLE, E.H. and TOOLE, V.K. (1954). *Bot. Gaz.*, **115**, 205

BUTLER, W.L., NORRIS, K.H., SIEGELMAN, H.W. and HENDRICKS, S.B. (1959). *Proc. Nat. Acad. Sci. U.S.A.*, **45**, 1703

BUTLER, W.L., SIEGELMAN, H.W. and HENDRICKS, S.B. (1964). *Photochem. Photobiol.*, **3**, 521

DOWNS, R.J. (1955). *Plant Physiol.*, **30**, 468

FONDEVILLE, J.C., BORTHWICK, H.A. and HENDRICKS, S.B. (1966). *Planta*, **69**, 357

HENDRICKS, S.B. and BORTHWICK, H.A. (1967). *Proc. Nat. Acad. Sci. U.S.A.*, **58**, 2125

HENDRICKS, S.B., BORTHWICK, H.A. and DOWNS, R.J. (1956). *Proc. Nat. Acad. Sci. U.S.A.*, **42**, 19

HENDRICKS, S.B. and TAYLORSON, R.B. (1975). *Proc. Nat. Acad. Sci. U.S.A.*, **72**, 306

PARKER, M.W., HENDRICKS, S.B. and BORTHWICK, H.A. (1950). *Bot. Gaz.*, **111**, 242

PARKER, M.W., HENDRICKS, S.B., BORTHWICK, H.A. and SCULLY, N.J. (1946). *Bot. Gaz.*, **108**, 1

PARKER, M.W., HENDRICKS, S.B., BORTHWICK, H.A. and WENT, F.W. (1949). *Am. J. Bot.*, **36**, 194

PIRINGER, A.A. and HEINZE, P.H. (1954). *Plant Physiol.*, **29**, 467

SIEGELMAN, H.W. and FIRER, E.M. (1964). *Biochemistry*, **3**, 418

SIEGELMAN, H.W. and HENDRICKS, S.B. (1965). *Fed. Proc. Fed. Am. Soc. Exp. Biol.*, **24**, 863

SIEGELMAN, H.W., TURNER, B.C. and HENDRICKS, S.B. (1966). *Plant Physiol.*, **41**, 1289

I

THE PERCEPTION OF LIGHT

2

THE NATURE OF THE BLUE LIGHT PHOTORECEPTOR IN HIGHER PLANTS AND FUNGI[1]

WINSLOW R. BRIGGS
Department of Plant Biology, Carnegie Institution of Washington, Stanford, California 94305, U.S.A.

Ever since the initial spectral detection and preliminary isolation of phytochrome (Butler *et al.*, 1959), workers interested in the regulatory influence of light on plant development have concentrated on the red and far-red regions of the spectrum in their studies. The extent of this concentration is revealed by a perusal of the recent review literature (*see* Briggs and Rice, 1972; Mohr, 1972). Great progress has been made in understanding the physical and chemical nature of this plant photoreceptor. There is now a substantial amount of literature addressed to its location and action in the cell, particularly its association with a variety of membrane systems. Indeed, most of the papers in this 22nd Easter School and in the following European Photomorphogenesis Symposium reflect the intensive central interest in phytochrome in the field of light and plant development.

There is, however, another photoreceptor system, both in higher and lower plants and in fungi, which should be considered to have equal importance. This pigment system absorbs light in the long-wavelength ultraviolet and blue regions of the spectrum, and fails to show the sort of photoreversibility that is characteristic of so many phytochrome responses. For convenience, the term 'blue light photoreceptor' will be used in this chapter. There exists in the literature a large number of action spectra for a fairly wide range of physiological and biochemical responses to light. Although the exact wavelengths for the various peaks and shoulders vary from organism to organism, the shape is always the same: a single broad action peak near 370 nm, a shoulder near 420 nm, an action maximum between 445 and 460 nm and an additional peak (or shoulder) between 470 and 490 nm. Such action spectra have been obtained for first positive phototropism of *Avena* coleoptiles (Curry, 1957, *see* Thimann and Curry, 1960; Shropshire and Withrow, 1958); second positive phototropism of *Avena* coleoptiles (Everett and Thimann, 1968); phototropism of *Phycomyces* sporangiophores (Curry and Gruen, 1959; Delbrück and Shropshire, 1960); the light-growth reaction in *Phycomyces* sporangiophores (Delbrück and Shropshire, 1960); carotenoid synthesis in the fungus

[1]CIW-DPB Publication No. 570

Fusarium (Rau, 1967); carotenoid synthesis in the fungus *Neurospora* (Zalokar, 1955); suppression of the expression of a circadian rhythm of conidiation in the fungus *Neurospora* (Sargent and Briggs, 1967); light-induced oxygen uptake by a carotenoidless mutant of the green alga *Chlorella* (Kowallik, 1967); polarotropism in germlings of the liverwort *Sphaerocarpos* and the fern *Dryopteris* (Steiner, 1967); and inhibition of hypocotyl elongation in lettuce (*Lactuca sativa*) (Hartmann, 1967). There is even an action spectrum for carotenogenesis in the bacterium *Mycobacterium* (Rilling, 1964; Howes and Batra, 1970) which shows the same general features.

Some of the above responses, for example first positive phototropism in *Avena* and phototropism in *Phycomyces*, require only very low irradiances to become saturated (low irradiance reaction, LIR). Others, such as the suppression of the conidiation rhythm in *Neurospora*, polarotropism in *Sphaerocarpos* and *Dryopteris* and inhibition of lettuce hypocotyl elongation, require high irradiances administered over periods of several hours or more. Some of the high-irradiance reactions (HIR) also have peaks in the far-red indicative of phytochrome action, for example lettuce hypocotyl inhibition, but others, for example *Neurospora* conidiation and *Sphaerocarpos* polarotropism, do not. Mohr (1972) has discussed in detail the evidence that such far-red action peaks are attributable to phytochrome. The present author agrees with Mohr, however, that where both blue and far-red action peaks are found, the blue peak must be attributable to some pigment other than phytochrome. Perhaps the most convincing evidence is the observation that no phytochrome-like photoresponses (nor, indeed, the measurable spectral photoreversibility characteristic of phytochrome) have ever been found in the fungi. The above list of photoresponses is not intended to be exhaustive; it should, however, remind the reader of the ubiquity of the shorter wavelength photoresponses in higher plants and fungi.

It is important to point out, with reference to the discussion to follow, that there are two known photoresponses in fungi which do not show the characteristic blue–ultraviolet sensitivity described above. The first of these responses is the phototropic response of conidio-phores of the fungus *Entomophthora coronata* (*Conidiobolus villosus*) (Page and Brungard, 1961). This response shows maximum sensitivity in the blue and red regions of the spectrum. The second response is the phototactic migration of pseudoplasmodia of the cellular slime mould *Dictyostelium discoideum* (Poff, Butler and Loomis, 1973) with clear action maxima near 430 and 560 nm. Since, as Page (1968) concedes, the *Entomophthora* system requires a more detailed action spectrum to obtain the precise action peaks, it is not possible to determine whether these two pigment systems are similar or not. Page and Brungard (1961) have suggested a porphyrin for *Entomophthora* on the basis of 20-fold greater sensitivity in the blue than in the red region and the presence of a substance in acetone extracts with a character-istic porphyrin absorption spectrum. As will become clear below, the evidence is strong that the photoreceptor for *Dictyostelium* phototaxis is also a porphyrin, a haem protein (Poff and Butler, 1974a). In any case, these two organisms contain photoreceptors which are neither phytochrome nor the blue light photoreceptor common to the other systems mentioned above.

Let us return for the moment to the predominating blue light photoreceptor. The first action spectrum for phototropism in *Avena* (Johnston, 1934) showed two peaks in the blue region (the ultraviolet region was not investigated). Wald and DuBuy (1936) found that *Avena* coleoptiles contained carotenoids, and therefore proposed that a carotenoid must be the photoreceptor for phototropism. Bünning (1937a,b) strengthened this notion with the discovery that carotenoids in the coleoptile were concentrated near the apex, also the site of highest photosensitivity. This view persisted until 1949, when Galston and Baker (1949) proposed an alternate and highly attractive candidate for the photoreceptor. They observed that riboflavin would sensitise the photo-oxidation of the plant auxin indole-3-acetic acid (IAA) as would a crude *brei* obtained from etiolated pea epicotyls. They determined action spectra for these two processes, and re-determined the action spectrum in the blue region for *Avena* phototropism. Within the resolution of their measurements, the three action spectra were very similar. Since Went (1928) had long before established that phototropic curvature of coleoptiles was a consequence of a light-induced difference in the amount of auxin in the lighted and shaded sides, Galston and Baker (1949) proposed a mechanism in which differential light-induced auxin inactivation, greater on the lighted than on the shaded side, caused the differential in auxin concentrations that ultimately caused the growth curvature. However, with the observation that there was no net loss of auxin as a consequence of phototropic induction in corn coleoptiles (Briggs, Tocher and Wilson, 1957; Briggs, 1963a), the mechanism proposed by Galston and Baker no longer appeared tenable. Many workers (the present author included, *see* Briggs, 1963b, 1964) thereupon rejected flavins as prime candidates for the photoreceptor as well.

The various action spectra obtained from the early 1960s onwards, all showing a broad action peak centred near 370 nm, did not resolve the photoreceptor identity although strong opinions were expressed. The problem was that flavins (but not carotenoids) were known to have a broad absorption band near 370 nm, while carotenoids (but not flavins) were considered to show the fine structure in the blue region that characterised the action spectrum. Then Hager (1970) demonstrated that carotenoids with a hydroxyl group, such as lutein, would indeed show absorption near 370 nm in a mixture of ethanol and water. By manipulating the solvent mixture, Hager could mimic precisely the phototropic action spectrum for *Avena* with the lutein absorption spectrum. However, Song and Moore (1974) carefully reconsidered Hager's system and concluded on theoretical grounds that carotenoids were singularly poor candidates for the photoreceptor.

They also noted that while energy transfer from carotenoids to a longer wavelength absorber such as chlorophyll was well documented, there was no evidence for a carotenoid–acceptor system in any of the action spectra mentioned above.

Several workers found evidence for photosensitivity in the region of protein absorption near 280 nm.   Curry, Thimann and Ray (1956) concluded that such action in *Avena* phototropism was a consequence of destruction of some auxin derivative (the action spectrum resembled the absorption spectrum of IAA shifted about 12 nm towards the visible region).   Equally plausible, however, would be absorption by an aromatic amino acid in a protein, with energy transfer to an associated chromophore.   Indeed, Pratt and Butler (1970) proposed just such a mechanism for phytochrome phototransformation by 280-nm light.   Both Curry and Gruen (1959) and Delbrück and Shropshire (1960) report a 280-nm peak in their *Phycomyces* action spectra, but neither mention the possibility that it might be attributable to absorption by the protein moiety of a chromoprotein.   Howes and Batra (1970) do make such a suggestion, concluding that the *Mycobacterium* photoreceptor must be a flavoprotein, although Batra (1971) concedes the possibility of a carotenoprotein complex as well.

Song and Moore (1974) point out that flavin molecules rigidly held, at low temperature for example, show band splitting to yield the type of fine structure required by the action spectra.   They also point out that the binding of a flavin to a protein at room temperature could confer sufficient rigidity to yield band splitting.   Rau (1967) even presents from the literature the absorption spectra between 340 and 510 nm for two flavoprotein enzymes.   These spectra show fine structure in the blue region which closely matches the action spectrum for carotogenesis in *Fusarium*.   Finally, Strittmatter (1966) presents the absorption spectrum of highly purified cytochrome b_5 reductase from liver, another flavoprotein which shows appropriate fine structure in the blue region of the spectrum.

About 3 years ago, Poff and his co-workers brought an entirely new approach to bear on the photoreceptor problem.   Poff, Butler and Loomis (1973) irradiated either dissociated pseudoplasmodia or cells from exponentially growing cultures of *Dictyostelium discoideum* with 560-nm light, a wavelength that is effective in inducing phototactic migration of the pseudoplasmodia.   Using extremely sensitive spectrophotometric techniques, they found a clear light-induced increase in absorbance at 411 nm.   The increase decayed in darkness with a half-time of about 7 s (presumably at room temperature).   They then obtained an action spectrum between 400 and 700 nm for phototactic migration, and showed that it matched the action spectrum for the light-induced change in absorbance between 515 and 700 nm (technical problems prevented measurement of the latter action spectrum below 515 nm).   The same signal could be observed in cell-free extracts provided that a reducing agent such as dithionite was present. The fraction containing photoactivity could be sedimented at 12 000 g for 30 min, and if this pellet were sonicated, could be solubilised, remaining in the supernatant following centrifugation at 100 000 g for 90 min. It is perhaps ironic that the first clear evidence from this spectral technique, implicating a pigment absorbing in the blue region, was obtained from a system that clearly contained a photoreceptor differing from that in the many systems which gave rise to the flavin *versus* carotenoid controversy.

In a further study, Poff, Loomis and Butler (1974) purified the pigment roughly 2 000-fold by ammonium sulphate fractionation and sucrose density gradient fractionation, obtaining a preparation which they estimated to be about 95 per cent pure. Gel exclusion chromatography indicated a molecular weight of about 240 000 (for a globular molecule in both cases). Photoactivity remained relatively stable throughout the purification procedure. The purified pigment was a chromoprotein with a strong Soret absorption band at 430 nm and a relatively weak and broad absorption band between 530 and 600 nm (in its dithionite-reduced form). On the basis of the similarity between this absorption spectrum and the phototaxis action spectrum, the authors proposed that the pigment is indeed the photoreceptor, and that the light-induced step is an oxidation. Dark reduction would follow photo-oxidation only in the presence of a suitable reductant.

Poff and Butler (1974a) next studied low-temperature spectra and pyridine haemochromagen formation, and concluded that the pigment was a haem protein. They also determined a crude action spectrum for the photo-oxidation, finding it similar to the action spectra for light-induced spectral changes *in vivo* and for phototactic migration. The evidence to date thus suggests that the photoreceptor for *Dictyostelium* phototaxis may be a reduced iron porphyrin protein.

The photo-oxidation described above was not the only light-induced absorbance change which Poff and Butler found, however. When they irradiated either mycelial mats of *Phycomyces* or cell suspensions of *Dictyostelium* with 470-nm light, they obtained a second absorbance change (Poff and Butler, 1974b). In freshly harvested samples, the light produced an absorbance increase near 430 nm, together with a decrease near 445 nm. Aged samples which had become anaerobic failed to show the 445-nm bleaching, although the 430-nm signal persisted. As was the case with the pigment system described above, the light-induced spectral changes decayed in the dark, although the half-time was somewhat longer (near 30 s). Poff and Butler succeeded in solubilising the *Dictyostelium* pigment, but were unsuccessful with *Phycomyces.* The spectral characteristics strongly suggested that light was bringing about the reduction of a *b*-type cytochrome. However, the cytochrome itself could not be the actual photoreceptor, since a crude action spectrum in the blue region for the solubilised pigment showed an action maximum near 465 nm for the photoreduction. The authors suggested the actual photoreceptor to be a flavin, and the spectral changes to be a flavin-mediated photoreduction of the cytochrome. They also suggested, on the basis of comparative action spectra, that the pigment might be the long-sought blue light photo-receptor. In this connection, there arose a second irony: *Dictyostelium*, from which they were able to solubilise the pigment, has no known photoresponse with the familiar blue–ultraviolet action spectrum; *Phycomyces,* for which such action spectra are available, failed to yield soluble photoactivity.

In further studies, Poff and Butler (1975) obtained light-minus-dark and oxidised-minus-reduced difference spectra for the solubilised *Dictyostelium* pigment. These spectra strongly suggested a *b*-type cytochrome. The authors could separate the *b*-type cytochrome from cytochrome *c* by brushite chromatography, but lost the photoactivity on doing so. The authors hypothesised that the chromophore itself, possibly a flavin, became lost during chromatography.

Muñoz, Brody and Butler (1974) were able to obtain this same blue-inducible absorbance change in mycelial mats of *Neurospora.* Once again, the light-minus-dark difference spectrum suggested photo-reduction of a *b*-type cytochrome. A preliminary action spectrum again showed an action peak near 460 nm, a wavelength at which the cytochrome did not absorb appreciably, and the same model was proposed: a flavin-mediated photoreduction of the cytochrome. As with *Dictyostelium* and *Phycomyces,* partially anaerobic samples were used.

Muñoz and Butler (1975) then characterised the *Neurospora* system in more detail. They obtained a light-minus-dark difference spectrum which showed, in addition to the obvious cytochrome reduction, a broad minimum near 460 nm, suggestive of flavin reduction as well. A carefully obtained action spectrum for the light-induced spectral changes showed the expected single broad band in the ultraviolet plus appropriate fine structure in the blue region, with no action at wave-lengths longer than 520 nm. Long-term irradiation with high-intensity light caused irreversible bleaching near 460 nm, plus several other spectral changes. As this bleaching progressed, there was a concomi-tant loss of the blue light-inducible absorbance change. Cell-free extracts failed to show the light-induced absorbance change, but addition of either FMN or FAD restored photosensitivity. In this case, cytochrome *c* was also photoreduced, but its dark decay was much slower (10 min half-time) than that for the *b*-type cytochrome (30 s half-time). Finally, in the presence of either FMN or FAD, Muñoz and Butler (1975) found a marked stimulation of oxygen uptake by blue light in the extracts. The effect could be inhibited by azide (concentration unspecified). The demonstration that spectral changes consistent with flavin reduction accompany the light-induced cytochrome reduction, and the observation that irreversible bleaching (consistent with the bleaching of a flavin) is accompanied by a loss of photoactivity strengthen the notion that a flavin of some sort is the primary photoreceptor for the observed light response. The strong similarity between the action spectrum for cytochrome photoreduction and the action spectra for suppression of expression of a circadian rhythm of conidiation (Sargent and Briggs, 1967) and for promotion of carotenogenesis (Zalokar, 1955) thus constitute the evidence that the pigment system being studied is indeed the blue light photo-receptor.

To provide convincing evidence that a particular pigment is the photoreceptor for a given biological process, one should make certain that several criteria are met. Firstly, one would like to obtain a pigment preparation *in vitro* with an absorption spectrum matching the action spectrum for the process. Screening pigments can cause the match to be less than perfect, and where photoreversibility of two pigment forms is involved and action depends upon a particular photoequilibrium, the match may fail altogether (Hartmann, 1966, has elegantly analysed this situation in the case of certain phytochrome responses). Secondly, when two different organisms have slightly shifted action spectra, the same pigment preparation from the two organisms should reflect the difference in the absorption spectra. Thirdly, conditions which alter the photosensitivity of the system might be expected to alter some property of the pigment in question. Fourthly, the pigment should undergo some meaningful photochemistry related to the measurable biological response to light. While there are obviously other criteria one might impose, this list suffices for the discussion that follows.

Several years ago, Dr I.A. Newman, in the author's laboratory, observed that within 1 s of the onset of blue light, there was a dramatic decrease in the electric potential difference between the *Avena* coleoptile tip and the base of the plant. The rapidity of the response suggests that the pigment might be membrane-associated, although it does not require it. However, if the photoreceptor system studied by Butler and his colleagues were membrane-associated and also involved in electron transport, and light in some way altered the rate of this electron transport, one might have a link between the photochemistry and the earliest biological response measured to date. Thus it seemed worthwhile to examine several plant membrane fractions with the other three criteria in mind.

Between July, 1973, and August, 1974, the author was privileged to spend his sabbatical year in the laboratory of Professor Rainer Hertel, of the University of Freiburg, West Germany. The experiments to be described below were carried out with Professor Hertel, Dr A. Jesaitis and Ms P. Heners, and will be published elsewhere in full with these workers.

Coleoptiles from etiolated corn seedlings were used in these studies since a large amount of physiological information was available about their phototropic responses (*see* Briggs, 1963b, 1964). The first experiments involved differential centrifugation to determine whether any particular fraction had the appropriate absorption spectrum. The coleoptiles were harvested under dim green light and extracted in MOPS (morpholinopropanesulphonic acid) buffer, 25 mM, with 0.1 mM $MgCl_2$, 3 mM EDTA, 14 mM 2-mercaptoethanol and 8 per cent (w/w) sucrose, adjusted to pH 7.4. Following chopping and grinding, the crude extract was centrifuged at 500, 9 000 and 21 000 g (15 min each) and 50 000 g for 1 h. The two final pellets (designated 21KP and 50KP) were then re-suspended in MOPS buffer similar to the extraction medium, but at pH 7.0, and without the mercaptoethanol. The re-suspended pellets were then examined in a Perkin-Elmer Model 356 dual-wavelength spectrophotometer, which permitted measurement of absorption and difference spectra in densely turbid samples. Both pellets showed a major absorption band near 420 nm, with smaller bands near 445 and 475 nm. End absorption by protein obscured any detail in the ultraviolet region. The peaks hence matched those in the action spectrum for *Avena* phototropism in position, but not in relative height. The 21KP was substantially enriched for a membrane fraction that binds naphthylphthalamic acid, and considered by Hertel, Thomson and Russo (1972) to be plasma membrane. The 50KP was enriched in cytochrome b_5 (as determined from the oxidised-minus-reduced difference spectrum) and the enzyme to reduce the cytochrome with NADH, considered to be a marker for endoplasmic reticulum (Strittmatter, 1966).

Oxidised-minus-reduced difference spectra of both fractions indicated that the 420–nm absorption band was the Soret band of a cytochrome, with a small alpha band near 555 nm. Both fractions contained bound flavin (between 1 and 2×10^{-7} M on a per gram fresh weight of tissue basis). The flavin was determined by converting it into lumiflavin by alkaline photolysis, re-acidifying with acetic acid and extracting the lumiflavin into chloroform for fluorescence assay. Riboflavin was used as a standard. The two cytochromes differed, however: that predominating in the 21KP could not be reduced by NADH, while that predominating in the 50KP could. Both were fully reducible by dithionite.

To determine with more precision the possible identity of the two membrane fractions, sucrose gradients (20–50 per cent) were run to equilibrium on the supernatant following the 9,000 g centrifugation. The remaining mitochondria were localised in the gradient by measuring cytochrome c oxidase, endoplasmic reticulum by measuring NADH-dependent cytochrome c reductase (Lord *et al.*, 1973), proplastids by measuring acetone-soluble carotenoids, Golgi by measuring glucan synthetase (Ray, Shininger and Ray, 1969) and presumptive plasma membrane by measuring naphthylphthalamic acid binding (Hertel, Thomson and Russo, 1972) (Dr P.M. Ray actually performed the glucan synthetase assays). The NADH-reducible cytochrome coincided precisely with the endoplasmic reticulum marker, as expected from the literature (Strittmatter, 1966; Lord *et al.*, 1973), while the non-NADH-reducible cytochrome followed the broad distribution of presumptive plasma membrane, peaking near 35 per cent sucrose.

Both fractions contained a complete electron transport chain from NADH to molecular oxygen. NADH oxidation was measured by following the loss of absorbance at 360 nm. It was cyanide resistant, inhibited by nitrogen (reversed by bubbling air through the sample) and inhibited by carbon monoxide in the dark, the carbon monoxide resistance being reversible only by bubbling air through the sample in the light.

Hertel (personal communication) has recently shown that both fractions will bind riboflavin. A small amount of radioactive riboflavin is added to samples, and then different amounts of cold riboflavin are added. The preparations are then centrifuged rapidly to pellet the membrane fraction, the pellet then being counted for radioactivity. A concentration of almost 10^{-4} M riboflavin is required in order to remove most of the radioactivity from the pellet, so the affinity for riboflavin is evidently not particularly high. However, when Hertel used the native soluble flavin from either corn or *Phycomyces,* it had an almost 10-fold greater affinity for the membranes than riboflavin. The identity of the flavin is at present unknown, although it is readily converted into lumiflavin, and hence can be assayed and quantified. The distribution of flavin binding in a sucrose gradient parallels that of the two cytochromes. Both fractions contain binding activity, although it may be higher in the endoplasmic reticulum fraction.

So far, nothing has been said that might link either fraction to the phototropic response. However, it is possible to alter the phototropic sensitivity of corn coleoptiles in two different ways. Briggs (1960) showed that exposure to light at an appropriate high intensity for a sufficiently brief duration, conditions under which the reciprocity law would fail for second positive curvature, made corn coleoptiles completely insensitive to normally effective phototropic doses of light. During roughly 30 min following the exposure to light, dark adaptation occurs and the coleoptiles recover their normal phototropic sensitivity. Thus it seemed reasonable to subject corn coleoptiles to such a light treatment and to examine the various pigmented fractions rather than measuring phototropic sensitivity. Surprisingly, light induced an increase in flavin absorption in a fraction pelleted between 10 000 and 50 000 g. Flavin was assayed as usual by the lumiflavin method, and protein by the Lowry method (Lowry *et al.,* 1951). The results are shown in *Table 2.1.* If the plants are extracted immediately following irradiation, the increase ranges from 19 to 29 per cent. However, if the coleoptiles are allowed to incubate in the dark for 60 min at 22 °C, the flavin approaches the control value again. Preliminary experiments indicate that the decay is gradual, being about half completed after 30 min. Thus, one is encouraged to think that the kinetics of flavin increase immediately following illumination, and the subsequent return to the dark level, may indeed parallel the light-induced alteration in phototropic sensitivity. The light-induced increase seems to be found predominantly in the 21KP and not in the 50KP.

Briggs (1963c) also showed that red light would decrease substantially the phototropic sensitivity of corn coleoptiles. However, the decrease was very slow, and required almost 1 h for completion. A preliminary experiment indicated that red light also increased the

Table 2.1   Alteration in flavin to protein ratio in membrane fraction by brief high-intensity irradiation

	Experiment No.		
	1	2	3
Dark control[1]	100	100	100
Irradiate, extract immediately	125	119	129
Irradiate, extract following 60–min darkness	–	91	106

[1] Arbitrarily set as 100 per cent

flavin to protein ratio in the isolated coleoptiles, and that the increase was gradual.   If the effect can be confirmed, then detailed kinetics should be considered in order to determine whether the time course for the red-induced change parallels that for the red-induced alteration in phototropic sensitivity.

The above experiments, showing a possible parallelism between alterations in phototropic sensitivity and alteration in the amount of bound flavin, are important in that they provide a link between the heavier of the two fractions, that enriched with presumptive plasma membrane, and the phototropic system as measured *in vivo*.   At the moment, however, it is premature to speculate what these light-induced changes mean.   Dose–response curves for their induction, and detailed kinetics for their appearance and decay, are needed at the very least.

Preliminary work with preparations obtained from *Phycomyces* sporangiophores yielded fractions with absorption spectra similar to those mentioned above for corn, but with the two longer wavelength peaks shifted to about 460 and 490 nm.   The shifts are consistent with the shift of these peaks in the *Phycomyces* action spectra.   The preparations also contained both cytochrome and flavin, but also sufficient carotenoid to make it unclear which pigment, flavin or carotenoid, was the major absorber.   Efforts to obtain such fractions from a carotenoidless mutant, C-2, met with failure, at least partially because of lack of membrane stability under the conditions used.

It is clear from the above discussion that substantial progress has been made during the past 3 years in identifying the blue light photoreceptor.   The best evidence comes from the studies initiated by Poff, Butler and Loomis (1973), identifying spectral changes as a consequence of light treatment. · The apparent photoreceptor in *Dictyostelium* is clearly a haeme protein which becomes oxidised directly upon light absorption.   The blue light photoreceptor itself is probably a flavoprotein which becomes photoreduced by light and thereupon reduces a *b*-type cytochrome.   The system becomes re-oxidised in following darkness.   In both cases, similarity between action spectra for a biological process and action spectra for the photoinduced absorbance changes make the case for the pigments as the legitimate photoreceptors strong.   In addition, our own rather preliminary evidence is completely consistent with the notion of a membrane-associated flavo-protein–cytochrome couple as the light-sensing system in corn, with the added information that conditions which alter phototropic sensitivity also alter the flavin to protein ratio of at least one membrane fraction. The first indications are that the kinetics of the two alterations mentioned are consistent with the kinetics of the physiological changes.

Clearly a great deal of work is still needed to explore the various systems before one can say with certainty that flavin absorption of light modifies electron transport across a specific membrane, leading ultimately to the measured biological change. It is this author's opinion, however, that the model proposed above is likely to turn out to be correct. Perhaps the blue light photoreceptor today is where phytochrome was in 1959 (Butler *et al.*, 1959). Given the same rapid progress with this photoreceptor as with phytochrome, we should know a great deal more about it within a relatively few years.

References

BATRA, P.P. (1971). In *Photophysiology* Vol.6, p.47. Ed. Giese, A.C. Academic Press, New York

BRIGGS, W.R., TOCHER, R.D. and WILSON, J.F. (1957). *Science*, **126**, 210

BRIGGS, W.R. (1960). *Plant Physiol.*, **35**, 951

BRIGGS, W.R. (1963a). *Plant Physiol.*, **38**, 237

BRIGGS, W.R. (1963b). *Annu. Rev. Plant Physiol.*, **14**, 311

BRIGGS, W.R. (1963c). *Am. J. Bot.*, **50**, 196

BRIGGS, W.R. (1964). In *Photophysiology*, **Vol.1**, p.223. Ed. Giese, A.C. Academic Press, New York

BRIGGS, W.R. and RICE, H.V. (1972). *Annu. Rev. Plant Physiol.*, **23**, 293

BÜNNING, E. (1937a). *Planta*, **27**, 148

BÜNNING, E. (1937b). *Planta*, **27**, 583

BUTLER, W.L., NORRIS, K.H., SIEGELMAN, H.W. and HENDRICKS, S.B. (1959). *Proc. Nat. Acad. Sci. U.S.A.*, **45**, 1703

CURRY, G.M., THIMANN, K.V. and RAY, P.M. (1956). *Physiol. Plant.*, **9**, 429

CURRY, G.M. (1957). *Studies on the Spectral Sensitivity of Phototropism*, Ph.D. Thesis, Harvard University

CURRY, G.M. and GRUEN, H.E. (1959). *Proc. Nat. Acad. Sci. U.S.A.*, **45**, 797

DELBRÜCK, M. and SHROPSHIRE, W. Jr. (1960). *Plant Physiol.*, **35**, 194

EVERETT, M. and THIMANN, K.V. (1968). *Plant Physiol.*, **43**, 1786

GALSTON, A.W. and BAKER, R.S. (1949). *Am. J. Bot.*, **36**, 773

HAGER, A. (1970). *Planta*, **91**, 38

HARTMANN, K.M. (1966). *Photochem. Photobiol.*, **5**, 349

HARTMANN, K.M. (1967). *Z. Naturforsch.*, **22b**, 1172

HERTEL, R., THOMSON, K.St. and RUSSO, V. (1972). *Planta*, **107**, 325

HOWES, C.D. and BATRA, P.P. (1970). *Arch. Biochem. Biophys.*, **137**, 175

JOHNSTON, E.S. (1934). *Smithsonian Misc. Collect.*, **92**, 1
KOWALLIK, W. (1967). *Plant Physiol.*, **42**, 672
LORD, J.M., TAGAWA, T., MOORE, T.S. and BEEVERS, H. (1973). *J. Cell Biol.*, **57**, 659
LOWRY, O.H., ROSEBROUGH, N.J., FARR, A.L. and RANDALL, J.R. (1951). *J. Biol. Chem.*, **193**, 265
MOHR, H. (1972). *Lectures on Photomorphogenesis*, Springer-Verlag, Berlin, 237pp
MUÑOZ, V., BRODY, S. and BUTLER, W.L. (1974). *Biochem. Biophys. Res. Commun.*, **58**, 322
MUÑOZ, V. and BUTLER, W.L. (1975). *Plant Physiol.*, **55**, 421
PAGE, R.M. and BRUNGARD, J. (1961). *Science*, **134**, 733
PAGE, R.M. (1968). In *Photophysiology*, Vol.3, p.65. Ed. Giese, A.C. Academic Press, New York
POFF, K.L., BUTLER, W.L. and LOOMIS, W.F. Jr. (1973). *Proc. Nat. Acad. Sci. U.S.A.*, **70**, 813
POFF, K.L., LOOMIS, W.F. Jr. and BUTLER, W.L. (1974). *J. Biol. Chem.*, **249**, 2164
POFF, K.L. and BUTLER, W.L. (1974a). *Photochem. Photobiol.*, **20**, 241
POFF, K.L. and BUTLER, W.L. (1974b). *Nature, Lond.*, **248**, 799
POFF, K.L. and BUTLER, W.L. (1975). *Plant Physiol.*, **55**, 427
PRATT, L.H. and BUTLER, W.L. (1970). *Photochem. Photobiol.*, **11**, 503
RAU, W. (1967). *Planta*, **72**, 14
RAY, P.M., SHININGER, T.L. and RAY, M.M. (1969). *Proc. Nat. Acad. Sci. U.S.A.*, **64**, 605
RILLING, H.C. (1964). *Biochim. Biophys. Acta*, **79**, 464
SARGENT, M.L. and BRIGGS, W.R. (1967). *Plant Physiol.*, **42**, 1504
SHROPSHIRE, W. Jr. and WITHROW, R.B. (1958). *Plant Physiol.*, **33**, 360
SONG, P.-S. and MOORE, T.A. (1974). *Photochem. Photobiol.*, **19**, 435
STEINER, A.M. (1967). *Naturwissenschaften*, **54**, 497
STRITTMATTER, P. (1966). In *Flavins and Flavoproteins*, Vol.8, p.325. Ed. Slater, E.C. Elsevier, Amsterdam
THIMANN, K.V. and CURRY, G.M. (1960). In *Comparative Biochemistry*, Vol.1, p.243. Ed. Florkin, M. and Mason, H.S. Academic Press, New York
WALD, G. and DuBUY, H.G. (1936). *Science*, **84**, 247
WENT, F.W. (1928). *Rec. Trav. Bot. Neerl.*, **25**, 1
ZALOKAR, M. (1955). *Arch. Biochem. Biophys.*, **56**, 318

3

RE-EXAMINATION OF PHOTOCHEMICAL PROPERTIES AND ABSORPTION CHARACTERISTICS OF PHYTOCHROME USING HIGH-MOLECULAR-WEIGHT PREPARATIONS[1]

LEE H. PRATT
Department of Biology, Vanderbilt University, Nashville, Tennessee 37235, U.S.A.

Introduction

Two recent discoveries have led to the necessity for re-examining the basic properties of the phytochrome molecule *in vitro*. Firstly, Gardner *et al.* (1971) demonstrated that high-molecular-weight phytochrome (large phytochrome) is unusually susceptible to proteolysis, yielding a chromopeptide (small phytochrome) which is resistant to further degradation. Secondly, Pike and Briggs (1972) reported that etiolated oat shoots yield a relatively high protease activity in crude extracts, explaining why purified oat phytochrome was always degraded (e.g. Mumford and Jenner, 1966; Hopkins, 1971; Rice and Briggs, 1973). Since most investigations of the basic photochemical properties and absorption characteristics of phytochrome utilised preparations from oats (e.g. Butler, Hendricks and Siegelman, 1964; Linschitz *et al.*, 1966; Mumford and Jenner, 1971), it becomes apparent that most of what we have learned about the molecular properties of phytochrome reflects, in fact, knowledge of a proteolytically derived product of the native molecule (*see* Briggs and Rice, 1972, for discussion).

All available biochemical evidence indicates that large phytochrome is the native molecule and that small phytochrome is either not present *in vivo* or is present only in small amounts (Gardner *et al.*, 1971; Pratt, 1973; Cundiff and Pratt, 1973, 1975a,b; Pratt, Kidd and Coleman, 1974). This conclusion is based largely upon observations that small phytochrome is detected in crude extracts only in proportion to the amount of proteolysis which takes place *in vitro*. Thus, many calculations which utilise the results of photochemical and absorption measurements on oat phytochrome *in vitro* (e.g. Hartmann, 1966; Oelze-Karow and Mohr, 1973) may need to be reconsidered.

[1] Paper presented at the Annual European Photomorphogenesis Symposium and invited by the Editor for publication in these Proceedings

Briggs and co-workers have re-investigated some of the photochemical properties and absorption characteristics of phytochrome using high-molecular-weight preparations from rye (e.g. Tobin and Briggs, 1973; Gardner and Briggs, 1974). However, because their large phytochrome was not purified from oats, the problem of possible interspecies differences in the properties being measured remains. In addition, they did not repeat the basic photochemical measurements of Butler, Hendricks and Siegelman (1964). Thus, we have been interested in studying the photochemical properties and absorption characteristics of large and small phytochrome from oats, as well as large phytochrome from rye, barley and peas. A summary of our observations and their significance is presented here.

Phytochrome Preparation

The original protocol with which we began to purify phytochrome is derived from that described by Hopkins and Butler (1970) and includes changes introduced by Rice, Briggs and Jackson-White (1973) (*Figure 3.1,* left). However, with oat seedlings this protocol yields only small phytochrome. Thus, modifications designed primarily to minimise proteolysis and consequently maximise the yield of large phytochrome were introduced, following in part suggestions made by Gardner *et al.* (1971). This modified protocol (*Figure 3.1,* right) provides relatively high yields of purified large phytochrome from a variety of plant sources, including oats (Pratt, 1973). In addition, it is a protocol designed for routine use by one person working alone.

When phytochrome is purified from oats using the modified protocol, and when the first ammonium sulphate cut is with 240 g l^{-1} (40 per cent) in order to co-purify small with large phytochrome (Pratt, 1973), the elution profile from the final P-300 column shows two well separated phytochrome peaks (*Figure 3.2*). The first is undegraded (large) phytochrome and the second is proteolytically degraded (small) phytochrome (Gardner *et al.,* 1971; Cundiff and Pratt, 1973, 1975b). Small phytochrome is commonly a 60 000–dalton monomer while large phytochrome is apparently a 240 000–dalton dimer of 120 000–dalton subunits. The latter conclusion is based upon combined results of sedimentation velocity, sedimentation equilibrium, gel exclusion chromatography and sodium dodecyl sulphate gel electrophoresis experiments (*see* Briggs and Rice, 1972, for discussion).

Phytochrome-containing fractions from the P-300 column are normally concentrated by ammonium sulphate precipitation. Precipitates are then dissolved in 60 mM potassium phosphate, pH 7.8, and stored in a liquid nitrogen freezer. At this temperature, preparations are stable for at least 3 years. All handling of phytochrome is carried out under dim green safelight (Pratt, 1973) except for specified red or far-red irradiations. Actinic light is filtered through appropriate Balzer B-40 interference filters. Phytochrome purity is expressed as its specific

PHYTOCHROME PURIFICATION

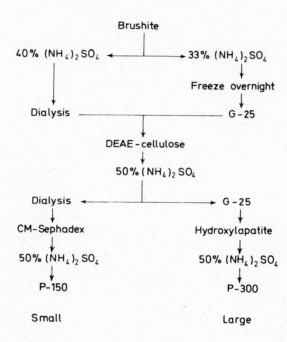

Figure 3.1 (left) Protocol for the purification of degraded (small) phytochrome and adaptations designed to maximise the yield of undegraded (large) phytochrome (right). Except when highest phytochrome purity was needed, either the diethyl-aminoethyl(DEAE)-cellulose or hydroxylapatite column was omitted. Other chromatographic steps include Sephadex G-25, carboxymethyl(CM)-Sephadex and Bio-Gel P-150 and P-300. Details of these protocols can be found in Hopkins and Butler (1970), Pratt (1973) and Rice, Briggs and Jackson-White (1973)

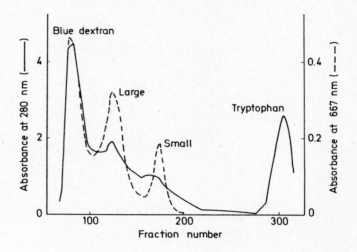

*Figure 3.2   Bio-Gel P-300 elution profile for a phytochrome preparation from Garry oats (*Avena sativa *L.) previously purified by brushite and DEAE-cellulose chromatography.   Blue dextran and tryptophan were added to the 5-ml phytochrome sample before application to the 2.4 × 58 cm column.   Blue dextran marks the void volume both by absorbance at 280 nm and at 667 nm;   tryptophan marks the total volume. Absorbances at 280 nm (solid line) and at 667 nm (broken line) are shown as a function of fraction number.   Each fraction contains about 1 ml.   The positions of the large and small phytochrome fractions are labelled.   (After Pratt, 1973)*

absorbance ratio ($A_{667\,nm}/A_{280\,nm}$ after saturating far-red irradiation). Absorbance spectra are recorded with a Shimadzu MPS-50L spectrophotometer.   Phototransformation measurements are made with a custom-built dual-wavelength spectrophotometer (Kidd and Pratt, 1973).

Absorption Spectra

Absorbance spectra characteristic of large and small Garry oat phytochrome co-purified and separated only by the final P-300 column (*Figure 3.2*) are given in *Figure 3.3* left and right.   The purification buffers contained no ethylenediaminetetraacetate (EDTA) and, after the brushite column, no 2-mercaptoethanol.   Spectra for small phytochrome after saturating 723–nm (a) or saturating 665–nm irradiation (b) are indistinguishable from those presented by other workers.   The Pr spectrum for large phytochrome (a) is similar to that for small phytochrome, but the

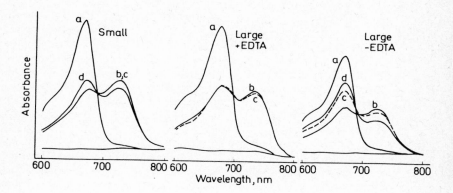

Figure 3.3 Absorption spectra of small oat phytochrome (left), large oat phyto-chrome in the presence of EDTA (centre) and large oat phytochrome in the absence of EDTA (right). All spectra were measured at 3 °C. (a) After saturat-ing irradiation with 723–nm light; (b) after saturating irradiation with 665–nm light; (c) same as (b) but after incubation for 30 min in darkness at 3 °C; (d) same as (c) but after incubation for 18 h in darkness at 4 °C. (After Pratt and Cundiff, 1975; Pratt, 1975a)

spectrum for large Pfr (b) is clearly different in that there is relatively less absorbance in the far-red region. Shorter wavelength red light, for example 620 nm, yields a spectrum for large Pfr that is indistinguishable from that shown in *Figure 3.3* (right, b).

If the two phytochrome fractions are allowed to remain in darkness for 30 min at 3 °C after red actinic irradiation, no appreciable change is seen in the spectrum for small Pfr (*Figure 3.3*, left, b → c). However, for large phytochrome, a large increase in red absorbance is observed which is accompanied only by a much smaller change in the far-red region (*Figure 3.3*, right, b → c). This rapid change in red absorbance is also seen with large phytochrome preparations from rye, barley and peas (Pratt and Cundiff, 1975) and exhibits first-order kinetics with a half-life at 3 °C of 122 s (Pratt, 1975a). Thus, it is essentially com-plete within 30 min. If the two red-irradiated samples are then left at 4 °C overnight after recording spectra c, both large and small phyto-chrome undergo reversion at approximately equal rates with changes of equivalent magnitude in the red and far-red regions (*Figure 3.3*, c → d).

A kinetic analysis of large phytochrome preparations with absorption properties comparable with those described above leads to the suggestion that they contain Pr in photoequilibrium with two kinds of Pfr (Pratt,

1975a). One form of Pfr would have normal far-red extinction while the other would have very little. This low-extinction form of Pfr, in darkness, would then revert rapidly to Pr (*Figure 3.3,* right, b → c), while the high-extinction form of Pfr would revert only slowly, as is the case for small phytochrome (*Figure 3.3,* right, c → d). However, an equivalent possibility is that only one type of Pfr is present but each Pfr molecule has one or more high far-red extinction chromophores and one or more low far-red extinction chromophores. This apparent difference in far-red extinction may reflect a reversible change in protein-chromophore interaction (Butler, Siegelman and Miller, 1964).

If either EDTA or 2-mercaptoethanol is added to large phytochrome preparations, this rapid dark increase in red absorbance after red actinic irradiation does not occur (*Figure 3.3,* centre, b → c) (Pratt and Cundiff, 1975). Thus, we suspected that this unusual spectral property may result from an interaction of Pfr with a divalent cation. However, we could never induce this apparent rapid reversion by addition of metal ions to samples containing 0.1 mM EDTA, although we did observe two other effects of cations on the spectral properties of large phytochrome (Pratt and Cundiff, 1975). Firstly, as previously reported by Briggs (personal communication) and more recently by Lisansky and Galston (1974), it is clear that some cations (here Cu^{2+}, Co^{2+} and Zn^{2+} at 1 mM) will bleach phytochrome and that this spectral denaturation (Butler, Siegelman and Miller, 1964) may be reversed by EDTA. Secondly, 1 mM Fe^{3+} accelerates the rate of reversion by the same order of magnitude as that seen by Mumford and Jenner (1971) with reductants such as reduced pyridine nucleotides. However, large phytochrome is relatively unresponsive to other cations (Mg^{2+}, Ca^{2+}, Mn^{2+}, Ni^{2+} and Fe^{2+} at 1 mM). Clearly, phytochrome reacts specifically and differentially to a variety of metal cations and it is possible that comparable interactions *in situ* could markedly and reversibly affect both spectral measurements of phytochrome and the photomorphogenic consequence of red and far-red irradiations.

Photochemical Measurements

Detailed treatments of the methodology and theoretical considerations involved in the determination of the basic photochemical properties (ϕ_r/ϕ_{fr}; $[Pfr]_\infty^{665}$; $[Pr]_\infty^{665}$) of phytochrome are presented elsewhere (Butler, Hendricks and Siegelman, 1964; Butler, 1972). The terms $[Pfr]_\infty^{665}$ and $[Pr]_\infty^{665}$ are the proportions of the total phytochrome pool present as Pfr and Pr, respectively, at photoequilibrium in 665–nm light, while ϕ_r/ϕ_{fr} is the ratio of quantum yields for the phototransformations of Pr to Pfr (ϕ_r) and Pfr to Pr (ϕ_{fr}). To determine $[Pfr]_\infty^{665}$ and $[Pr]_\infty^{665}$, one can use the following relationship:

$$1 - [Pfr]_\infty^{665} = [Pr]_\infty^{665} = (A_{665_{min}}/A_{665_{max}})/(1 + [\phi_r/\phi_{fr}]) \qquad (3.1)$$

where A_{665min} (A_{665max}) is the absorbance of the sample at 665 nm at equilibrium under 665(723)-nm light. Absorbance values are taken directly from corresponding spectra such as those presented in *Figure 3.3* while the ratio of the quantum yields is determined as follows:

$$\phi_r/\phi_{fr} = [E_{723}A_{723max}(dA/dt)_{665}]/[E_{665}A_{665max}(dA/dt)_{723}] \tag{3.2}$$

where E_{723} (E_{665}) is the actinic quantum flux rate at 723 (665) nm, A_{723max} is the absorbance at 723 nm at photoequilibrium under 665-nm light and $(dA/dt)_{665}$ $[(dA/dt)_{723}]$ is the initial rate of change of absorbance at 727 nm under 665(723)-nm actinic irradiation. Again, A_{723max} is taken from the appropriate absorption spectrum. Intial rates of change of absorbance were measured with a dual-wavelength spectrophotometer with 798 nm as the reference wavelength. Actinic quantum fluxes were measured with a YSI Radiometer (Pratt, 1975b) and corrected for absorption by the sample to obtain effective incident quantum fluxes as follows:

$$E = E_0(1 - 10^{-A})/2.3A \tag{3.3}$$

where E_0 is the measured quantum flux and A is the absorbance of the sample in the actinic light path at the actinic wavelength (A is always less than 0.2).

Using small oat phytochrome, presumably comparable to that studied by Butler, Hendricks and Siegelman (1964), I confirmed their measurements of 0.81 for $[Pfr]_\infty^{665}$ and about 1.5 for ϕ_r/ϕ_{fr} (*Table 3.1*). Hence, the protocol used here cannot be the source of the difference between their reported values and those I obtained for large phytochrome. In contrast, however, $[Pfr]_\infty^{665}$ for large phytochrome from the same oat cultivar and in the presence of EDTA is only 0.75 while the ratio of quantum yields is 1.0 (*Table 3.1*).

In the absence of EDTA, there is evidence that at least two different photoequilibria are established (Pratt, 1975a). Hence, it is not possible to use the above equations since one of the assumptions in their derivation — that only one photoequilibrium is involved (Butler, 1972) — is not met.

Absolute quantum yields for the two opposing phototransformations can also be determined from the data used above as described by Butler (1972):

$$\phi_r = (k_{665}[Pfr]_\infty^{665})/(E_{665}\epsilon_{r665}) \tag{3.4}$$

$$\phi_{fr} = (k_{723}[Pr]_\infty^{723})/(E_{723}\epsilon_{fr723}) \tag{3.5}$$

where k_{665} (k_{723}) is the first-order rate constant for the phototransformation using 665(723)-nm actinic light of the given actinic quantum flux rate (E_{665} and E_{723} respectively) and ϵ_{r665} (ϵ_{fr723}) is the extinction coefficient of Pr (Pfr) at 665 (723) nm. The rate constants are

Table 3.1   Photochemical characteristics of small and large phytochrome from Garry oats
    Each entry represents an average of five determinations with an independently purified sample

Phytochrome preparation	Specific absorbance ratio	ϕ_r/ϕ_{fr}	$[Pr]_\infty^{665}$	$[Pfr]_\infty^{665}$
Small	0.13	1.64	0.174	0.826
	0.13	1.29	0.198	0.802
Large (+ EDTA)	0.22	0.92	0.259	0.741
	0.13	0.96	0.244	0.756
	0.028	1.07	0.247	0.753
	0.14	0.98	0.251	0.749

obtained from the same data used to obtain initial rates of absorbance change for equation 3.2 (Pratt, 1975b) while extinction coefficients, which are needed only for these calculations of absolute quantum yields, are taken from the paper by Tobin and Briggs (1973).   While their extinction coefficients are for large rye phytochrome, large oat phytochrome has almost identical absorption characteristics and, as will be evident below, the estimated quantum yield values for large oat and rye phytochrome are equal.
    Absolute quantum yields ($\phi_r = \phi_{fr}$) are then estimated to be 0.17 for large oat phytochrome (Pratt, 1975b).   While it might at first appear that the values of 0.28 and 0.20 reported by Gardner and Briggs (1974) for large rye phytochrome suggest a different result, this apparent difference instead derives from error in calculation.   When their data are used and when

(1)   their actinic quantum flux measurements are corrected by equation 3.3
(2)   rate constants to the base 10 rather than to the base e are used
(3)   a value of 0.75 rather than 0.81 is used for $[Pfr]_\infty^{665}$,

a value of 0.17 is again obtained indicating complete agreement between our measurements (Pratt, 1975b).

Significance of this Re-examination

Three points argue that one should utilise the photochemical parameters measured for large phytochrome in the presence of EDTA (*Table 3.1*; $[Pfr]_\infty^{665} = 0.75$; $\phi_r/\phi_{fr} = 1.0$) rather than the significantly different values which have long been accepted (Butler, Hendricks and Siegelman, 1964; $[Pfr]_\infty^{665} = 0.81$; $\phi_r/\phi_{fr} = 1.5$) but which we now know to have been determined with proteolytically degraded phytochrome.   Firstly, as pointed out above, phytochrome *in situ* is almost certainly present in the high-molecular-weight form. Secondly, the absorbance characteristics of large phytochrome more closely match *in vivo* measurements than do the characteristics of small phytochrome (Pratt and Cundiff, 1975). Thirdly, the ratio of quantum yields for large phytochrome *in vitro* agrees well with the only such estimate *in vivo* (Pratt and Briggs, 1966) while the same ratio for small phytochrome is distinctly different.

Values of $[Pfr]_\infty$ have often been calculated for phytochrome *in vivo* as a function of wavelength (e.g. Hartmann, 1966). However, these calculations must inevitably begin with an assumption regarding $[Pfr]_\infty$ at some wavelength in the red spectral region. Thus, one must always use a primary value for $[Pfr]_\infty^{red}$ which, as discussed by Butler, Hendricks and Siegelman (1964) and Butler (1972), can *only* be determined with phytochrome *in vitro* even though it is (and always has been) possible that the *in vivo* value is changed upon extraction of phytochrome. While there is no reason as yet to suspect such a change in $[Pfr]_\infty^{red}$ upon extraction, there is also, unfortunately, no method yet available for carefully testing this latter possibility.

The new photochemical parameters for large phytochrome reported here (*Table 3.1*) are most important whenever calculations are made from spectral data to determine the amounts of Pr and Pfr present (Butler and Lane, 1965). Such calculations are commonly made either when estimating the amount of reversion which occurs *in vivo* (e.g. Butler and Lane, 1965; Marmé, Marchal and Schäfer, 1971) or when deriving precise values regarding the amount of Pfr present at photoequilibrium as a function of wavelength or present as a function of varying actinic light doses (e.g. Hartmann, 1966). In most cases the differences between 0.81 and 0.75 for $[Pfr]_\infty^{red}$ and 1.5 and 1.0 for ϕ_r/ϕ_{fr} will be within the limits of error of the measurement and thus are of relatively little concern. However, it may be necessary to reconsider some calculations which lead to conclusions that might be dependent upon values (i.e. $[Pfr]_\infty^{red}$, ϕ_r/ϕ_{fr}) which are now known to have been measured with a degradation product of the native molecule (e.g. Oelze-Karow and Mohr, 1973).

The reversible changes in the spectral properties of large phytochrome which have been summarised here (*Figure 3.3* and effects of cations) and presented in detail elsewhere (Pratt and Cundiff, 1975; Pratt, 1975a) provide further reason to interpret spectral measurements with caution, if not sometimes with scepticism. It is, for example, often possible to explain unusual spectral data (e.g. Marmé, Marchal and Schäfer, 1971) on the basis of reversible phytochrome extinction changes which clearly occur at least *in vitro*. It is thus not necessary to postulate complex models to describe such behaviour in the absence of independent evidence derived from another assay method. In addition, since reversible extinction changes are seen almost exclusively in the far-red region of the spectrum and since many investigators assay for phytochrome at 730 *versus* 800 nm, it may often be necessary to confirm observations by also measuring at 665 *versus* 800 nm.

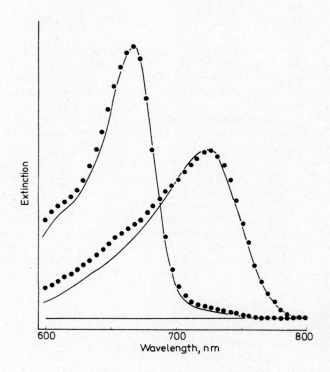

Figure 3.4 Normalised spectra for small (solid line) and large (dotted line) oat phytochrome (the latter in the presence of EDTA) calculated as described in the text. Absorption maxima for Pr = 667 nm and for Pfr = 724 nm. Extinction values for Pfr have been adjusted as though 100 per cent of the Pr were transformed to Pfr so that they can be directly compared with those for Pr

With the photochemical data and absorption characteristics presented here, it is possible to compare directly large and small phytochrome spectra in the Pfr as well as in the Pr form. Extinction values for Pfr, normalised to the corresponding spectra for Pr, can be calculated as follows:

$$\epsilon_{fr}^{\lambda} \propto \{(A_{Pfr}^{\lambda,m}) - ([Pr]_{\infty}^{665})(A_{Pr}^{\lambda,m})\} \, ([Pr]_{\infty}^{723}/[Pfr]_{\infty}^{665}) \tag{3.6}$$

where ϵ_{fr}^{λ} is the normalised extinction of Pfr at wavelength λ; $A_{Pfr}^{\lambda,m}$ $(A_{Pr}^{\lambda,m})$ is the measured absorbance of Pfr (Pr) at wavelength λ and at photoequilibrium under 665(723)-nm actinic light; and $[Pr]_{\infty}^{723}$ is the proportion of the total phytochrome pool present as Pr at photoequilibrium under 723-nm actinic light. While it is assumed that $[Pr]_{\infty}^{723} = 1$, the use of another value (e.g. 0.95) would not have a significant effect on the result. For purposes of comparison with the small phytochrome extinction values, the large phytochrome spectra, which were measured with EDTA present, can be further normalised by multiplying by the ratio of A_{665max} for the small phytochrome spectra used to A_{665max} for the large phytochrome spectra used. Such normalised and calculated spectra are presented in *Figure 3.4*.

Both large and small Pr and Pfr have identical absorption maxima at 667 and 724 nm, respectively. The calculated Pfr spectra clearly indicate that the second Pfr peak in the red region is almost entirely (if not exclusively) a reflection of the Pr still present at photoequilibrium in red light. In addition, it becomes apparent that the difference in measured Pfr spectra between small and large phytochrome (*Figure 3.2*, left and centre, spectra b), as well as the difference in $[Pfr]^{665}$ values (*Table 3.1*), is more a reflection of a difference in quantum yield ratio (*Table 3.1*) than a difference in relative extinction (*Figure 3.4*). The minor difference between small and large phytochrome observed in the red region of these spectra is variable, depending upon the original spectra chosen for comparison, and should not be considered significant.

Acknowledgements

This research programme was supported by National Science Foundation Grant GB-17057 and by grants from the Vanderbilt University Research Council.

References

BRIGGS, W.R. and RICE, H.V. (1972). *Annu. Rev. Plant Physiol.*, **23**, 293
BUTLER, W.L. (1972). In *Phytochrome*, p.185. Ed. Mitrakos, K. and Shropshire, W. Jr. Academic Press, New York
BUTLER, W.L., HENDRICKS, S.B. and SIEGELMAN, H.W. (1964). *Photochem. Photobiol.*, **3**, 521

BUTLER, W.L. and LANE, H.C. (1965). *Plant Physiol.*, **40**, 13
BUTLER, W.L., SIEGELMAN, H.W. and MILLER, C.O. (1964). *Biochemistry*, **3**, 851
CUNDIFF, S.C. and PRATT, L.H. (1973). *Plant Physiol.*, **51**, 210
CUNDIFF, S.C. and PRATT, L.H. (1975a). *Plant Physiol.*, **55**, 207
CUNDIFF, S.C. and PRATT, L.H. (1975b). *Plant Physiol.*, **55**, 212
GARDNER, G. and BRIGGS, W.R. (1974). *Photochem. Photobiol.*, **19**, 367
GARDNER, G., PIKE, C.S., RICE, H.V. and BRIGGS, W.R. (1971). *Plant Physiol.*, **48**, 686
HARTMANN, K.M. (1966). *Photochem. Photobiol.*, **5**, 349
HOPKINS, D.W. (1971). *Protein Conformational Changes of Phytochrome*, Ph.D. Thesis, University of California
HOPKINS, D.W. and BUTLER, W.L. (1970). *Plant Physiol.*, **45**, 567
KIDD, G.H. and PRATT, L.H. (1973). *Plant Physiol.*, **52**, 309
LINSCHITZ, H., KASCHE, V., BUTLER, W.L. and SIEGELMAN, H.W. (1966). *J. Biol. Chem.*, **241**, 3395
LISANSKY, S.G. and GALSTON, A.W. (1974). *Plant Physiol.*, **53**, 352
MARMÉ, D., MARCHAL, B. and SCHÄFER, E. (1971). *Planta*, **100**, 331
MUMFORD, F.E. and JENNER, E.L. (1966). *Biochemistry*, **5**, 3657
MUMFORD, F.E. and JENNER, E.L. (1971). *Biochemistry*, **10**, 98
OELZE-KAROW, H. and MOHR, H. (1973). *Photochem. Photobiol.*, **18**, 319
PIKE, C.S. and BRIGGS, W.R. (1972). *Plant Physiol.*, **49**, 521
PRATT, L.H. (1973). *Plant Physiol.*, **51**, 203
PRATT, L.H. (1975a). *Photochem. Photobiol.*, **21**, 99
PRATT, L.H. (1975b). *Photochem. Photobiol.*, **22**, 33
PRATT, L.H. and BRIGGS, W.R. (1966). *Plant Physiol.*, **41**, 467
PRATT, L.H. and CUNDIFF, S.C. (1975). *Photochem. Photobiol.*, **21**, 91
PRATT, L.H., KIDD, G.H. and COLEMAN, R.A. (1974). *Biochim. Biophys. Acta*, **365**, 93
RICE, H.V. and BRIGGS, W.R. (1973). *Plant Physiol.*, **51**, 927
RICE, H.V., BRIGGS, W.R. and JACKSON-WHITE, C.J. (1973). *Plant Physiol.*, **51**, 917
TOBIN, E.M. and BRIGGS, W.R. (1973). *Photochem. Photobiol.*, **18**, 487

4

INTERMEDIATES IN THE PHOTOCONVERSION OF PHYTOCHROME

R.E. KENDRICK
Department of Plant Biology, The University, Newcastle-upon-Tyne NE1 7RU, U.K.
C.J.P. SPRUIT
Laboratory of Plant Physiological Research, Agricultural University, Wageningen, The Netherlands

Introduction

The mechanism of phytochrome photoconversion is of great interest since it is the production of Pfr from Pr that initiates the wide range of phytochrome mediated responses. Butler, Hendricks and Siegelman (1964) reported that the photoconversions of Pr to Pfr and of Pfr to Pr were apparently simple first-order photoreactions. However, in 1966 the application of two different techniques demonstrated that the reactions were not simple, each involving several intermediates. Linschitz *et al.* (1966), using flash photolysis *in vitro*, and Spruit (1966a–d), using low-temperature techniques both *in vivo* and *in vitro*, observed a number of pigment forms that were unstable at room temperature. Although similarities between the observed intermediates *in vivo* and *in vitro* were obvious, a direct comparison presents difficulties. This has led to a confusing range of different terminologies in the literature. Some of the problems encountered are:

(1) Absorbance maxima of phytochrome are shifted upon extraction, making comparison of *in vitro* and *in vivo* data difficult (Everett and Briggs, 1970).
(2) Use of low temperatures shifts the absorbance maxima of phytochrome to longer wavelengths (Spruit and Kendrick, 1973).
(3) Flash photolysis work presumably has been carried out with the 60 000-dalton pigment form, a product of proteolytic degradation of the 120 000-dalton native pigment (Gardner *et al.*, 1971).
(4) The adopted terminology overlaps with the existing well established usage for photosynthetic pigments.

31

In this chapter, we wish to collate the available data and to propose a new system of terminology.

Techniques

FLASH PHOTOLYSIS

Flash photolysis has been used to determine the presence of intermediates *in vitro* at 0 °C (Linschitz *et al.*, 1966). This technique enables rapid absorbance measurements to be made after a short intense actinic flash. Absorbance changes during the time interval from 0.2 ms to several minutes following the flash provided evidence for intermediates in the Pr to Pfr and Pfr to Pr pathways. A double-flash technique (Linschitz and Kasche, 1967), involving the use of a carefully timed second flash of far-red light, has been used to investigate the behaviour of intermediates formed from Pr by an initial red flash.

LOW TEMPERATURE

At the temperature of liquid nitrogen (77 K, −196 °C), the photoconversion of Pr to Pfr and Pfr to Pr is not possible. Intermediates stable at this temperature are formed from Pr and Pfr by exposure to actinic light. Other intermediates, presumably formed from these initial photoproducts by dark relaxation reactions, are observed after irradiation at temperatures greater than 77 K as well as by warming the photoproducts produced at 77 K. Low-temperature techniques have been used both *in vivo* (Spruit, 1966a–d; Spruit and Kendrick, 1973) and *in vitro* (Cross *et al.*, 1968; Pratt and Butler, 1968; Kroes, 1970; Kendrick and Spruit, 1973c). The kinetics of intermediate reactions have been studied using a quasi-continuous measuring spectrophotometer (Spruit, 1971; Kendrick and Spruit, 1973a,b). So far, studies of phytochrome phototransformation at the temperature of liquid helium (4 K) have not been reported.

PIGMENT CYCLING

Briggs and Fork (1969a,b) demonstrated that under conditions of pigment cycling when Pr and Pfr are both excited (e.g. with mixed red and far-red light), intermediates between Pr and Pfr accumulate. A limitation, inherent in their technique, is that the actinic light used to maintain the intermediates has to be strictly separated from the beams measuring the absorption changes. To this end, in the instruments used by Briggs and co-workers (Briggs and Fork, 1969a,b;

Everett and Briggs, 1970; Gardner and Briggs, 1974), complementary wavelength regions were used for the actinic and measuring beams.
A cut-off filter in front of the photomultiplier prevented interference by the high-intensity actinic light. This procedure, however, precluded measurement in the red region of the spectrum and meant that only the minor absorption bands of phytochrome in the blue region of the spectrum could be studied. Using this technique, intermediates that decay in darkness to Pfr were demonstrated both *in vivo* and *in vitro*, although in the latter case readings were too small for detailed analysis.

A quasi-continuous measuring spectrophotometer (Spruit, 1971; Kendrick and Spruit, 1972a,b; 1973a,c) enabled intermediates that accumulate under conditions of pigment cycling to be measured at any wavelength in the visible range. Although this instrument measures intermediates that accumulate under cycling conditions the absorbance changes observed may be underestimated since the instrument measures them during a short dark interval starting about 0.2 ms after the end of each actinic flash. The level of intermediates measured is therefore an integration of absorbance values during this time interval and only reaction intermediates with lifetimes considerably in excess of 0.2 ms will be observed. However, this limitation is offset to a large extent by the possibility of measuring at temperatures down to 77 K.

DEHYDRATION

It has been known for some time that photoconversion of Pr to Pfr is not possible in dehydrated tissue (Tobin and Briggs, 1969; Kendrick, Spruit and Frankland, 1969; McArthur and Briggs, 1970; Grill and Spruit, 1972), in *in vitro* samples dried on to gelatin film (Tobin, Briggs and Brown, 1973) and in the presence of high glycerol concentrations (Balangé, 1974). Rehydration of such samples restores the photoreversibility. Recently, it has been demonstrated that in freeze-dried tissue, phytochrome intermediates are formed from Pr and Pfr even though complete phototransformation is restricted (Kendrick, 1974; Kendrick and Spruit, 1974). In many respects, the effect of dehydration on the phytochrome phototransformations appears to be similar to that of low temperature.

Terminology

We propose here a new terminology for phytochrome intermediates observed during phototransformation of both Pr and Pfr. Since the spectral characteristics of these intermediates are not completely known and also for the reasons given in the Introduction, any new terminology must be independent of the spectral data of the intermediates and allow for accommodation of additional intermediates if they are found in the future. The terminology proposed is an adaptation of that worked out for the phototransformation of the visual pigments. All pigment forms originating from Pr will be indicated as R and those originating from Pfr as F. *Table 4.1* compares the new terminology with that used in our previous publications (e.g. Kendrick and Spruit, 1973b,c).

Table 4.1   Comparison of new and old terminology

Pigment form	Old terminology	New terminology
Stable	Pr Pfr	Pr Pfr
First photoproducts	P698 P650	lumi-R lumi-F
Dark relaxation products	P710, Pbl P690, Px	meta-Ra, meta-Rb meta-Fa, meta-Fb

Photoconversion of Pr

Pr TO lumi-R

The first photoproduct to be observed upon irradiation of Pr will be called lumi-R. It has been detected by flash photolysis (Linschitz *et al.*, 1966) and low-temperature studies (Cross *et al.*, 1968; Pratt and Butler, 1968; Kroes, 1970) *in vitro*, and low-temperature studies (Spruit, 1966a–c; Spruit and Kendrick, 1973; Kendrick and Spruit, 1973b) and dehydration (Kendrick, 1974; Kendrick and Spruit, 1974) *in vivo*. Its presence can be observed in difference spectra at 77 K as a peak at about 698 nm. Since there should be considerable negative absorption at the short-wavelength side, the position of the maximum in the absorption spectrum of lumi-R could be lower than 698 nm. Difference spectra for the reaction of Pr to lumi-R at two temperatures are shown in *Figure 4.1*.

At 77 K, lumi-R is thermostable, but can be photoconverted back into Pr. A photoequilibrium therefore exists at low temperature between Pr and lumi-R. At 203 K, *in vivo* lumi-R formed from Pr by red light is not only in photoequilibrium with Pr, but also undergoes dark reversion to Pr (Kendrick and Spruit, 1973b). This reaction is also seen in freeze-dried tissue at 0 °C (Kendrick, 1974; Kendrick and Spruit, 1974) and *in vitro* at 203 K (Cross *et al.*, 1968).

$$
\begin{array}{c}
d \\
\downarrow \quad hv \\
\text{Pr} \;\underset{hv}{\overset{}{\rightleftharpoons}}\; \text{lumi-R}
\end{array}
$$

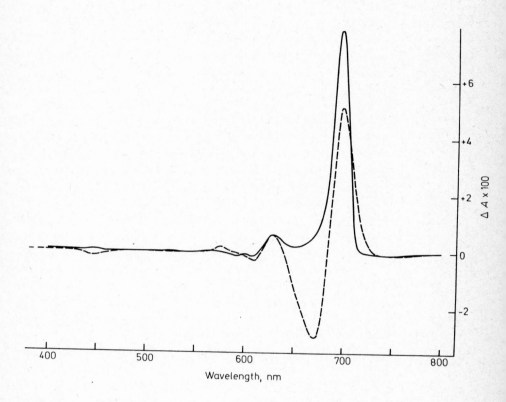

Figure 4.1 Difference spectra for the phototransformation of Pr to lumi-R in pea hook tissue: solid line, at 77 K; broken line, at 203 K

RELAXATION REACTIONS OF lumi-R

At temperatures above 203 K and in tissue sufficiently hydrated at 0 °C, lumi-R undergoes a series of dark relaxation reactions to form Pfr. Several intermediates have been identified by their absorption bands in difference spectra and are called meta-Rn. It must again be emphasised that the absorbance maximum of the same intermediate may vary *in vivo* and *in vitro* as well as with temperature. Flash photolysis (Linschitz *et al.*, 1966) and photoconversion of Pr at 228 K *in vitro* (Cross *et al.*, 1968) and 254 K *in vivo* (Spruit and Kendrick, 1973) suggest that lumi-R transforms into meta-Ra with an absorption

maximum of about 710 nm.   *Figure 4.2* shows the difference spectrum
of the red/far-red reversible reaction that occurs with Pr *in vivo* at
254 K.   It demonstrates photoreversibility between Pr and meta-Ra:

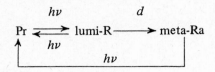

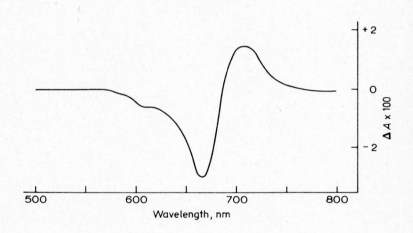

*Figure 4.2   Difference spectrum for the phototransformation of Pr to meta-Ra
in peak hook tissue at 254 K*

Both flash photolysis (Linschitz *et al.*, 1966) and low-temperature
studies (Cross *et al.*, 1968) *in vitro* as well as investigations under
conditions of pigment cycling *in vitro* and *in vivo* (Kendrick and Spruit,
1972a,b; 1973a–c) suggest that a relatively weakly absorbing intermediate
directly precedes Pfr.   This intermediate we will call meta-Rb.   *Figure
4.3* shows a difference spectrum for the reactions of intermediates
which form Pfr in darkness after the pigment *in vitro* has been cycled
under mixed red/far-red light.   Since the meta-Rb to Pfr reaction is
the slowest dark reaction in the Pr to Pfr and Pfr to Pr pathways, it
is this intermediate which accumulates in greatest concentration when
the pigment cycles and *Figure 4.3* is essentially a difference spectrum
for the reaction meta-Rb to Pfr.

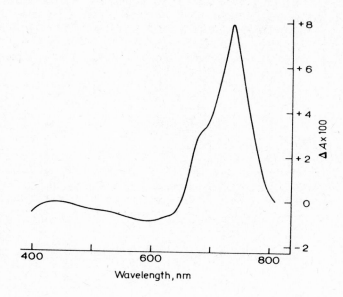

Figure 4.3   Difference spectrum for the dark relaxation reactions of the inter-mediates (mainly meta-Rb) accumulating during irradiation at 0 °C with high-intensity incandescent light.   Phytochrome in glycerol buffer (3:1)

Photoconversion of Pfr

Pfr TO lumi-F

At 77 K Pfr is photoconverted into lumi-F.   A maximum of about 650 nm in the difference spectrum points to an absorbance maximum of this form in the red region of the spectrum (Spruit, 1966a; Spruit and Kendrick, 1973).   Evidence for a similar photoproduct was found by using flash photolysis (Linschitz *et al.*, 1966).   It must be pointed out that when studying the Pfr to Pr pathway, results are difficult to interpret since the starting material is not 100 per cent Pfr, but a mixture of about 80 per cent Pfr and 20 per cent Pr, the photo-stationary equilibrium maintained by red light.   *Figure 4.4* shows the difference spectrum for the Pfr to lumi-F reaction at 77 K *in vivo*. This reaction is photoreversible (Kendrick and Spruit, 1973b) and a photostationary equilbrium is reached when Pfr is irradiated at this temperature. At 203 K, lumi-F undergoes a slow dark reversion to Pfr (Kendrick and Spruit, 1973b). This reaction can also be observed in freeze-dried tissue at 0 °C (Kendrick and Spruit, 1974). Freeze-dried tissue is a potentially useful material for studying the Pfr to Pr photoconversion since in this material lumi-F is a relatively stable photoproduct. Any Pr present forms lumi-R, which, however, quickly disappears at temperatures of 0 °C and higher by a rapid dark reversion to Pr (Kendrick, 1974; Kendrick and Spruit, 1974):

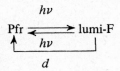

This means that one can effectively study the reactions of lumi-F in isolation using this system and it is at present being used to investigate the Pfr to Pr pathway (Spruit, Kendrick and Cooke, 1975).

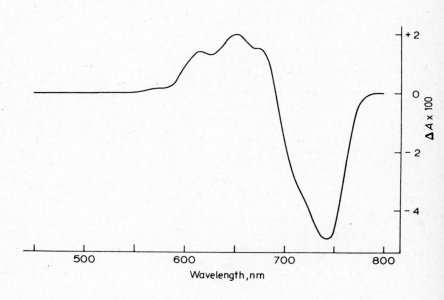

Figure 4.4   Difference spectrum for the phototransformation of Pfr to lumi-F in pea hook tissue at 77 K

RELAXATION REACTIONS OF lumi-F

Warming lumi-F formed at 77 K results in two reactions: firstly the dark reversion to Pfr (Kendrick and Spruit, 1973b) and secondly the formation of an intermediate, showing up in the difference spectra *in vivo* as a maximum at about 690 nm (Spruit and Kendrick, 1973; Kendrick and Spruit, 1973b). We will call this form meta-Fa. Freeze-dried tissue irradiated with far-red light at 0 °C shows that a mixture of lumi-F and meta-Fa is formed from Pfr. If, on the other hand,, Pfr is irradiated in freeze-dried tissue at 77 K, only lumi-F can be detected. The difference spectrum for warming of lumi-F formed at 77 K in freeze-dried tissue is shown in *Figure 4.5* and demonstrates the formation of meta-Fa. At 0 °C, meta-Fa and lumi-F appear to be in a dark equilibrium (Spruit, Kendrick and Cooke, 1975).
Kinetic studies at low temperature also indicate that the reaction sequence Pfr → lumi-F → meta-Fa is complex (Kendrick and Spruit, 1973b). These three pigment forms appear to be interconvertible by light as well as being connected by dark equilibrium reactions:

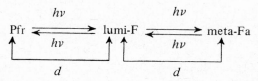

Studies are being carried out with freeze-dried tissue to characterise these reactions more clearly.

Warming of meta-Fa formed at low temperature results in the formation of Pr (Spruit and Kendrick, 1973). However, several experiments with *in vivo* material cooled to low temperatures under conditions of pigment cycling provide evidence for the accumulation of a relatively weakly absorbing intermediate meta-Fb, which forms Pr upon warming (Kendrick and Spruit, 1973b; Spruit and Kendrick, 1973). This intermediate appears to be a direct precursor of Pr and is presumably formed from meta-Fa:

$$\text{meta-Fb} \xrightarrow{d} \text{Pr}$$

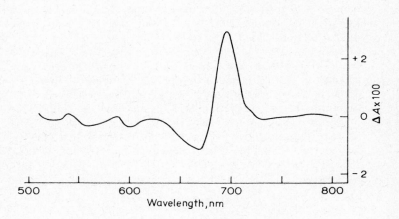

Figure 4.5   Difference spectrum for the dark relaxation of lumi-F to meta-Fa.
Freeze-dried pea epicotyl tissue

Discussion

The data available can be summarised in the following scheme for
phototransformation:

$$\text{Pr} \underset{h\nu}{\overset{h\nu}{\rightleftarrows}} \text{lumi-R} \xrightarrow{d} \text{meta-Ra} \xrightarrow{d} \text{meta-Rb} \xrightarrow{d} \text{Pfr}$$

$$\text{Pr} \xleftarrow{d} \text{meta-Fb} \xleftarrow{d} \text{meta-Fa} \underset{h\nu}{\overset{h\nu}{\rightleftarrows}} \text{lumi-F} \underset{h\nu}{\overset{h\nu}{\rightleftarrows}} \text{Pfr}$$

Similarities can be seen in both Pr to Pfr and Pfr to Pr photoconversions.   At physiological temperatures, these intermediates may only be
transients, except for those directly preceding the slowest dark reactions
in the cycle.   This has been demonstrated for the intermediate meta-
Rb, which accumulates under conditions of pigment cycling and forms
Pfr in darkness (Kendrick and Spruit, 1973c).

At low temperatures, Pr undergoes photoconversion to lumi-R, a reaction which is photoreversible. On warming, lumi-R undergoes dark reversion to Pr. These reactions are interpreted as being events restricted to the chromophore. At low temperatures, protein–chromophore interactions are restricted, thus preventing the formation of the relaxation intermediate meta-Ra. Similarly, the photoreversible reaction of Pfr to lumi-F is interpreted as being restricted to the chromophore and again, on warming, dark reversion of lumi-F to Pfr occurs preferentially, when protein–chromophore interactions are restricted. Dehydration appears to have an effect similar to low temperature and in the freeze-dried material events appear to be predominantly restricted to the chromophore.

The rate of the dark reaction meta-Rb to Pfr has been demonstrated to be strongly influenced by the molecular environment. The rate of this reaction is increased by reducing agents such as sodium dithionite *in vitro* (Kendrick and Spruit, 1973c) and anaerobic conditions *in vivo* (Kendrick and Spruit, 1973a). On the other hand, it is slowed *in vitro* by glycerol (Briggs and Fork, 1969b; Everett and Briggs, 1970; Kendrick and Spruit, 1973c); substitution of D_2O for H_2O (Kendrick and Spruit, 1973c) and *in vivo* by partial dehydration (Kendrick and Spruit, 1974).

Interestingly, the dark reversion of Pfr to Pr observed *in vivo* (Butler, Lane and Siegelman, 1963; Butler and Lane, 1965) and *in vitro* (Mumford, 1966; Mumford and Jenner, 1971; Pike and Briggs, 1972) is likewise strongly influenced by the molecular environment. The Pfr form of phytochrome can be regarded as metastable and under favourable conditions the requirement of far-red light to form Pr can be by-passed. It is therefore attractive to speculate that a dark equilibrium exists between Pfr and lumi-F, permitting the formation of Pr. A protonated form of Pfr (PfrH) formed at low pH and under reduced conditions (Mumford and Jenner, 1971) accumulates under conditions which favour dark reversion of Pfr to Pr. This product appears spectrally to be similar to lumi-F. If it is related to lumi-F or meta-Fa, this might form a possible mechanism by which Pr can be formed from Pfr in darkness. If this hypothesis can be confirmed, it will also give information about the chemistry of the Pfr to lumi-F reaction.

Since the difference in chromophore structure of Pr and Pfr has not yet been completely worked out, little can be said about the exact nature of the intermediate reactions. Theoretical analysis of C.D. spectra has enabled the three-dimensional structure of the chromophore to be predicted in the cases of Pr, Pfr and meta-Rb (Burke, Pratt and Moscowitz, 1972). Pr and Pfr both appear to conform to an extended form of the tetrapyrrole chromophore, in contrast to a tightly folded form for meta-Rb. Such evidence points to large changes in the three-dimensional structure of the chromophore during the intermediate relaxation reactions. However, investigation of the protein of Pr and Pfr suggests that little three-dimensional structural change occurs (Tobin and Briggs, 1973; Gardner, Thompson and Briggs, 1974). The photo-induced change between Pr and Pfr therefore appears to represent only minor changes in surface residues of the protein, presumably in the vicinity of the chromophore. Nevertheless, these changes could bring about the steric interaction with membranes that presumably initiates the wide display of phytochrome-controlled responses. The intermediates discussed here may be important in understanding the mechanism of phytochrome action, particularly under conditions of pigment cycling. Until now, only the physiological consequences of a proportion of total phytochrome being present as intermediates under continuous irradiation have been considered, and only further investigation can reveal if intermediates themselves are physiologically active.

References

BALANGÉ, A.P. (1974). *Physiol. Vég.,* **12,** 95
BRIGGS, W.R. and FORK, D.C. (1969a). *Plant Physiol.,* **44,** 1081
BRIGGS, W.R. and FORK, D.C. (1969b). *Plant Physiol.,* **44,** 1089
BURKE, M.J., PRATT, D.C. and MOSCOWITZ, A. (1972). *Biochemistry,* **11,** 4025
BUTLER, W.L., HENDRICKS, S.B. and SIEGELMAN, H.W. (1964). *Photochem. Photobiol.,* **3,** 521
BUTLER, W.L. and LANE, H.C. (1965). *Plant Physiol.,* **40,** 13
BUTLER, W.L., LANE, H.C. and SIEGELMAN, H.W. (1963). *Plant Physiol.,* **38,** 514
CROSS, D.R., LINSCHITZ, H., KASHE, V. and TENENBAUM, J. (1968). *Proc. Nat. Acad. Sci. U.S.A.,* **61,** 1095
EVERETT, M.S. and BRIGGS, W.R. (1970). *Plant Physiol.,* **48,** 679
GARDNER, G. and BRIGGS, W.R. (1974). *Photochem. Photobiol.,* **19,** 367
GARDNER, G., PIKE, C.S., RICE, H.V. and BRIGGS, W.R. (1971). *Plant Physiol.,* **48,** 686
GARDNER, G., THOMPSON, W.F. and BRIGGS, W.R. (1974). *Planta,* **117,** 367
GRILL, R. and SPRUIT, C.J.P. (1972). *Planta,* **108,** 203
KENDRICK, R.E. (1974). *Nature, Lond.,* **250,** 159
KENDRICK, R.E. and SPRUIT, C.J.P. (1972a). *Nature New Biol.,* **237,** 281
KENDRICK, R.E. and SPRUIT, C.J.P. (1972b). *Planta,* **107,** 341
KENDRICK, R.E. and SPRUIT, C.J.P. (1973a). *Photochem. Photobiol.,* **18,** 139
KENDRICK, R.E. and SPRUIT, C.J.P. (1973b). *Photochem. Photobiol.,* **18,** 153
KENDRICK, R.E. and SPRUIT, C.J.P. (1973c). *Plant Physiol.,* **52,** 327
KENDRICK, R.E. and SPRUIT, C.J.P. (1974). *Planta,* **120,** 265
KENDRICK, R.E., SPRUIT, C.J.P. and FRANKLAND, B. (1969). *Planta,* **88,** 293
KROES, H.H. (1970). *Meded. Landbouwhogesch. Wageningen,* **70,** 1

LINSCHITZ, H. and KASCHE, V. (1967). *Proc. Nat. Acad. Sci. U.S.A.,* **58**, 1059

LINSCHITZ, H., KASCHE, V., BUTLER, W.L. and SIEGELMAN, H.W. (1966). *J. Biol. Chem.,* **241**, 3395

McARTHUR, J.A. and BRIGGS, W.R. (1970). *Planta,* **91**, 146

MUMFORD, F.E. (1966). *Biochemistry,* **5**, 522

MUMFORD, F.E. and JENNER, E.L. (1971). *Biochemistry,* **10**, 98

PIKE, C.S. and BRIGGS, W.R. (1972). *Plant Physiol.,* **49**, 514

PRATT, L.H. and BUTLER, W.L. (1968). *Photochem. Photobiol.,* **8**, 477

SPRUIT, C.J.P. (1966a). In *Currents in Photosynthesis,* p.67. Ed. Thomas, J.B. and Goedheer, J.C. Donker, Rotterdam

SPRUIT, C.J.P. (1966b). *Meded. Landbouwhogesch. Wageningen,* **66**, 1

SPRUIT, C.J.P. (1966c). *Biochim. Biophys. Acta,* **120**, 186

SPRUIT, C.J.P. (1966d). *Biochim. Biophys. Acta,* **120**, 454

SPRUIT, C.J.P. (1971). *Meded. Landbouwhogesch. Wageningen,* **71**, 1

SPRUIT, C.J.P. and KENDRICK, R.E. (1973). *Photochem. Photobiol.,* **18**, 145

SPRUIT, C.J.P., KENDRICK, R.E. and COOKE, R.J. (1975). *Planta,* **127**, 121

TOBIN, E.M. and BRIGGS, W.R. (1969). *Plant Physiol.,* **44**, 148

TOBIN, E.M. and BRIGGS, W.R. (1973). *Photochem. Photobiol.,* **18**, 487

TOBIN, E.M., BRIGGS, W.R. and BROWN, P.K. (1973). *Photochem. Photobiol.,* **18**, 497

5

THE 'HIGH IRRADIANCE REACTION'

EBERHARD SCHÄFER

Institut für Biologie II, Universität Freiburg, D-78 Freiburg i. Br., Schänzlestrasse 9-11, West Germany

On the basis of different irradiation conditions, it is possible to distinguish two types of reaction in photomorphogenesis: induction-reversion reactions and high-irradiance reactions (cf. Mohr, 1972). The established operational criteria for the involvement of phytochrome in a light-mediated response require that an induction effected by a red-light pulse can be fully reversed by a subsequent pulse of far-red light. Implicitly, it is assumed that the law of reciprocity holds. Under these irradiation conditions ($t < 5$ min), the action spectrum for the induction of a light-mediated response shows a peak at 660 nm and reflects the absorption spectrum of Pr (*Figure 5.1a*). Under prolonged irradiation, the action spectra looks completely different when compared with the

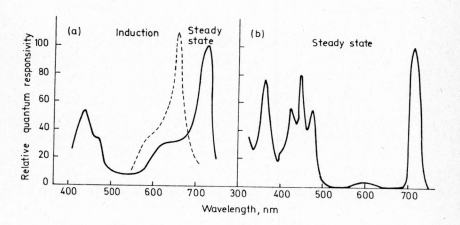

Figure 5.1   (a) Action spectra for control of anthocyanin synthesis in mustard seedlings (Sinapis alba L.) *under inductive and steady-state conditions (after Wagner and Mohr, 1966).   (b) Action spectrum for control of hypocotyl lengthening in lettuce seedlings* (Lactuca sativa, *cv. Grand Rapids) under steady-state conditions. (After Hartmann, 1967a)*

induction spectrum (Siegelmann and Hendricks, 1957; Mohr, 1957). The action spectra obtained always show a sharp peak in the far-red region of the spectrum (between 700 and 730 nm) and several smaller peaks in the blue region of the spectrum (*Figure 5.1a,b*).

In the past, attempts have been made to develop models of the phytochrome system to explain the action of phytochrome under induction and HIR conditions (Hartmann, 1966; Schäfer, 1971).

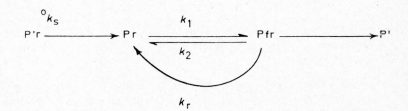

Scheme 5.1   Diagram and rate constants of an open phytochrome model

Scheme 5.1 shows a model which accounts for the major features of the phytochrome system as it occurs in mustard cotyledons in darkness and under continuous irradiation.   This model is based on direct photometric measurements of Pr and Pfr (and Ptot) in the cotyledons and hypocotylar hook of the mustard seedling (Schäfer, Marchal and Marmé, 1972; Schäfer, Schmidt and Mohr, 1973).   It is characterised by a zero-order *de novo* synthesis,

$$P'r \xrightarrow{^{o}k_{s}} Pr$$

(Schäfer, Marchal and Marmé, 1972; Quail, Schäfer and Marmé, 1973a,b), a first-order destruction,

$$Pfr \xrightarrow{k_{d}} P'fr$$

(Marmé, Marchal and Schäfer, 1971; Schäfer, Marchal and Marmé, 1972; Schäfer, Schmidt and Mohr, 1973), a first-order dark reversion,

$$Pfr \xrightarrow{k_{r}} Pr$$

(Marmé, Marchal and Schäfer, 1971) and a first-order light back-and-forth reaction,

$$Pr \underset{k_{2}}{\overset{k_{1}}{\rightleftharpoons}} Pfr$$

(Butler, Siegelmann and Miller, 1964; Schmidt *et al.*, 1973).

This model accounts for the action of phytochrome under induction conditions (Drumm and Mohr, 1974) and for the fact that the phytochrome system tries to approach a steady state under continuous irradiation. This is in agreement with the experimental data.

Irrespective of the phytochrome content at the beginning of the irradiation, a constant level of total phytochrome will be established (*Figure 5.2*). The model predicts that the steady-state level of Pr is dependent not only on wavelength but also on irradiance, whereas the level of Pfr remains constant under all irradiation conditions, indicating

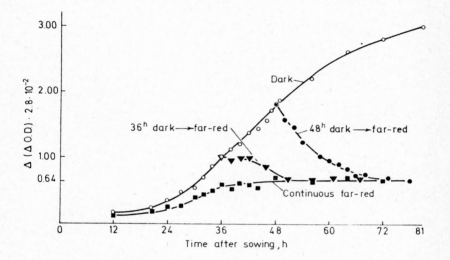

Figure 5.2 Time course of change of the total phytochrome in the mustard cotyledons for various times of onset of continuous far-red light. (After Schäfer, Marchal and Marmé, 1972)

that the 'photostationary' state (i.e. the ratio Pfr to Pr + Pfr) is irradiance dependent. A specific prediction from the model is that the term Ptot − 1 under steady-state conditions is proportional to the reciprocal of the irradiance:

$$\text{Ptot} - 1 = \frac{1}{N_\lambda} \cdot \frac{k_r + k_d}{k_1 + k_2}$$

(equation 8 in Schäfer and Mohr, 1974), where $k_1 = N_\lambda \, \epsilon_{r\lambda} \, \phi_{r\lambda}$, $k_2 = N_\lambda \, \epsilon_{fr\lambda} \, \phi_{fr\lambda}$, ϵ_r and ϵ_{fr} are the extinction coefficients of Pr and Pfr at the wavelength λ, and ϕ_r and ϕ_{fr} are quantum yields of Pr, Pfr phototransformations at the wavelength λ. This prediction was verified by direct photometric measurements using three different dark-light programmes (*Figure 5.3*). This indicates that the phytochrome system has to be an open system showing aequifinality (von Bertalanffy, 1968).

This simple model (*Scheme 5.1*) is completely satisfactory as long as the law of reciprocity is valid. Under these inductive conditions the photomorphogenic responses are correlated to the Pfr established by the light pulse and Pfr can be regarded as the effector/effector element of the phytochrome system (Oelze-Karow, Schopfer and Mohr, 1970; Drumm and Mohr, 1974). Beyond that, the model indeed yields an irradiance dependence of Ptot and the ratio Pfr/Ptot, whereas Pfr has to be independent of irradiance, and therefore Pfr (as defined in *Scheme 5.1*) cannot be the effector of the HIR.

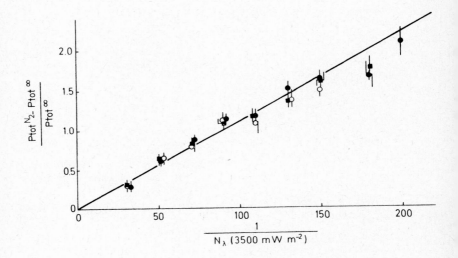

Figure 5.3   Irradiance dependence of the steady-state level of total phytochrome (Ptot) in mustard cotyledons under continuous far-red light.   o, 48 h far-red; ■ , 36 h dark + 24 h far-red; ● , 48 h dark + 24 h far-red.   (After Schäfer and Mohr, 1974)

The most precise action spectrum for an HIR was published by Hartmann (1967a) for the response 'inhibition of hypocotyl lengthening' in lettuce seedings (*Figure 5.1b*). Hartmann (1966, 1967b) has obtained convincing experimental evidence, using dichromatic irradiation, that the far-red peak of action (*Figure 5.1*) is due in some way to the phytochrome system (*Figure 5.4*). The same technique (simultaneous irradiation with two wavelengths) has been used by other investigators to show that photomorphogenic responses obtained under continuous far-red light are mediated in some way by phytochrome (Blondon and Jacques, 1970; Hachtel, 1972).

However, in spite of some effort (Hartmann, 1966), the manner in which phytochrome is involved in the HIR remains unclear. The major

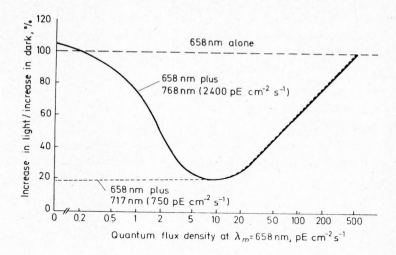

Figure 5.4 Hypocotyl lengthening in lettuce seedlings under continuous simultaneous irradiation with different quantum flux densities in the red (658 nm) at a constant high quantum flux density of different far-red bands (717, 768 nm). (After Hartmann, 1967b)

problem is that any explanation of the action spectrum of the HIR in terms of phytochrome seems to require the assumption that the optimum effectiveness of phytochrome under continuous irradiation is already reached at low Pfr/Ptot ratios, that the photoequilibrium for the response maxima is not always the same and that the effectiveness decreases with increasing Pfr/Ptot ratios (cf. Hartmann, 1966). Such an optimum effectiveness of low Pfr/Ptot ratios cannot be deduced from the phytochrome model presented in *Scheme 5.1*.

Recently, Quail, Marmé and Schäfer (1973) and Quail and Schäfer (1974a,b) have advanced a phytochrome model which ignores *de novo* synthesis of Pfr but includes an interaction of Pfr and Pr with a receptor X or X' (*Scheme 5.2*).

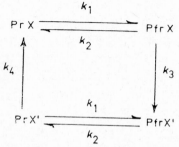

Scheme 5.2 Diagram and rate constants of a closed phytochrome receptor model

This model was developed in order to account for the binding properties to an operationally defined 'membrane' fraction in corn coleoptiles. The major finding was that Pr in the dark does not bind to the membrane, irradiation with red light (forming Pfr) enhances the binding and subsequent irradiation with far-red light does not reduce this level to that observed prior to red-light irradiation. These data were taken to indicate that the receptor (or receptor site) X, to which phytochrome binds, exists in two forms (X and X') and that the transition will be mediated by Pfr.

The rate constant for this transition (k_3) was determined by measuring the irradiance dependence of the binding and was found to be about ln 2 (Schäfer, Lassig and Schopfer, 1975). The enhanced level of Pr binding decays exponentially in the dark, re-approaching the control level with a half-life of 50 min [$k_4 = (\ln 2)/50$] in maize. This cycle is repeatable (Quail, Marmé and Schäfer, 1973). A similar reaction chain was also obtained for a squash phytochrome receptor system (Quail, Marmé and Schäfer, 1973; Boisard, Marmé and Briggs, 1974).

Recently Quail (1975) has established that Pfr binds *in vitro* to degraded ribonucleoprotein (RNP) material and that subsequent irradiation with far-red light does not reduce the binding level to that observed prior to red-light irradiation. These data were taken to indicate that all measurements analysing the binding of phytochrome to pelletable fractions do not represent the *in vivo* situation (Quail, Marmé and Schäfer, 1973; Marmé, Boisard and Briggs, 1973; Quail and Schäfer, 1974a,b; Marmé *et al.*, 1974; Boisard, Marmé and Briggs, 1974; Schäfer, 1974,1975; Lehmann and Schäfer, 1975).

In contrast, I wish to draw attention to the fact that the closed phytochrome receptor model is based on measurements of the binding of phytochrome to a 'membrane' fraction after different light–dark programmes given *in vivo*. The enhanced Pr binding cannot be explained by assuming an unspecific binding of Pr *in vitro* to degraded RNP material obtained after the homogenisation of the tissue. Different amounts of pelletable phytochrome were measured, although in dark-grown seedlings and seedlings that were irradiated with red followed by far-red light Pr was always present during the extraction procedure. This result can only be explained by assuming a Pfr-induced induction *in vivo*.

A mathematical treatment of *Scheme 5.2* under steady-state conditions yields the following results. For a given wavelength (k_1, k_2), the amounts of PrX and PfrX' decrease with increasing irradiance, whereas PfrX and PrX' increase with increasing irradiance up to the saturation level. This level is a function of the wavelength and the maximum is obtained for $k_1/k_2 = \sqrt{k_4/k_3}$. This corresponds approximately to a wavelength of 710 nm (Butler, Siegelmann and Miller, 1964). The system element PfrX and the flux $k_3 \cdot$PfrX can be related to the HIR, because both show an irradiance dependence and an optimum in the far-red region in reasonably good agreement with the observed position of the far-red peak of action in several 'high irradiance responses.'

Further optima of these parameters will be obtained whenever k_2 becomes larger than k_1, which occurs at least twice in the blue region of the visible spectrum, as indicated by the absorption spectra of Pr and Pfr (Butler, Siegelmann and Miller, 1964).

This closed (cyclic) photochrome receptor model (*Scheme 5.2*) shows in principle the most important features of the HIR (the wavelength and irradiance dependence) but it predicts that the total amount of phytochrome (Ptot) is independent of wavelength and irradiance, which is, however, not the case (Schäfer and Mohr, 1974). Obviously, one has to consider *de novo* synthesis of Pr and destruction of Pfr, which will transform the closed system of *Scheme 5.2* into an open system.

Kinetic studies of Pfr destruction in squash cotyledons indicated that the destruction process was always preceded by a fast transition between two unknown Pfr states (Schäfer and Schmidt, 1974; Schmidt and Schäfer, 1974). The Arrhenius activation energy of the destruction reaction is independent of temperature, whereas that of the fast transition between the two unknown Pfr states shows a jump at 20 °C. This effect indicates that the destruction reaction is always preceded by a binding of Pfr to a membrane (Schäfer and Schmidt, 1974). For maize coleoptile, oat coleoptile and mesocotyl tissue and squash cotyledons, a correlation between binding and rate of destruction was demonstrated with respect to wavelength and irradiance dependence (Schäfer, Lassig and Schopfer, 1975,1976). It seems that the system 'counts' the total number of Pfr molecules it has seen during irradiation, so that an irradiation with low intensity at high photostationary state can be substituted by an irradiation with high intensity at low photostationary state (Quail and Schäfer, 1974a,b; Schäfer, Lassig and Schopfer, 1975, 1976). These data suggest that the destruction of Pfr always starts from a 'relaxed' Pfr receptor state, PfrX'.

On the other hand, it is well known that only Pr is synthesised in the dark as well as in the light (Quail, Schäfer and Marmé, 1973a,b). Because no significant binding of Pr is detectable in dark-grown seedlings (Quail, Marmé and Schäfer, 1973), it is concluded that Pr will be synthesised in an unbound form. For further calculations, it is assumed that the *de novo* synthesis is a zero-order reaction and that the rate constant is independent of the pool size of Ptot. This holds true at least for cases where the amount of Ptot is decreased by continuous irradiation with far-red light (Schäfer, Marchal and Marmé, 1972) or short irradiation with red light (Quail, Schäfer and Marmé, 1973a,b). The rate constants for the phototransformation were chosen so as to be the same in the bound and unbound phytochrome. It was shown by Quail (1974a,b) that this assumption indeed holds true in the red/far-red region of the spectrum.

In the following discussion, dark reversion will be disregarded because it seems to be negligible under continuous irradiation of appreciable irradiance. Based on these premises, *Scheme 5.2* can be modified in the form shown in *Scheme 5.3*.

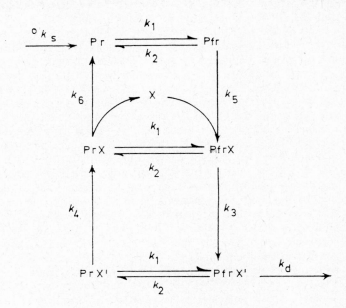

Scheme 5.3   Diagram and rate constants of an open phytochrome–receptor model

This model was tested by analysing the amount of pelletable phyto-chrome from maize coleoptiles as a function of the delay time between short pulses of red and far-red light given *in vivo* (*Figure 5.5*). The time course of the kinetics can be split into three phases which were interpreted in accordance with *Scheme 5.3*: (1) the relaxation of the slowest phytochrome intermediate (*ca.* 4 s); (2) the binding reaction, Pfr + X $\xrightarrow{k5}$ PfrX (*ca.* 40 s); and (3) the transition of the receptor site, PfrX $\rightarrow$ PfrX$'$ (*ca.* 2 min) (Lehmann and Schäfer, 1975).

Since the amount of the receptor sites is not limiting (Schäfer, 1974, 1975a), the binding reactions are fast when compared with the other reactions (Lehmann and Schäfer, 1975) and the cooperativity of the binding (Quail and Schäfer, 1974a,b; Schäfer, 1974, 1975a) does not come into play under steady-state conditions where the total Pfr con-centration is very low, the model can be reduced to the form shown in *Scheme 5.4* under steady-state conditions.

For both models (*Schemes 5.3* and *5.4*), which are open systems, the stationary solution of PfrX$'$ is independent of irradiance and wave-length (influx = efflux; $^{o}k_s = k_d \cdot$ PfrX$'$). Therefore, PfrX$'$ cannot be the effector of the HIR. On the other hand, PfrX and PrX$'$ show a

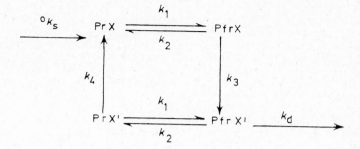

Scheme 5.4 *Diagram and rate constants of an open phytochrome–receptor model under steady-state conditions*

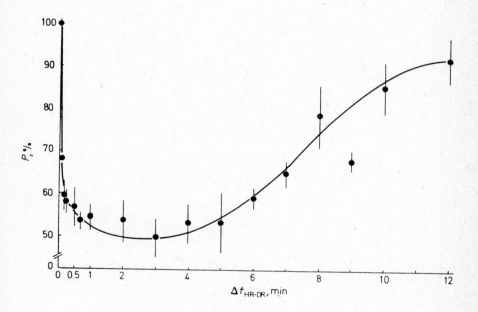

Figure 5.5 *Phytochrome pelletability as a function of the time delay between the red-light induction and the reversion with far-red light. The red-light induction level was normalised to 100 per cent. (After Lehmann and Schäfer, 1975)*

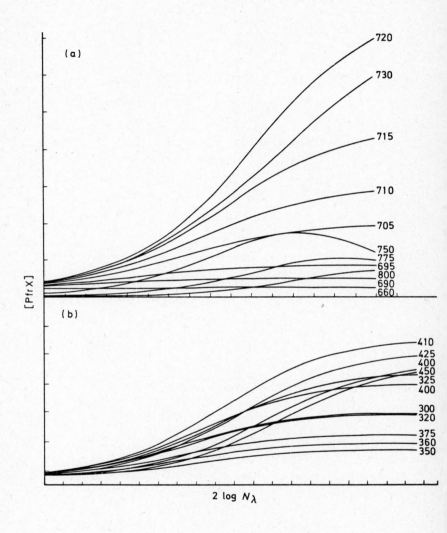

Figure 5.6 Semi-logarithmic plot of some calculated irradiance effect curves: (a) in the red/far-red range; (b) in the blue range of the visible spectrum. (After Schäfer, 1975b)

similar wavelength and irradiance dependence as described for *Scheme 5.2,* but one has to take into account that for wavelengths longer than 730 nm and very low irradiances, no steady-state conditions can be established because of the limiting rate of *de novo* synthesis (Schäfer, 1975b). *Figure 5.6a,b* shows some irradiance effect curves using values for k_1 and k_2 from a representative *in vitro* absorption spectrum (Butler, Siegelmann and Miller, 1964) and $k_3 = \ln 2$ and $k_4 = (\ln 2)/50$. The calculated dose effect curves show that the Weber–Fechners law (proportionality between response and the logarithm of the quantum flux density) is valid over a wide range of quantum flux densities, although the dose effect curves are not parallel. The value for the maximum slope of the irradiance effect curves is theoretically independent of k_3 but strongly dependent on k_4 and on the absorption spectra of Pr and Pfr and will, therefore, differ from plant to plant.

An action spectrum for the amount of PfrX or for the flux under steady-state conditions can be obtained from *Figure 5.6a,b* by plotting $N_{720}/N\lambda$ for constant PfrX as a function of Pfr/Ptot. A broad optimum curve with a half bandwidth of two orders of magnitude is obtained. This is typical for most of the optimum curves reported for physiological responses (Blaauw and Blaauw-Jansen, 1970; Hild and Hertel, 1972; Hild, 1974). A plot of the same data *versus* wavelength rather than ligand concentration results in an action spectrum which is typical for all 'high irradiance responses'. (*Figure 5.7*). Whereas the far-red peak

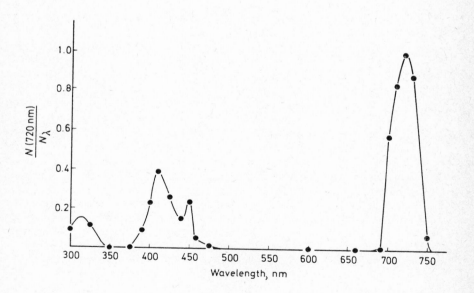

Figure 5.7   Action spectrum for the amount of PfrX, calculated on the basis of the data in Figure 5.6a,b. The relative quantum responsivity is plotted as a function of wavelength. (After Schäfer, 1975b)

of the theoretical action spectrum fits most of the known HIR action spectra very well (Wagner and Mohr, 1966; Hartmann, 1967a), the peaks in the blue region of the spectrum show significant deviations from the empirical values. This is probably caused by at least two factors: firstly, the *in vitro* absorption spectra of Pr and Pfr as used in the present argument (Butler, Siegelmann and Miller, 1964) differ considerably from the *in vivo* spectra in the blue region of the spectrum, whereas the red and far-red region of the spectrum is not affected significantly by extraction of the pigment (Everett and Briggs, 1970); secondly, the measured action spectra in the blue region are probably affected by the simultaneous action of a blue photoreceptor in addition to phytochrome (Mohr, 1972).

It should be mentioned that the open phytochrome 'receptor' model can also be predicted purely on the basis of a theoretical analysis of the action spectra obtained under HIR conditions. To obtain a phytochrome responsivity curve with an optimum for low Pfr concentrations and an irradiance dependence, one has to predict a cyclic phytochrome system including light reactions and competing slow–dark reactions.

Irrespective of these restrictions, the conclusion seems to be justified that the open phytochrome–receptor model (*Schemes 5.3* and *5.4*), which is based on *in vivo* spectroscopic measurements, as well as on the analysis of phytochrome–receptor interactions after irradiations performed *in vivo,* allows one to explain the regulation under induction and HIR conditions. One has to predict a regulation by two distinct effector elements. The PfrX pool, which is a transient pool and is irradiance and wavelength dependent, is the effector element under HIR conditions. Under induction conditions, the regulation will be from the PfrX$'$ pool. In *Schemes 5.3* and *5.4,* the pool size of PfrX and the flux of the system after short light pulses are not affected by the destruction and therefore the PfrX pool cannot be the effector under induction conditions.

This interpretation indicates that under HIR conditions one has to expect a simultaneous action of the PfrX and PfrX$'$ pool. At very low irradiances, the induction response should dominate and therefore red light should be more effective than far-red or blue light. With increasing irradiances, far-red and blue light should become more and more effective (by action of PfrX), whereas the effectiveness of red light should remain constant. This principal type of irradiance effect curve should be obtained wherever induction and HIR are acting simultaneously (*Figure 5.1a*). This results in a crossing of the irradiance effect curves as reported by Wagner and Mohr (1966) for the phytochrome-mediated control of the anthocyanin synthesis in mustard cotyledons (*Figure 5.8*).

Whenever a threshold reaction is discussed on the basis of *Schemes 5.3* and *5.4,* one has to predict that this response shows no HIR reaction and no destruction below the threshold. This behaviour is the conclusion of the assumption that the

$$\text{PfrX} \xrightarrow{k_3} \text{PfrX}'$$

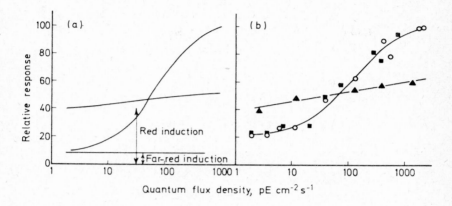

Figure 5.8   Semi-logarithmic plot of some irradiance effect curves: (a) theoretical calculations (after Schäfer, 1975b); (b) data derived from Wagner and Mohr (1966). ▲, Irradiation with red light; ○, irradiation with far-red light; ■, irradiation with blue light

transition shows a threshold phenomenon.   Such a reaction type was reported in detail for the phytochrome-mediated control of lipoxygenase synthesis in the mustard seedling (Oelze-Karow, Schopfer and Mohr, 1970).   Some more consequences based on this assumption are reported in more detail by Schäfer (1975b) and Mohr and Oelze-Karow (Chapter 17).

A consequence of this concept is that one cannot expect 'normal' HIR action spectra in monocotyledon systems.   In these tissues, the destruction is irradiance and wavelength dependent and saturated at very low Pfr concentrations (Schäfer, Lassig and Schopfer, 1975). Therefore, the far-red peak at 720 nm should be shifted towards longer wavelengths, which induce Pfr concentration levels below the saturation level for the destruction.

A further consequence of this model is that phytochrome action always starts from a relaxed protein form, because the action is always preceded by a binding reaction, which is much slower than the relaxations from the photochemical intermediates.   This is in contrast to the action of rhodopsin and chlorophyll, which are constituents of the receptor membranes.

Acknowledgement

This work was supported by the Deutsche Forschungsgemeinschaft (SFB 46).

References

BERTALANFFY, L. von.   (1968). *General System Theory*, Braziller, New York
BLAAUW, O.H. and BLAAUW-JANSEN, G. (1970). *Acta Bot. Neerl.*, **19**, 764
BLONDON, F. and JACQUES, R. (1970). *C.R. Acad. Sci., Paris*, **270**, 947
BOISARD, J., MARMÉ, D. and BRIGGS, W.R. (1974). *Plant Physiol.*, **54**, 272
BUTLER, W.L., SIEGELMANN, H.W. and MILLER, C.O. (1964). *Biochemistry*, **3**, 851
DRUMM, H. and MOHR, H. (1974). *Photochem. Photobiol.*, **20**, 151
EVERETT, M.S. and BRIGGS, W.R. (1970). *Plant Physiol.*, **45**, 679
HACHTEL, W. (1972). *Planta*, **102**, 247
HADELER, K.P. (1974). *Mathematik für Biologen*. Springer, Berlin, Heidelberg, New York
HARTMANN, K.M. (1966). *Photochem. Photobiol.*, **5**, 349
HARTMANN, K.M. (1967a). *Z. Naturforsch.*, **22b**, 1172
HARTMANN, K.M. (1967b). In *Book of Abstracts, European Photobiology Symposium, Hvar (Yugoslavia)*, 29
HILD, V. and HERTEL, R. (1972). *Planta*, **108**, 245
HILD, V. (1974). *Dissertation*, University of Freiburg
LEHMANN, U. and SCHÄFER, E. (1975). In preparation
MARMÉ, D., MARCHAL, B. and SCHÄFER, E. (1971). *Planta*, **100**, 331
MARMÉ, D., BOISARD, J. and BRIGGS, W.R. (1973). *Proc. Nat. Acad. Sci. U.S.A.*, **70**, 3861
MARMÉ, D., MACKENZIE, J.M. Jr., BOISARD, J. and BRIGGS, W.R. (1974). *Plant Physiol.*, **54**, 263
MOHR, H. (1957). *Planta*, **49**, 389
MOHR, H. (1972). *Lectures on Photomorphogenesis*, Springer, Berlin, Heidelberg, New York
MUMFORD, F.E. and JENNER, E.L. (1966). *Biochemistry*, **5**, 3657
OELZE-KAROW, H., SCHOPFER, P. and MOHR, H. (1970). *Proc. Nat. Acad. Sci. U.S.A.*, **65**, 51
OELZE-KAROW, H. and MOHR, H. (1973). *Photochem. Photobiol.*, **18**, 319
QUAIL, P.H. (1975). *Planta*, **123**, 223
QUAIL, P.H., MARMÉ, D. and SCHÄFER, E. (1973). *Nature New Biol.*, **245**, 189
QUAIL, P.H., SCHÄFER, E. and MARMÉ, D. (1973a). *Plant Physiol.*, **52**, 124
QUAIL, P.H., SCHÄFER, E. and MARMÉ, D. (1973b). *Plant Physiol.*, **52**, 128
QUAIL, P.H. and SCHÄFER, E. (1974a). *J. Membr. Biol.*, **15**, 393
QUAIL, P.H. and SCHÄFER, E. (1974b). In *Mechanisms of Regulation of Plant Growth*. Ed. Bieleski, R.L., Ferguson, A.C. and Cresswell, M.M. *R. Soc. N.Z. Bull.*, **12**, 351
QUAIL, P.H. (1974a). *Planta*, **118**, 345
QUAIL, P.H. (1974b). *Planta*, **118**, 357
SCHÄFER, E. (1971). *Dissertation*, University of Freiburg

SCHÄFER, E., MARCHAL, B. and MARMÉ, D. (1972). *Photochem. Photobiol.,* **15,** 457

SCHÄFER, E., SCHMIDT, W. and MOHR, H. (1973). *Photochem. Photobiol.,* **18,** 331

SCHÄFER, E. and SCHMIDT, W. (1974). *Planta,* **116,** 257

SCHÄFER, E. and MOHR, H. (1974). *J. Math. Biol.,* **1,** 9

SCHÄFER, E. (1974). In *Membrane Transport in Plants.* Ed. Zimmermann, U. and Dainty, J. Springer, Berlin, Heidelberg, New York

SCHÄFER, E. (1975a). *Photochem. Photobiol.,* **21,** 189

SCHÄFER, E. (1975b). *J. Math. Biol.,* **2,** 41

SCHÄFER, E., LASSIG, U. and SCHOPFER, P. (1975). *Photochem. Photobiol.,* **22,** 193

SCHÄFER, E., LASSIG, U. and SCHOPFER, P. (1976). *Photochem. Photobiol.,* submitted for publication

SCHMIDT, W., MARMÉ, D., QUAIL, P. and SCHÄFER, E. (1973). *Planta,* **111,** 329

SCHMIDT, W. and SCHÄFER, E. (1974). *Planta,* **116,** 267

SIEGELMANN, H.W. and HENDRICKS, S.B. (1957). *Plant Physiol.,* **32,** 393

WAGNER, E. and MOHR, H. (1966). *Photochem. Photobiol.,* **5,** 397

II

THE SITE OF PHYTOCHROME ACTION

6

PHYSIOLOGICAL EVIDENCE AND SOME THOUGHTS ON LOCALISED RESPONSES, INTRACELLULAR LOCALISATION AND ACTION OF PHYTOCHROME

W. HAUPT
M.H. WEISENSEEL
Botanisches Institut, Universität Erlangen-Nürnberg, D-852 Erlangen, West Germany

Introduction

There are several ways to demonstrate phytochrome being bound to membranes:

(1) In cell extracts, part of the phytochrome is found in a particulate fraction which probably is principally plasma lemma material. Rubinstein, Drury and Park (1969) seem to have been the first to demonstrate this fact. Subsequently, the laboratories at Freiburg and Harvard thoroughly pursued investigations along this line. Marmé, Quail and co-workers, in their contributions to this book, deal with the most important results of these investigations (*see* Chapters 8 and 9).

(2) Cytochemical investigations on whole cells seem to be capable of demonstrating distinct localisations of phytochrome. Up to now, the most important results have been obtained from immuno-chemical studies. These are dealt with in the paper by Pratt, Coleman and Mackenzie (*see* Chapter 7).

(3) Several unique physiological responses can be directly related to membrane-bound phytochrome. In fact, the first evidence that active phytochrome is probably bound to the plasma membrane came from physiological experiments. Those experiments, their interpretation and some new thoughts about the possible mode of action of phytochrome are described in this chapter. Firstly, let us consider a few questions on the subject:

  (a) What are the specific physiological experiments from which a distinct intracellular localisation of phytochrome can be inferred?

63

(b) Is there evidence for an effect of phytochrome on the localisation site?

(c) Does phytochrome show conformational changes when bound
to the plasma membrane?

(d) Are there physiological effects pointing towards unbound
phytochrome *in vivo*?

(e) What might be the primary effect of phytochrome?

To answer these questions by experiments, we can hardly use cells
or tissues which respond to a phytochrome photoconversion as a whole.
Instead, we need systems and responses where spatial differences in the
state of phytochrome within a single cell result in a localised response.
So far we know only two types of systems and responses which meet
these requirements: phototropism in the filamentous gametophyte of
ferns and mosses (the so-called protonema), and the orientation movement of the chloroplast in the green alga *Mougeotia*.   Most of the
information referred to in this chapter comes from the latter.

Phytochrome is Bound to the Plasma Membrane and it Affects the State of the Membrane

The cylindrical cell of *Mougeotia* is unique in its structure, having a
single flat chloroplast which divides the cell into two half-cylinders
(cf. *Figure 6.1*) and which is able to turn in response to the direction
of light.   When unilateral red light is applied to *Mougeotia,* the chloroplast turns so as to show its face to the light.   This response can occur
in complete darkness after a short red induction, and this induction is
far-red reversible.   Thus, we are dealing with a phytochrome effect
(Haupt, 1959).   By irradiating small areas of a cell with micro-beams of
red light, it was found that in order to be orientated by the light the
chloroplast itself does not need to be hit by the light (Haupt, 1970,
1972).   Thus, phytochrome is localised outside the chloroplast.

Still more important was the discovery of an action dichroism (*Figure
6.1*).   In linearly polarised light, the response strongly depends on the
orientation of the electric vector.   The chloroplast movement is induced
if the electric vector and the long axis of the cell are perpendicular to
each other.   By combination of the micro-beam method with polarisation, we arrived at the conclusion that the phytochrome molecules, Pr,
are dichroic and oriented in a definite pattern; their absorption vectors
(or transition moments) are oriented parallel to the cell surface and
parallel to an invisible screw structure of the cytoplasm (*Figures 6.1*
and *6.4a*)

Such a strongly defined pattern requires a structural base.   By far the
most probable site is the plasma membrane (plasmalemma).   Chloroplast
orientation as the physiological response, then, is induced and controlled
by an intracellular gradient of Pfr in such a way that the edges of the
chloroplast are 'pushed away' by high Pfr levels (*see* Haupt, 1972, for a
review and references to earlier papers).

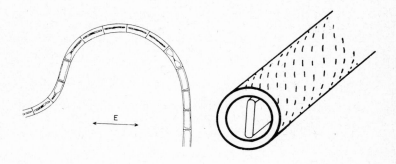

*Figure 6.1   Action dichroism (left) and inferred dichroic orientation of phyto-
chrome Pr in the* Mougeotia *cell (right).   The filament, originally with the chloroplasts in profile position, has been irradiated with polarised red light; the double
arrow shows the plane of the electric vector;   the dashes represent the direction
of main absorption.   (After Haupt, 1972)*

A similar conclusion about the localisation of phytochrome has been
drawn from experiments with polarised red and far-red light in the fern
protonema cell (Etzold, 1965).   Here the response to Pfr is the determination of the growth site at the cell tip.   A change in the direction
of the polarised light results in a new direction of growth (*Figure 6.2*).
    Knowing that phytochrome is most likely associated with the plasma-
lemma does not necessarily mean that this fact is important for the
response.   We therefore were looking for a membrane effect of phytochrome in *Mougeotia*.   It is true that some electrical phytochrome
effects have already been reported in roots and coleoptiles of higher
plants (Tanada, 1968; Jaffe, 1968; Racusen and Miller, 1972; Newman
and Briggs, 1972).   No such response, however, is known so far for
Mougeotia,   mainly for two reasons which became apparent during our
research.   Firstly, the micro-electrodes used for intracellular recording
do not easily penetrate the plasmalemma, or the injury caused by the
micro-electrode is too serious (Weisenseel and Dorn, work in progress).
Meanwhile, Weisenseel and Smeibidl (1973) looked for other effects and
discovered that phytochrome controls plasmolysis in *Mougeotia*.   The
most striking facts of these experiments are that Pfr greatly reduces the
time for the start of plasmolysis and deplasmolysis and Pfr increases the
degree of protoplast contraction (*Figure 6.3*).   These results are attributed mainly to a change in the permeability of the membrane to water.
Since this effect becomes visible a few seconds after irradiation, a direct
phytochrome effect on the membrane is suggested.   This would fit very
well the physiological importance of membrane-bound phytochrome.
However, at the moment we do not see a causal relationship between

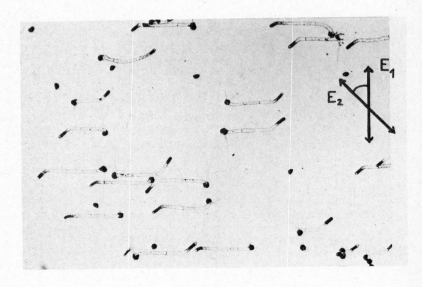

Figure 6.2   *The polarotropic curvature of fern protonema.   The protonema were grown in polarised red light for 9 days (E1), then the direction of the polarised light was rotated by 50° for 8 h (E2).   After an additional 16 h dark period, the protonema   were photographed.   (After Etzold, 1965)*

the permeability of the plasmalemma to water and the movement of the chloroplast. We shall return to this point later.

Phytochrome Changes its Conformation upon Photoconversion

In the fern protonema, Etzold (1965) suggested that only the Pr form is orientated parallel to the cell surface, but that Pfr is orientated normal to it (*Figure 6.4b*). This assumption turned out to be very stimulating for experiments with *Mougeotia* and has since been proved conclusively (Haupt, 1970, 1972). Only one experiment will be given here to demonstrate this conclusion (*Figure 6.5*). As stated above, the chloroplast edge always moves away from that region where the highest Pfr concentration is produced. We can consequently induce a movement by localised phototransformation of Pr to Pfr near the edge. This is achieved with a red micro-beam in cells which otherwise have a low level of Pfr. At the beginning of the experiment, the chloroplast is in the face position; the red micro-beam is placed to the flank. If, in addition, this micro-beam is polarised, the result should depend on the orientation of the electric vector. Indeed, this action dichroism is

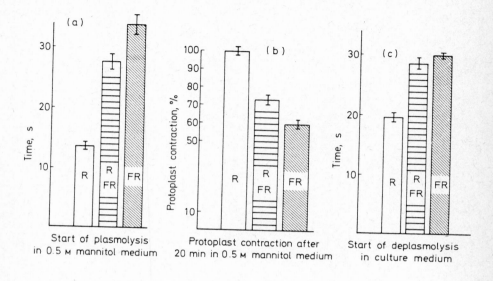

Figure 6.3   Start of plasmolysis (a), deplasmolysis (c) and amount of protoplast length contraction (b) in Mougeotia *after various light pre-treatments (± S.E.M.). In (b) the strongest chloroplast contraction (R) is taken as 100 per cent. R = red light; FR = far-red light.   (After Weisenseel and Smeibidl, 1973)*

as expected: localised response, that is, movement of the chloroplast away from the flank, is obtained only with the electric vector parallel to the surface. Thus, surface parallel orientation of Pr is confirmed. Yet the most important part of the experiment is when post-irradiation with a far-red micro-beam is applied in order to cancel the red light effect. By means of polarised far-red light we can test the dichroic orientation of Pfr at this site. We found that far-red light is much more effective if the electric vector is normal to the surface than parallel to it (*Figure 6.5*).

Thus, the chromophore of phytochrome has different orientations in space in the Pr and Pfr forms. The protein moiety, on the other hand, which is thought to be associated with the plasmalemma, is assumed not to change its orientation in space. Thus, upon photoconversion, the chromophore changes its orientation not only relative to the cell but also relative to the phytochrome protein. This may result in small but significantly different conformations of the protein moiety associated with the plasma membrane. Such a conformational change of a membrane-bound protein might be the basis of the above-

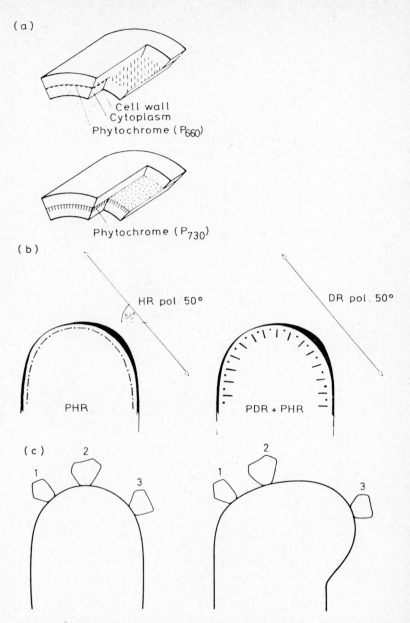

(a)

Cell wall
Cytoplasm
Phytochrome (P_{660})

Phytochrome (P_{730})

(b)

HR pol. 50°

DR pol. 50°

PHR

PDR + PHR

(c)

Figure 6.4 Schematic diagram of the dichroic phytochrome (dashes) parallel (P_{660} = P_{HR} = Pr) or normal to the surface (P_{730} = P_{DR} = Pfr). (a) Part of a cylindrical Mougeotia cell. (After Haupt, 1970.) (b) Chloronema tips after dark adaptation (P_{HR}) or red light (P_{DR} + P_{HR}); both polarised red light (HR polarisation 50°) and polarised far-red light (DR polarisation 50°) vibrating 50° to a line normal to the cell length axis can cause the same Pfr gradient in the tip as denoted by the thickness of the tip lining. (c) Position of markers (starch granules) before and after tip growth in response to a Pfr gradient according to (c). (After Etzold, 1965)

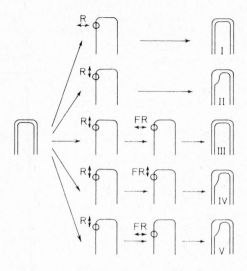

Figure 6.5   Partial response of the Mougeotia *chloroplast as a result of micro-beam irradiations with polarised red and far-red light.   From left to right: starting position (chloroplast fully in face position); positioning of the micro-beam (chloroplast omitted); result of the irradiation.   The double arrows denote the direction of the electrical vector.   The lower row (V) shows a control where R and FR were placed at different sites, hence no reversion of R effect by FR. (After Haupt, 1972)*

mentioned membrane effect of phytochrome in *Mougeotia,* namely control of water permeability.   Besides the possible importance for the mechanism of phytochrome action, the well defined change of dichroic orientation upon photoconversion once again confirms our conclusion about well defined localisation and a strong association of phytochrome to the cell membrane,

Is There Also Physiological Evidence for Unbound Phytochrome in the Cell?

When Boisard, Briggs, Marmé, Quail and Schäfer (e.g. Schäfer, 1974) found particle-bound phytochrome in cell extracts, this seemed to be an excellent confirmation of our physiological findings.   However, on more detailed analysis, a serious discrepancy seemed to arise.   In those extracts, only Pfr is found in the pellet in high concentrations, while Pr is found mainly in the supernatant.   This raises the important question of whether these *in vitro* results may find some parallelism in our physiological system, i.e. whether there is also some difference *in vivo* in the binding properties of Pr and Pfr, especially whether there is also physiological evidence for unbound Pr.

Some recent results with *Mougeotia* indeed point in this direction. They will be referred to briefly in spite of their very preliminary character (Haupt and Bretz, work in progress). We start with a *Mougeotia* cell which has the chloroplast in the profile position and which has been pre-irradiated with saturating far-red light, thus having phytochrome mainly in the Pr form. We try to induce the movement, i.e. orientation into the face position, with one or two flashes of polarised red light lasting about 1 ms each. Under these conditions, with a single flash, we obtain only a weak response, i.e. a low percentage of cells orienting the chloroplast, but 100 per cent response is obtained if the first flash is followed, after 15 s, with a second flash. Now, two aspects seem most important:

(1)   two flashes at the same time are much less effective than two flashes with an interval of 15 s between them;
(2)   two flashes with a 15–s interval between them induce full response even if their intensity is reduced.

This means that the first flash somehow 'conditions' the system for the second flash. To relate this to our problem of unbound phytochrome, we have to consider the absorption pattern in the cell (Haupt, 1970). Let us assume first a dichroic orientation of phytochrome parallel to the cell surface and perpendicular to the long axis of the cell. Polarised light, vibrating perpendicular to the long axis, will be absorbed only in the median of the cell, i.e. at top and bottom but not at the flanks. Thus, a strong absorption gradient is obtained. Now, let phytochrome be oriented parallel to the surface and along helical lines, as we have pointed out earlier (*Figure 6.1*). Such molecules would also have absorption vectors only near the median and, with perpendicular vibrating light, a strong absorption gradient will arise and therefore also a gradient of Pfr.

If, however, Pr were distributed at random throughout the cell, absorption would be possible all round the cell with equal probability, and thus no absorption gradient would be established. This would hold true for Pr which is not bound to the plasma membrane. (We assume that phytochrome has an inherent polar structure which precludes a random association with membranes.) Going back to our experiment with the flashes and its tentative explanation: the first flash may find only a small percentage of Pr bound to the membrane and therefore only a small Pfr gradient is established. The Pfr formed from free Pr by the first flash may bind uniformly to the membrane. The next flash then acts on bound Pfr but, because of the peculiar dichroic situation, red light can establish a Pfr gradient even if phytochrome is present mainly as Pfr. Alternatively, Pfr formed by the first flash may 'condition' the membrane to bind more Pr near the Pfr sites; the second flash then acts on more bound Pr and thus can establish a strong Pfr gradient. Clearly, much more work is needed, and is already in progress.

A possible turnover of free and bound phytochrome in the *Mougeotia* cell (as implied in these considerations) should have another interesting consequence: whenever differential photoconversion of Pr and Pfr is established, the remaining gradient of free Pr can be expected to level within a few minutes, resulting in a gradient of Ptot. This might be tested by appropriate experiments.

Some Ideas About Phytochrome Localisation and its Primary Effect

This section is mainly speculative. Let us return to the plasmolysis experiments in *Mougeotia*. The most puzzling result is the large and persistent difference in the degree of protoplast contraction between cells with high and low concentrations of Pfr (*Figure 6.3b*). This difference remains at least up to 1 h (unpublished results). In a previous discussion (Weisenseel and Haupt, 1974), we made some calculations about the necessary permeability of the plasmalemma to mannitol in far-red irradiated cells in order to account for this difference. We found it improbably high (10^{-7} cm s^{-1}). Such a high permeability should necessarily lead to some obvious deplasmolysis within 20 min, but none occurs.

One point we had not considered so far, however, is the possibility of a high permeability for mannitol and other solutes in far-red irradiated cells at the start of plasmolysis only. This high permeability could later have been obviated by the shrinkage of the protoplast and by probable concomitant changes in the membrane. In other words, it is conceivable that the reflection coefficient for mannitol and solutes is much less than unity in cells with a low level of Pfr and becomes almost unity in cells with a high level of Pfr. It is also possible that there is even a narrow band of Pfr concentrations which triggers such a transition in the membrane state: we may call it a threshold and relate it to cooperativity phenomena (cf. Mohr and Oelze-Karow, Chapter 17). In this regard, it is interesting to remember the work of Fondeville, Borthwick and Hendricks (1966) and many others on *Mimosa pudica*, where it was found that the leaflets close after one transfers the plant from light to dark; this response is under phytochrome control, and the transition from 'no Pfr effect' to 'full Pfr effect' has been shown to occur within a narrow concentration band of Pfr. We can assume here that the sufficiently low Pfr level triggers a large increase in the solute permeability of the pulvinule motor cell membrane, which in turn might explain the rapid loss of turgor in the ventral parts of the motor tissue (Dainty, 1969). Accordingly, Satter, Marinoff and Galston (1970) observed a large efflux of K$^+$ from these cells.

Summarising our present knowledge, there are several unique effects such as permeability to water (and presumably also to solutes) in *Mougeotia*, K^+ fluxes in *Mimosa*, surface charges and electrical potentials in roots and coleoptiles, all of which seem to be controlled by phytochrome and which point toward a Pfr-mediated change in the plasma membrane. Something seems to happen to the membrane – but what? Is there a common mechanism to all of these effects? Is there a so-called coupling factor? We can imagine the existence of such a factor: the *calcium ion*. From Rasmussen (1970) '... calcium is a key component of the membrane. Changes in calcium binding alter many of the physical properties of the membrane – for example, its permeability to water, other ions and solutes, and its deformability.' (*See* Rasmussen, 1970, for further literature.) Also, from the work of Miledi (1973) and Miller (1974), we know that calcium acts on the inside of the membrane and that the influx of calcium can increase the level of calcium ions in the cytoplasm by stimulating its release from intracellular stores.

We shall now describe a simple model of phytochrome action which is meant mainly to stimulate further research in a direction we think might be most rewarding and to aid in our understanding of phytochrome physiology. The model is based on our present knowledge of the structure and properties of cytoplasmic membranes. According to Singer and Nicholson (1972), the cytoplasmic membrane is comparable to a mosaic where proteins are inserted in a lipid-double-layer matrix. These proteins are free to move, as has been demonstrated for rhodopsin in retina disc membranes (Poo and Cone, 1974). Also important is the fact that biological membranes normally have a very low permeability towards divalent ions.

The basic assumption in our model is: *Pfr molecules in membranes function there as Ca^{2+} carriers.* It should be emphasised that the term 'carrier' does not necessarily imply that it is an active mechanism which consumes cell energy. For calcium ions, it can safely be assumed that under normal conditions, for example, the flux from outside to inside is energetically downhill (Rasmussen, 1970; Baker, 1972; Robinson and Jaffe, 1973). Such a carrier can be effective, transporting up to 10^4 ions per second (Bamberg *et al.*, 1974), which makes it possible that a low concentration of Pfr might have a large effect. Once the calcium ions have traversed the membrane, they are probably bound to membrane proteins and negatively charged membrane lipids, to contractile proteins and to enzyme systems and taken up into calcium stores such as mitochondria and perhaps into the vacuole.

Such binding to the membrane constituents could change the membrane state and could change its reflection coefficient and permselectivity to ions. It could also provide sites where Pr could become bound to the membrane.

Binding of calcium to intracellular fixed proteins could establish a calcium gradient and thereby a sufficiently strong intracellular electric field to segregate charged entities, thus leading to cell polarity and localised growth (Jaffe, Robinson and Nuccitelli, 1974). This attractive hypothesis and our model would explain the Pfr-controlled localised growth in fern protonemata (*Figure 6.4*). When calcium ions interact with contractile proteins, the result may be the movement of intra-cellular components such as the chloroplast in *Mougeotia*. Even the binding of calcium ions to enzymes and their consequent regulation could be considered as a result of phytochrome photoconversion. It is recognised that in this simple model some important elements may be missing, for instance secondary messengers, the release of calcium from intracellular pools and the deactivation of Pfr or the specificity of Pfr for calcium ions. It is conceivable that under certain circumstances, such as in black lipid membranes (Roux and Yguerabide, 1973), Pfr may conduct ions other than calcium. In general, and in essence, what we are suggesting is that phytochrome affects the concentration of free calcium ions in the cytoplasm.

In conclusion, we would like to make a few suggestions on how to check the basic assumptions of this model experimentally. There are mainly three types of experiments which might be suitable: tracer experiments, electrical measurements and cytochemical experiments. If Pfr increases the permeability of the plasmalemma for calcium, we can expect at least a transient higher influx of $^{45}Ca^{2+}$ in red light. In addition to standard procedures for flux measurements, the 'steady flow method' of Baker and Mason (1974) seems well suited. With very large cells such as *Nitella,* it might even be possible to measure a trans-cellular flux of calcium when part of the cell is illuminated with red and part with far-red light. The method of performing such an experi-ment may be similar to that used for transcellular osmosis (Dainty, 1969; House, 1974). Low-temperature (below $-70\ ^{\circ}C$) autoradiography may be applicable to demonstrate local $^{45}Ca^{2+}$ influx when localised Pfr gradients are established. Electrical measurements of the membrane potential and membrane conductance should be able to detect a sudden change in calcium flux. A transcellular calcium flux or a flux where calcium is taken up at one site and other cations are extruded else-where could be detected, even in small cells, with a newly developed vibrating electrode system (Jaffe and Nuccitelli, 1974).

Cytochemically, a calcium ion influx might be measured with the calcium-sensitive extract from certain jelly fishes, aequorin (e.g. Stinnakre and Tauc, 1973). Injected into the cells, it would detect calcium ions since it becomes luminescent when in contact with free calcium, the degree of luminescence being proportional to the concen-tration of calcium ions.

We would like to emphasise the advantage of those systems where localised phytochrome responses can be observed within one cell, i.e. differential responses of different cell regions as a result of differential phytochrome photoconversion. It seems desirable, therefore, to look for phytochrome responses in as large cells as possible, e.g. *Nitella, Chara* or *Acetabularia.*

References

BAKER, P.F. (1972). *Prog. Biophys. Mol. Biol.*, **24**, 177

BAKER, P.F. and MASON, W.T. (1974). *J. Physiol.*, **242**, 50P

BAMBERG, E., BENZ, R., LÄUGER, P. and STARK, G. (1974). *Chem. Unserer Zeit*, **8**, 33

DAINTY, J. (1969). In *The Physiology of Plant Growth and Development*, p.420. Ed. Wilkins, M.B. McGraw-Hill, Maidenhead

ETZOLD, H. (1965). *Planta*, **64**, 254

FONDEVILLE, J.C., BORTHWICK, H.A. and HENDRICKS, S.B. (1966). *Planta*, **69**, 357

HAUPT, W. (1959). *Planta*, **53**, 484

HAUPT, W. (1970). *Physiol. Veg.*, **8**, 551

HAUPT, W. (1972). In *Phytochrome*, p.554. Ed. Mitrakos, K. and Shropshire, W. Jr. Academic Press, London, New York

HOUSE, C.R. (1974). *Water Transport in Cells and Tissues*, p.182. Edward Arnold, London

JAFFE, L.F. and NUCCITELLI, R. (1974). *J. Cell Biol.*, **63**, 614

JAFFE, L.F., ROBINSON, K.R. and NUCCITELLI, R. (1974). *Ann. N.Y. Acad. Sci.*, **238**, 372

JAFFE, M.J. (1968). *Science, N.Y.*, **162**, 1016

MILEDI, R. (1973). *Proc. R. Soc. Lond.*, B, **183**, 421

MILLER, D.J. (1974). *J. Physiol.*, **242**, 93P

NEWMAN, I.A. and BRIGGS, W.R. (1972). *Plant Physiol.*, **50**, 687

POO, M. and CONE, R.A. (1974). *Nature, Lond.*, **247**, 438

RACUSEN, R. and MILLER, K. (1972). *Plant Physiol.*, **49**, 654

RASMUSSEN, H. (1970). *Science, N.Y.*, **170**, 404

ROBINSON, K.R. and JAFFE, L.F. (1973). *Dev. Biol.*, **35**, 349

ROUX, S.J. and YGUERABIDE, J. (1973). *Proc. Nat. Acad. Sci. U.S.A.*, **70**, 762

RUBINSTEIN, B., DRURY, K.S. and PARK, R.B. (1969). *Plant Physiol.*, **44**, 105

SATTER, R.L., MARINOFF, P. and GALSTON, A.W. (1970). *Am. J. Bot.*, **57**, 916

SCHÄFER, E. (1974). In *Membrane Transport in Plants*, p.435. Ed. Zimmermann, U. and Dainty, J. Springer, Berlin, Heidelberg, New York

SINGER, S.J. and NICHOLSON, G.L. (1972). *Science, N.Y.*, **175**, 720

STINNAKRE, J. and TAUC, L. (1973). *Nature New Biol.*, **242**, 113

TANADA, T. (1968). *Plant Physiol.*, **43**, 2070

WEISENSEEL, M.H. and SMEIBIDL, E. (1973). *Z. Pflanzenphysiol.*, **70**, 420

WEISENSEEL, M. and HAUPT, W. (1974). In *Membrane Transport in Plants*. p.427. Ed. Zimmermann, U. and Dainty, J. Springer, Berlin, Heidelberg, New York

7

IMMUNOLOGICAL VISUALISATION OF PHYTOCHROME

LEE H. PRATT
RICHARD A. COLEMAN
JOHN M. MACKENZIE, Jr.
Department of Biology, Vanderbilt University, Nashville, Tennessee 37235, U.S.A.

Immunology — An Alternative Assay for Phytochrome

The assay used to isolate, purify and characterise phytochrome is derived from the unique, photoreversible spectral properties of this chromoprotein (Butler and Lane, 1965). However, it has long been apparent that this spectral assay suffers from a number of limitations. Firstly, the assay is not suited for quantitative use when large amounts of chlorophylls are present and thus is limited primarily to work with etiolated tissues and tissue extracts. Secondly, the assay is relatively insensitive and thus requires large samples, which in turn limits spatial resolution when making *in vivo* measurements. The spectral assay is therefore not suitable for studies of phytochrome within single, defined cells. Thirdly, the assay requires that phytochrome itself remain undisturbed since a wide variety of conditions are known to reduce or eliminate almost entirely its absorption of visible light (Butler, Siegelman and Miller, 1964; Lisansky and Galston, 1974; Marmé, 1975; Pratt and Cundiff, 1975). Fourthly, the most serious problem is that it is not possible to relate spectral measurements directly to the absolute amount of phytochrome protein present.

Hence, additional means of assay for phytochrome are clearly desirable. One possibility as an assay system is the use of antisera specific for phytochrome. While available immunochemical assays, like the spectral assay, provide no direct information concerning biological activity of the phytochrome being measured, the immunochemical assays, unlike the spectral assay, are neither dependent upon the presence of the chromophore nor on its extinction coefficient and thus may be used as a direct measure of the amount of phytochrome protein present. Hence, by using both assay methods, an investigator may obtain a much broader range of information regarding the biochemical properties and physiological role of phytochrome than by using either method alone.

One area in which immunochemical assays may provide information supplementary to the spectral assay is in a study of the subcellular localisation of phytochrome. Other complementary approaches to this problem are discussed elsewhere in these proceedings and thus will not be reviewed here (Haupt and Weisenseel, Chapter 6; Marmé, Bianco and Gross, Chapter 8; Quail and Gressel, Chapter 9; Evans, Chapter 10; Furuya and Manabe, Chapter 11). However, as will be evident below, the immunocytochemical **assay** simultaneously provides additional information about phytochrome and is not limited to answering the question of its subcellular localisation.

In addition to the work described here, phytochrome-specific antisera have been used for other purposes such as: (1) to measure directly and quantitatively the amount of phytochrome protein present in samples of unknown phytochrome content (Pratt, Kidd and Coleman, 1974); (2) to compare phytochromes isolated from different plants (Cundiff and Pratt, 1975a; Pratt, 1973; Rice and Briggs, 1973); (3) to search for differences between Pr and Pfr (Cundiff and Pratt, 1975a; Hopkins and Butler, 1970; Pratt, 1973); and (4) to analyse the products of phytochrome proteolysis (Cundiff and Pratt, 1973, 1975b).

Immunocytochemical Protocols

Since it is not possible to observe directly phytochrome *in situ* with either the light or electron microscope, the principle of the immunocytochemical assay is to stain phytochrome specifically with a readily visible label. From the several immunocytochemical approaches available, we have chosen a double indirect method (*Figure 7.1*) because it is potentially the most sensitive (Moriarty, 1973; Sternberger *et al.*, 1970). The first step in this, as in any other immunocytochemical procedure, is to fix the tissue histologically so as to prevent movement during localisation of the antigen being studied while simultaneously preserving structural and ultrastructural features (Hökfelt *et al.*, 1973). The fixed tissue must then be sectioned and prepared so that it is freely accessible to all of the reagents. Finally, the fixation and sectioning processes must be sufficiently mild that antigenic activity of phytochrome is retained to a maximal extent in order to have suitable sensitivity in the assay. Development of an appropriate protocol must therefore involve an extensive search for the best compromise among these three competing requirements. In addition, when interpreting micrographs, especially those made with the electron microscope, one must remember that conditions which yield the best ultrastructural detail either prevent penetration of the large immunochemical reagents or inactivate phytochrome as an antigen (Coleman and Pratt, 1974b). Thus, one should not anticipate the same anatomical detail as is conventionally obtained in electron microscopy. Since little work had previously been done on methodology for the intracellular localisation of plant antigens, much of our effort has involved development of suitable protocols for such

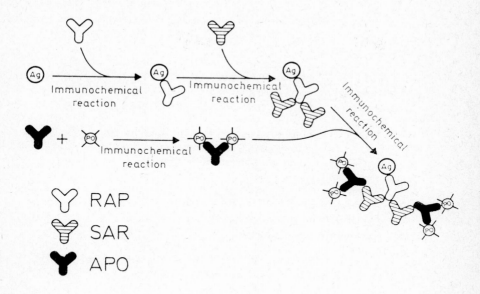

Figure 7.1 A double indirect technique for the visualisation of an antigen. Components of the labelling procedure are: phytochrome, Ag; rabbit antiphytochrome serum, RAP; sheep antirabbit immunoglobulin serum, SAR; rabbit antiperoxidase immunoglobulins, APO; and peroxidase, PO

experiments. With further methodological advances, it should become possible to define the intracellular localisation of phytochrome with even greater resolution than is now possible.

An outline of the basic protocol developed for the visualisation of phytochrome with the light microscope is as follows (Pratt and Coleman, 1974). Tissue is fixed in darkness for 24 h, normally at 25 °C, in 4 per cent formaldehyde (from paraformaldehyde) in 0.1 M sodium phosphate, pH 7.6. The fixed tissue is dehydrated through an ethanol series, transferred through an ethanol-xylene series into xylene, embedded in paraffin and sectioned, with the 8-12 μm sections then being attached to glass microscope slides. The paraffin is then removed from the sections with xylene and the sections are finally rehydrated by transfer first into ethanol and then through an ethanol-water series back into water. Reagents are applied to the sections directly on the microscope slides. Unless otherwise noted, tissue is exposed only to green safelight until fixation is complete.

For electron microscopy (Coleman and Pratt, 1974b), tissue which has been fixed as for light microscopy is first transferred to ethanol and then embedded in polyethylene glycol (PEG). The PEG-embedded tissue is sectioned 3 μm thick and the sections are immersed directly in saline (0.15 N NaCl). Since the sections are free in suspension, reagents have access to them from both sides, thus enhancing penetration. Sections are collected for transfer from one reagent or wash solution to the next by centrifugation at low speed (300 g) in 1-ml glass conical-tip centrifuge tubes.

The duration and sequence of reagent and wash treatments are given in *Table 7.1.* Basically, as depicted in *Figure 7.1*, the tissue sections are first incubated in specific rabbit antiphytochrome serum (RAP) which contains immunoglobulins that immunospecifically bind to antigenically active phytochrome in the tissue. After washing away excess of RAP, immunoglobulins from sheep antirabbit immunoglobulin serum (SAR) are immunospecifically bound to RAP. This second step is possible because immunoglobulins from one animal (here rabbit) may serve as antigen in another animal (here sheep). Because all immunoglobulins from one animal, regardless of their different immunospecificities, possess a region of common structure, SAR is capable of binding with any rabbit immunoglobulin. After washing away excess of SAR, rabbit antiperoxidase–peroxidase complex (PAP) is then immunospecifically cross-linked to RAP by SAR. This last step is possible because every immunoglobulin has two immunospecific binding sites and in this protocol only one of the SAR binding sites is used when reacting with RAP. Peroxidase is thus used as the label for phytochrome, being attached to phytochrome through an intervening 'pyramid' of three specific immunoglobulins. Peroxidase is made visible by incubation with the substrate, 3,3$^{/}$-diaminobenzidine (DAB) in the presence of hydrogen peroxide. The insoluble orange reaction product is directly visible with the light microscope and can be made electron-dense for electron microscopy by subsequent incubation in osmium tetroxide. For observation with the electron microscope, the stained sections are re-embedded in any medium suitable for ultra-thin sectioning.

RAP was prepared in a routine fashion by injecting highly purified low-molecular-weight Garry oat (*Avena sativa* L.) phytochrome into a rabbit (Pratt, 1973). RAP should not contain immunoglobulins which bind with other antigens in the tissue being studied and thus its specificity must be verified by extensive *in vitro* investigations (Pratt, 1973; Cundiff and Pratt, 1973, 1975a,b). SAR can be obtained commercially from a number of sources. However, since SAR will cross-link any rabbit immunoglobulin to RAP, PAP must be highly purified so that it contains nothing but peroxidase and antiperoxidase immunoglobulins. A simple immunospecific purification of this complex, which simultaneously provides the complex in the needed soluble form, is described in detail by Sternberger *et al.* (1970). Peroxidase was chosen as the label in preference to the more commonly used fluorescent dyes or ferritin for a variety of reasons:

Table 7.1 Immunocytochemical protocols

Treatment[1]	Duration of treatment	
	Light microscopy	Electron microscopy
Non-immune sheep serum[2] or penetration test solution[3]	30 min	2 h
RAP, penetration test solution or non-immune rabbit serum	30 min (1/40)[4]	2 h (1/60)
Saline[5] rinse	15 s (running)	2 rinses
Saline wash	40 min	3 h
Non-immune sheep serum	15 min	1 h
SAR	30 min (1/80)	2 h (1/120)
Saline rinse	15 s (running)	2 rinses
Saline wash	40 min	3 h
Non-immune sheep serum	15 min	1 h
PAP	40 min (1/60)	3 h (1/90)
Saline rinse	15 s (running)	2 rinses
Saline wash	50 min	10 h (2 °C)
DAB (0.02% in 50 mM Tris-Cl, pH 7.6)	—	1 h (0 °C)
DAB (0.02% in 50 mM Tris-Cl, pH 7.6) + 0.01% H_2O_2	—	5 min
DAB (0.05% in 50 mM Tris-Cl, pH 7.6) + 0.01% H_2O_2	7 min	—
Saline rinse	—	2 rinses
OsO_4 (1% in saline)	—	45 min

[1] All treatments at room temperature unless otherwise specified
[2] All sera and washing solutions for electron microscope protocol contain 0.02 per cent of sodium azide to eliminate bacterial contamination
[3] Penetration test solution is composed of 1 part sheep serum, 1 part rabbit serum and 2 parts saline
[4] All dilutions are made with non-immune sheep serum
[5] Saline for electron microscope protocol contains 10 per cent (by volume) of non-immune sheep serum

(1) As an enzyme it provides high sensitivity.
(2) It yields stain visible with both the light and electron microscopes.
(3) Its activity is only marginally decreased during the preparation of the PAP complex.

Immunocytochemical Controls

Two different controls are required. The first is designed to detect either:

(1) non-specific adsorption of any of the reagents to the tissue sections
(2) endogenous DAB-oxidising activity.

This localisation control is the most conservative possible in that it utilises sections immediately adjacent to the experimental sections and follows exactly the same protocol except for the substitution of

non-immune rabbit serum (i.e. serum which does not immunospecifically bind with any tissue component) for RAP in the first step of
the procedure (*Table 7.1*). For electron microscopy, it is necessary to
determine whether all reagents have penetrated all parts of the tissue
section, and for this purpose we developed a penetration control.
Since we found that rabbit immunoglobulins would non-specifically bind
with the tissue (see below), we first obtained such non-specific binding
and then used precisely the same protocol as for the experimental
sections but in this instance the purpose was to localise non-specifically-
bound rabbit immunoglobulins rather than specifically-bound RAP
(*Table 7.1*).

Our first attempts to stain phytochrome produced experimental and
localisation control sections which were indistinguishable (*Figure 7.2d*).
We found the difficulty to be non-specific binding of immunoglobulins
and were therefore able to reduce localisation control activity by
saturating non-specific binding sites with non-immune sheep immunoglobulins which are not active reagents in the assay. We accomplished
this goal in two ways. Firstly, the tissue sections were exposed to
concentrated non-immune sheep serum prior to treatment with each of
the three immunochemical reagents. Secondly, since the immunochemical reagents are highly diluted, we performed this dilution with
non-immune sheep serum rather than saline as is more commonly done.
Either modification alone greatly reduces control activity (*Figure 7.2b,c*),
while both modifications combined eliminate it entirely (*Figure 7.2a*).
These modifications also eliminate localisation control activity at the
electron microscope level (compare *Figure 7.3a* and *7.3b*). Since
localisation controls are virtually invisible they are generally not presented
even though they are always prepared.

Successful penetration controls should have a uniform field of stain
with no differential staining of organelles and cytoplasm. When sections
are sufficiently thin (3 μm) and incubation times (*Table 7.1*) sufficiently
long, it is clear that thorough penetration by reagents is achieved
(*Figure 7.3c*). Again, although penetration controls are always prepared
and evaluated, they are not shown for each experiment since they provide no other information. With this control, it is possible to exclude
the possibility that the absence of stain results only from the inability
of one or more reagents to reach phytochrome. Thus, one can conclude
that if antigenic phytochrome were present, it would be detected. The
possibility that phytochrome antigenic activity has been eliminated during
the preparation of the tissue sections unavoidably remains. However,
this limitation in interpretation is minimised by the observation that,
whenever we have anticipated the presence of phytochrome based upon
spectral measurements, it has been observed immunocytochemically.
Thus, one can be reasonably confident that if phytochrome were present
above the limit of detection it would be observed.

An indication of the degree of ultrastructural preservation is given by
a localisation control section which is post-stained with lead citrate and
uranium acetate (Coleman and Pratt, 1974b). It is evident that

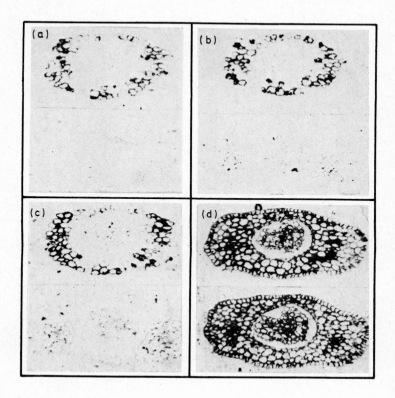

Figure 7.2 Development of light-level localisation protocol. Top section in each pair is an experimental section; bottom section in each pair is a localisation control. (a) Active sera or immunoglobulins are diluted with non-immune sheep serum and the sections are pre-incubated in non-immune sheep serum prior to treatment with each of the three active reagents as described in Table 7.1. (b) Active sera or immunoglobulins are diluted with non-immune sheep serum but no pre-incubations are used. (c) Sections are pre-incubated as in (a) but active sera or immunoglobulins are diluted with saline rather than non-immune sheep serum. (d) Active sera are diluted with saline and no pre-incubations are used. Micrographs are bright-field views of 3-day-old (25 °C) etiolated Garry oat (Avena sativa L.) coleoptiles about 0.4 mm behind the coleoptile tip. Magnification 70X. (After Pratt and Coleman, 1974)

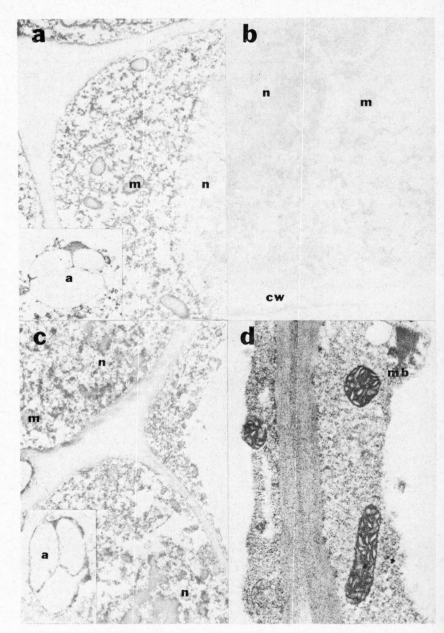

Figure 7.3   Electron micrographs of Garry oat coleoptile parenchyma cells comparable to those presented in Figure 7.2 *but prepared by the electron microscope protocol.   (a) Micrograph of experimental section showing phytochrome-specific electron density.   Mitochondrion, m; nucleus, n.   Magnification 13,000X.   Inset: amyloplast, a.   Magnification 11,000X.   (b) Micrograph of localisation control section.   Cell wall, cw.   Magnification 17,000X.   (c) Micrograph of penetration control section.   Magnification 8,000X.   Inset: amyloplast.   Magnification 11,000X. (d) Localisation control section stained on the observation grid with uranyl acetate and lead citrate.   Microbody, mb.   Magnification 28,000X.   (By courtesy of Coleman and Pratt, 1974b)*

organelles are generally well preserved and that the cytoplasm, not evident in the localisation control micrograph (*Figure 7.3b*), is still present (*Figure 7.3d*).

Thus, by utilising the above protocols with appropriate controls, it is possible to stain specifically tissue sections for phytochrome. Since the localisation controls are unstained and since the tissue is otherwise unstained except by these protocols, all stain or electron density in the experimental micrographs is that associated with phytochrome. Organelles or regions which do not contain phytochrome are thus essentially invisible. Representative results of some applications of the above immunocytochemical protocols are outlined below.

Phytochrome Distribution

Using the light microscope protocol, one can readily determine the distribution of phytochrome throughout a plant with a high degree of sensitivity and resolution (Pratt and Coleman, 1971, 1974). Thus, one can better correlate phytochrome distribution with both plant photosensitivity and with the distribution and function of other morphogenic agents. While the spectral assay gives limited information regarding phytochrome distribution (Briggs and Siegelman, 1965), the immunochemical assay provides information on a cell-by-cell basis.

The immunocytochemically observed presence of relatively high levels of phytochrome in the tip of etiolated grass coleoptiles and in the region of the coleoptilar nodes (e.g. *Figure 7.4a*) agrees well with earlier spectrophotometric observations (Briggs and Siegelman, 1965). However, the immunocytochemical method makes it possible to determine specifically which cells contain phytochrome. Thus, for example, one may observe that tips of young adventitious roots, still within the shoot, contain phytochrome (e.g. *Figure 7.4b*) and root caps of mature roots also contain high levels (e.g. *Figure 7.5*). Both of the latter observations agree well with physiological experiments which imply the presence of phytochrome in these regions (e.g. Tepfer and Bonnett, 1972).

The immunocytochemical protocol is in no way limited, at least in theory, to the study of grass seedlings or etiolated tissue. We have detected phytochrome in *Pinus* embryos, Alaska pea and Zucchini squash shoots, *Lemna* fronds and orchid embryos, the latter grown under continuous red light (*Figure 7.6*). Thus, possibilities for future work are extensive and include the potential ability to visualise phytochrome in fully green tissue.

Even though specific observations vary from one plant to the next, it is always apparent that phytochrome is found primarily in morphogenically active regions of a plant (Pratt and Coleman, 1971, 1974). Further, phytochrome possesses a very discrete distribution in a plant even within what appears otherwise to be a homogeneous tissue. Again, however, the pattern seems to be different from one plant to another (Pratt and Coleman, 1974).

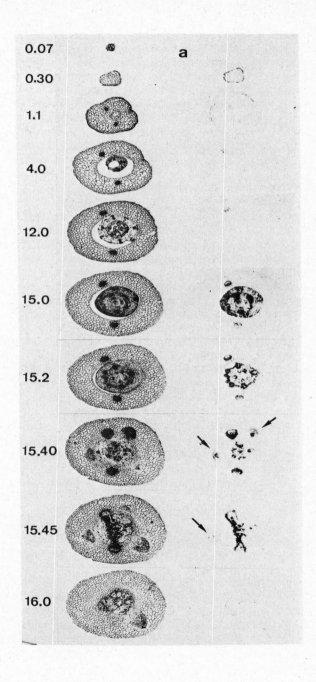

0.07

0.30

1.1

4.0

12.0

15.0

15.2

15.40

15.45

16.0

a

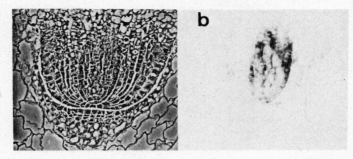

b

Figure 7.4 Survey of phytochrome distribution in a 5-day-old (25°C) etiolated rice (Oryza sativa L.) shoot. (a) Each experimental section was photographed with both dark-field illumination to show structure (left) and bright-field illumination to show phytochrome-specific stain (right). Numbers indicate distance in millimetres behind the coleoptile tip of the section presented. Arrows indicate locations of developing adventitious roots. Magnification 25X. (b) Higher magnification view of adventitious root from 15.4–mm section presented both as seen by phase contrast to indicate structure and by bright-field to indicate phytochrome-specific stain. Magnification 60X. (After Pratt and Coleman, 1974)

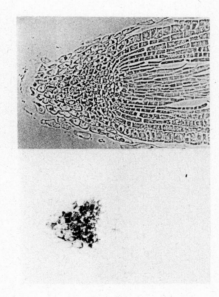

Figure 7.5 Phytochrome distribution in a 3-day-old (25°C) dark-grown Garry oat root tip. An experimental section was photographed both with phase contrast optics to indicate structure (top) and with bright-field optics to indicate phytochrome-specific stain (bottom). Magnification 140X

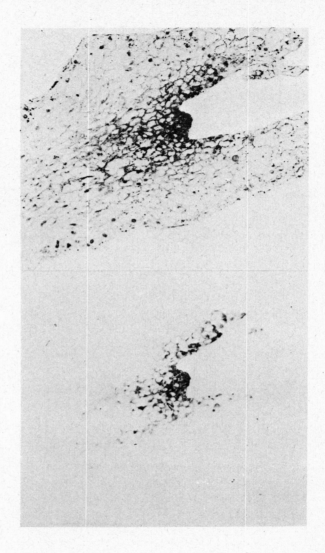

Figure 7.6   Distribution of phytochrome in a sterile-cultured orchid embryo (Cypripedium harbor X Cyp. merlin) grown under continuous red light (Balzers Filtraflex K6).   An experimental section was photographed both with dark-field optics to indicate structure (top) and with bright-field optics to indicate phytochrome-specific stain.   Magnification 45X

Phytochrome Synthesis

Use of the spectral assay for detection of phytochrome does not allow one to distinguish between two possibilities regarding the appearance of phytochrome during the germination and initial growth of seedlings. Firstly, the appearance of spectrophotometrically detectable phytochrome may represent only a final step in the maturation of the complete phytochrome molecule such as, for example, the attachment of the chromophore, or it may represent a *de novo* synthesis of the protein moiety following the initiation of germination.

When using the immunocytochemical assay to detect phytochrome protein, we have observed (Coleman and Pratt, 1974a) with both Garry oat and Balbo rye grains that antigenically detectable phytochrome does not appear until after more than at least 4 h of germination at 25 °C (*Figure 7.7*). Hydration at 0 °C for 24 h does not result in the appearance of antigenic activity, indicating that detectability is not simply a matter of hydration of the protein. Since antiphytochrome immunoglobulins readily detect not only small fragments of the native molecule (Cundiff and Pratt, 1973, 1975b) but also denatured forms of the chromoprotein (Pratt, Kidd and Coleman, 1974), one can conclude that the appearance of spectrally detectable phytochrome in these two cases results from *de novo* synthesis. Examination of tissue sections of all regions of 24-h hydrated seedlings indicates that the pattern of distribution at this early age is indistinguishable from that observed in older seedlings, thus leading to the further conclusion that phytochrome is probably synthesised in the cells in which it is later found. Thus, the immunocytochemical assay supplements information obtained by other approaches such as the use of heavy isotopes for density labelling (Quail, Schäfer and Marmé, 1973) or radioactive labelling (Correll *et al.*, 1968).

Phytochrome Destruction

Since the spectrophotometric assay tells one only that phytochrome photoreversibility or extinction is being lost during the destruction process (Butler and Lane, 1965), it is not possible to conclude anything about the products or mechanism of the destruction reaction based upon the results of this assay. However, since the immunological assay is a direct measure of phytochrome protein, it is possible for the first time to examine the fate of phytochrome protein during the course of the destruction process. In addition, since the immunocytochemical assay requires the presence of only a single antigenic determinant of phytochrome, it is capable of detecting only a fragment of the remaining molecule (Cundiff and Pratt, 1973, 1975b). Thus, it is possible to conclude from the absence of antigenic activity following extensive phytochrome destruction (*Figure 7.8*) that the protein moiety is almost certainly being largely, if not completely, degraded. These

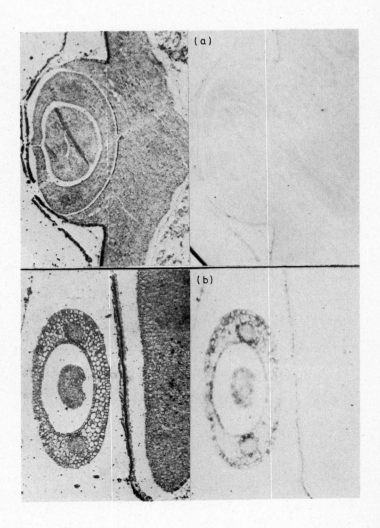

Figure 7.7 *Appearance of antigenically detectable phytochrome in germinating Garry oat grains prepared by the light microscope protocol. (a) Dark-field micrograph to indicate structure (left) and bright-field micrograph to indicate phytochrome-specific stain (right) of cross-section through a 4–h hydrated 25 °C grain. (b) As in (a) but a 24–h hydrated grain. Magnification 45✕. (After Coleman and Pratt, 1974a)*

Figure 7.8   Electron micrographs of Garry oat coleoptile parenchyma cells
similar to those presented in Figure 7.2 except prepared as experimental sections
by the electron microscope protocol.   Dark control (left).   From a shoot exposed
to white light at 20 °C for 5.5 h (right).   Mitochondrion, m; nucleus, n.   All
sections in each case were collected at the bottom of a 15–ml centrifuge tube
after OsO₄ incubation and photographed to show the general level of phytochrome-
specific stain observed (insets).   Magnification 15,000X.   (After Pratt, Kidd and
Coleman, 1974)

immunocytochemical observations are supported by a variety of complementary *in vitro* immunochemical assays (Pratt, Kidd and Coleman,
1974) which lead to the same conclusion.

Subcellular Distribution of Pr

Examination of a number of cell types which contain phytochrome
and have not been exposed to other than a green safelight before
fixation indicates that phytochrome, as Pr, possesses no discernible,
unique association with a *single* organelle or membrane type (Coleman
and Pratt, 1974b,c).   It is apparent, however, that stain for phytochrome
is observed in association with mitochondria (*Figure 7.9*, inset) and plastids (*Figure 7.3a*) as well as the cytoplasm in general but not nuclei
and vacuoles (*Figures 7.3a* and *7.9*).   Hence, while not conclusive, the
observations are nevertheless consistent with the possibility that phytochrome is present largely as a soluble protein in the cytoplasm of
etiolated plant tissue prior to its photoconversion to Pfr.   One may

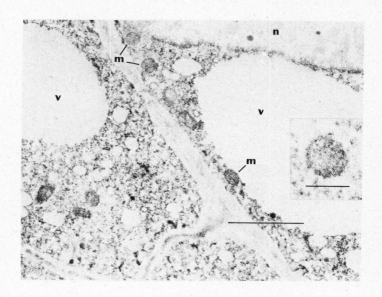

Figure 7.9   Electron micrograph of experimental section from the coleoptilar node region of a 3-day-old (25 °C) etiolated rye seedling.   Cells presented are parenchyma cells from the base of a primary leaf.   Mitochondrion, m; nucleus, n; vacuole, v. Bar represents 2 μm.   Inset: high-magnification view of a mitochondrion from this section.   Bar represents 0.3 μm.   (By courtesy of Coleman and Pratt, 1974c)

still not conclude, however, that there is no biologically significant association of a small but active pool of Pr with some unidentified subcellular component.

Subcellular Distribution of Pfr

Recent observations that extracts of plant tissue contain a high percentage of phytochrome in a pelletable fraction only after conversion to Pfr (e.g. Marmé, Bianco and Gross, Chapter 8; Quail and Gressel, Chapter 9) indicate, as one possibility, that Pfr may have a different subcellular localisation than Pr.   At least in Garry oat (*Figure 7.10*) and rice seedlings this appears to be the case (Mackenzie *et al.*, 1975).   At low magnification, and hence low resolution, phytochrome localised after transformation to Pfr appears to disappear (compare *Figure 7.10a* and *b*).   That this observation is not merely a function of fixation as Pfr is clear from the observation that fixation at 0 °C following immediate phototransformation back to Pr does not change the result (*Figure 7.10c*).   However, if fixation is delayed for 2 h at room temperature following the red–far-red irradiation sequence, then stain for phytochrome reappears (*Figure 7.10d*), indicating that this change induced by red light is reversible. Examination of the same cells at higher magnification, and thus higher resolution, leads to the conclusion that the antigenic activity of phytochrome need not have been lost as a result of red exposure, as might first have been assumed. Rather, it appears as though Pfr has migrated from a uniform distribution (*Figure 7.10e*) to discrete areas (*Figure 7.10f,g*) which are not resolved (and hence not seen) at low magnification. We have found that the red-induced movement of phytochrome is very rapid, occurring within 1 min at 3 °C. The return to the original distribution following subsequent far-red irradiation by contrast is very slow. The discrete areas of activity seen in *Figure 7.10g* appear to enlarge in size over a 1–h period at 25 °C and only after about 2 h at this temperature is the original condition obtained (*Figure 7.10h;* Mackenzie *et al.*, 1975).

It was recently observed that as much as 70 per cent of the phytochrome in red-irradiated Garry oat shoots will pellet at 20 000 *g* in 25 mM MOPS (*N*-morpholino-3-propanesulphonic acid) buffer, pH 7.5, that is 10 mM in $MgCl_2$ (Grombein *et al.*, 1975). However, there is as yet no evidence that the immunocytochemically observed redistribution of phytochrome is related to this preferential pelletability of Pfr *in vitro* (Marmé, Bianco and Gross, Chapter 8; Quail and Gressel, Chapter 9). Also, while the observations summarised here provide clear evidence for the intracellular movement of phytochrome as a function of its conformation, there is also no evidence yet which permits the conclusion that this movement is related to the morphogenic activity of phytochrome. However, it is tempting to postulate that this movement may represent a binding of Pfr to specific, biologically active receptor sites within the cell. One may then further postulate that phytochrome as Pr, following a red–far-red light sequence, would only slowly relax from these receptor sites as indicated both here (*Figure 7.10*) and by the pelletability assays of Quail, Marmé and Schäfer (1973). In this way, one could explain the so-called *Zea* paradox (Briggs and Chon, 1966) without postulating the existence of two pools of phytochrome in an etiolated plant, one active and one inactive, with different photochemical properties (Hillman, 1967). As originally suggested by Hendricks, if a cell contained a large amount of phytochrome but only a small number of active receptor sites, a very low dose of red light would saturate the response. However, the response would still be reversed by far-red light (which produces even more Pfr than the red light did and thus gives rise to the paradox), since the unbound Pfr produced by the far-red light could not bind to the receptor sites which would already be filled (albeit now with inactive Pr).

In order to test the hypothesis of Hendricks, as well as others proposed elsewhere in this volume, future experimentation should clearly be designed to answer three questions:

(1) Is enhanced pelletability of Pfr a consequence of the same phenomenon observed by immunocytochemical means?

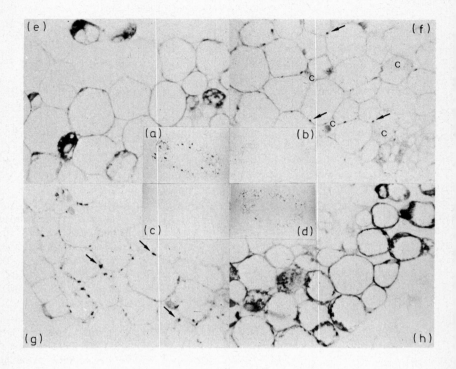

Figure 7.10   Bright-field micrographs showing phytochrome-specific stain in Garry oat coleoptile parenchyma cells from 3–μm experimental sections similar to those shown in Figure 7.2.   Sections were prepared by the electron microscope protocol except that the OsO$_4$ incubation was omitted and the sections were photographed without re-embedding for ultrathin sectioning.   All tissue was fixed at 0 °C immediately following the indicated treatment.   (a,e) Fixation prior to any light exposure; (b,f) fixation after 8 min exposure to Sylvania Gro-Lux lamps ('red') at 25 °C; (c,g) fixation after 8 min red followed immediately by 5 min far-red exposure provided by filtration of incandescent light through FRF–700 Plexiglas; (d,h), as in (c,g) except that a 2–h dark period follows the red–far-red sequence and precedes fixation.   Arrows identify some areas of discrete phytochrome-specific stain.   Following the red light-induced movement, unstained cytoplasm, c, indicates that phytochrome is no longer uniformly distributed throughout the cytosol.   Magnification of (a–d) 50×; magnification of (e–h) 600×.   (After Mackenzie et al., 1975).

(2) Is either phenomenon related to the photomorphogenic function of phytochrome?

(3) With what cellular and/or molecular component(s) is Pfr associating

These questions are presently under intense investigation in several laboratories and it is likely that at least partial answers to them will have been obtained by the time this book appears.

Conclusion

Immunocytochemistry is but one example of the wide range of assays for phytochrome which are made possible by its activity as an antigen. As indicated above, immunochemical assays do not replace the spectral assay but rather supplement it and thus allow one to answer questions which cannot be answered by using only the spectral assay. A further example of the potential of the immunological assay system is demonstrated by calculations which indicate that picogram levels of phytochrome should be detectable by means of a radioimmunoassay. Thus, one has a potential assay at least 1000-fold more sensitive than even the most sensitive spectral assay. One could then begin to ask, and hopefully to answer, an even wider range of questions for which the spectral assay alone is inadequate.

Acknowledgements

This research programme was supported by National Science Foundation Grant GB-17057 to L.H.P. and by grants from the Vanderbilt University Research Council.

References

BRIGGS, W.R. and CHON, H.P. (1966). *Plant Physiol.*, **41**, 1159

BRIGGS, W.R. and SIEGELMAN, H.W. (1965). *Plant Physiol.*, **40**, 934

BUTLER, W.L. and LANE, H.C. (1965). *Plant Physiol.*, **40**, 13

BUTLER, W.L., SIEGELMAN, H.W. and MILLER, C.O. (1964). *Biochemistry*, **3**, 851

COLEMAN, R.A. and PRATT, L.H. (1974a). *Planta*, **119**, 221

COLEMAN, R.A. and PRATT, L.H. (1974b). *J. Histochem. Cytochem.*, **22**, 1039

COLEMAN, R.A. and PRATT, L.H. (1974c). *Planta*, **121**, 119

CORRELL, D.L., EDWARDS, J.L., KLEIN, W.H. and SHROPSHIRE, W. Jr. (1968). *Biochim. Biophys. Acta*, **168**, 36

CUNDIFF, S.C. and PRATT, L.H. (1973). *Plant Physiol.*, **51**, 210

CUNDIFF, S.C. and PRATT, L.H. (1975a). *Plant Physiol.*, **55**, 207

CUNDIFF, S.C. and PRATT, L.H. (1975b). *Plant Physiol.*, **55**, 212

GROMBEIN, S., RUDIGER, W., PRATT, L. and MARMÉ, D. (1975). *Plant Sci. Lett.*, **5**, 275

HILLMAN, W.S. (1967). *Annu. Rev. Plant Physiol.*, **18**, 301
HÖKFELT, T., FUXE, K., GOLDSTEIN, M. and JOH, T.H. (1973). *Histochemie*, **33**, 231
HOPKINS, D.W. and BUTLER, W.L. (1970). *Plant Physiol.*, **45**, 567
LISANSKY, S.G. and GALSTON, A.W. (1974). *Plant Physiol.*, **53**, 352
MACKENZIE, J.M. Jr., COLEMAN, R.A., BRIGGS, W.R. and PRATT, L.H. (1975). *Proc. Nat. Acad. Sci. U.S.A.*, **72**, 799
MARMÉ, D. (1975). In *Proceedings of Annual European Photomorphogenesis Symposium.* Ed. Smith, H. University of Nottingham School of Agriculture, Sutton Bonington, Loughborough, Leics., U.K.
MORIARTY, G. (1973). *J. Histochem. Cytochem.*, **21**, 855
PRATT, L.H. (1973). *Plant Physiol.*, **51**, 203
PRATT, L.H. and COLEMAN, R.A. (1971). *Proc. Nat. Acad. Sci. U.S.A.*, **68**, 2431
PRATT, L.H. and COLEMAN, R.A. (1974). *Am. J. Bot.*, **61**, 195
PRATT, L.H. and CUNDIFF, S.C. (1975). *Photochem. Photobiol.*, **21**, 91
PRATT, L.H., KIDD, G.H. and COLEMAN, R.A. (1974). *Biochim. Biophys. Acta*, **365**, 93
QUAIL, P.H., MARMÉ, D. and SCHÄFER, E. (1973). *Nature New Biol.*, **245**, 189
QUAIL, P.H., SCHÄFER, E. and MARMÉ, D. (1973). *Plant Physiol.*, **52**, 124
RICE, H.V. and BRIGGS, W.R. (1973). *Plant Physiol.*, **51**, 939
STERNBERGER, L.A., HARDY, P.H. Jr., CUCULIS, J.J. and MEYER, H.G. (1970). *J. Histochem. Cytochem.*, **18**, 315
TEPFER, D.A. and BONNETT, H.T. (1972). *Planta*, **106**, 311

8

EVIDENCE FOR PHYTOCHROME BINDING TO PLASMA MEMBRANE AND ENDOPLASMIC RETICULUM

DIETER MARMÉ
JACQUELINE BIANCO[1]
JOACHIM GROSS
Institut für Biologie III, Universität Freiburg, D-78 Freiburg i. Br., Schänzlestrasse 9-11, West Germany

Introduction

A number of studies during the past few years have suggested that an early consequence of the phototransformation of phytochrome (P) from the red light-absorbing form (Pr) to the far-red light-absorbing form (Pfr) might be some change in the functional properties of plant membranes. For example:

(1)   Red light causes isolated root tips of barley (Tanada, 1968a) or mung bean (Tanada, 1968b) to adhere to a charged glass surface, and subsequent far-red irradiation causes their release;

(2)   There are clear phytochrome-mediated alternations in the bioelectric potentials of oat coleoptiles (Newman and Briggs, 1972), and both effects can be observed within a few seconds;

(3)   Phytochrome-controlled turgor changes in the motor cells of *Albizzia julibrissin* can be directly related to massive fluxes of potassium ions through the membrane (Satter, Marinoff and Galston, 1970).

Because of these considerations, the question of where phytochrome is localised at the cellular and subcellular level is highly correlated with the search for the primary action mechanism. For a long time, phytochrome was believed to be a soluble chromoprotein and association of phytochrome with membranes or organelles was not evident from early biochemical work. However, recent physiological investigations have indicated that at least some of the phytochrome has physiologically significant associations with subcellular structures. Experiments on the localisation of phytochrome in the green alga *Mougeotia* suggest

[1]Present address: Laboratoire de Physiologie Végétale, Université de Nice, 28 Avenue Valrose, 06-Nice, France

that a fraction of the pigment must be located and oriented at or very near to the plasma membrane (Haupt, Mörtel and Winkelnkemper, 1969). In addition, physiological activity has been reported for isolated oat etioplasts; structural development of the etioplast can be regulated by red and far-red light, indicating that phytochrome is associated with this organelle (Wellburn and Wellburn, 1973). Phytochrome-dependent reduction of NADP in the mitochondrial fraction, isolated from etiolated pea epicotyls, indicates the occurrence of the pigment in mitochondria (Manabe and Furuya, 1974). Irradiation of corn coleoptile cells with polarised red and subsequently with far-red light reveals significant differences in the amount of phytochrome phototransformed, depending upon whether the plane of polarisation is parallel or perpendicular to the longitudinal axis of the cells (Marmé and Schäfer, 1972). Fixed orientation of phytochrome in the plasma membrane seems to be a probable explanation of this result. Immunochemical assay at the subcellular level reveals that most of the phytochrome in etiolated oats and rye is distributed as Pr more or less uniformly throughout the cytoplasm and might be associated with the nuclear envelope, amyloplasts and mitochondria (Coleman and Pratt, 1974). Any other association of either Pr or Pfr with a membrane has not been demonstrated by this technique. Finally, phytochrome-induced conductance changes in a model membrane were interpreted as Pfr binding to the oxidised cholesterol membrane (Roux and Yguerabide, 1973).

Thus, many different considerations have led to the hypothesis that phytochrome is functionally associated, at least as Pfr, with one or more cell membranes. The first evidence for particulate phytochrome after cell fractionation was published by Solon Gordon in 1961 (Gordon, 1961). The next paper dealing with the pelletability of phytochrome from extracts appeared 8 years later (Rubinstein, Drury and Park, 1969). More recent studies have demonstrated that phytochrome binds specifically in its far-red absorbing form, Pfr, to particulate material *in vivo* and *in vitro* (Quail, Marmé and Schäfer, 1973; Marmé, Boisard and Briggs, 1973; Marmé *et al.*, 1974; Grombein *et al.*, 1975). In this paper, binding of phytochrome *in vivo* and *in vitro* to particulate structures from corn and zucchini is described and the differences between *in vivo* and *in vitro* induction of binding and between corn and zucchini are discussed.

Materials and Methods

Seeds of zucchini squash (*Cucurbita pepo* L. cv. Black Beauty) and corn grains (*Zea mays* L., WF 9 × 38 from Bear) were germinated in darkness for 6 days at 25 °C. The hypocotyl hooks (about 1 cm in length) from squash seedlings and corn coleoptiles with primary leaves (about 2 cm in length) were harvested in dim green safelight and immediately placed on ice. All manipulations — irradiation *in vivo* and *in vitro* and extraction — were carried out on ice or at 1 °C. Actinic red light was obtained from a 660-nm interference filter of about 15-nm half bandwidth.

The methods used to obtain particulate phytochrome have been published in detail elsewhere (Quail, Marmé and Schäfer, 1973; Marmé, Boisard and Briggs, 1973; Grombein *et al.*, 1975) and are summarised below. The extraction buffer was normally made by mixing a 25 mM *N*-morpholino-3-propanesulphonic acid (MOPS) solution with a 25 mM trishydroxymethylaminomethane (Tris) solution to give the pH indicated for the different experiments. The buffers contained 0.1 per cent of 2-mercaptoethanol. Magnesium chloride and ethylenediaminetetraacetic acid (EDTA) were added where indicated. For the experiments represented in *Figures 8.1–8.4*, 15-s extractions in a homogeniser (Bühler, Germany) were used. In all other experiments, the tissue was chopped in the extraction buffer with a razor blade and then ground in a mortar. The pellets, containing the bound phytochrome, were re-suspended in half of the extraction volume. Phytochrome was assayed using calcium carbonate as a scattering agent (Butler and Norris, 1960) and a Perkin-Elmer 356 spectrophotometer modified for automatic phytochrome measurements (Schmidt, 1974). The sample thickness was 3 mm. The ratio of calcium carbonate to test volume was always 0.5 g to 0.5 ml. Such a sample, when prepared from a dark homogenate after homogenisation of 1 g of zucchini hypocotyl hooks in 4 ml of extraction buffer, which was the normal ratio of tissue to buffer, gave an average signal of 1.5×10^{-2} absorbance units. The measuring wavelengths were 660 and 730 nm.

Phytochrome Binding in Zucchini

It has been documented that phytochrome pelletability from crude extracts can be enhanced by irradiation with red light either of the tissue prior to extraction (*in vivo*) or of the crude extract (*in vitro*) (Marmé, Boisard and Briggs, 1973; Marmé *et al.*, 1974; Boisard, Marmé and Briggs, 1974). Results have been published indicating differences between *in vivo* and *in vitro* induced binding:

(1) Phytochrome binding induced *in vivo* by red light could be partially reversed in this system by subsequent far-red irradiation *in vivo* (Marmé, Boisard and Briggs, 1973; Boisard, Marmé and Briggs, 1974), whereas binding induced by red irradiation of the extract is not reversed but enhanced by subsequent exposure to far-red light (Marmé, Boisard and Briggs, 1973).
(2) A difference in the pH dependence of pelletability with respect to *in vivo* or *in vitro* irradiation has also been reported (Marmé, 1975).

In this chapter, these differences between *in vivo* and *in vitro* binding are examined further. Specifically, the Mg^{2+} and pH dependences

of the phytochrome pelletability from zucchini hypocotyl hooks irradiated *in vivo* are compared with those obtained after irradiation of the
crude extract.

Zucchini hypocotyl hooks are irradiated on ice for 3 min with red
light *in vivo* and the tissue is extracted with MOPS-Tris buffer at pH
7.3 containing various concentrations of Mg^{2+}. *Figure 8.1* (open
triangles) shows that even in the absence of Mg^{2+} more than 50 per
cent of the total extractable pigment is pelletable at 20 000 g for 30
min. At 10 mM Mg^{2+}, the percentage of bound phytochrome increases
to over 80 per cent and decreases again at higher Mg^{2+} concentrations.
Even at 100 mM Mg^{2+}, more than 30% remains bound. The Mg^{2+}
dependence after irradiation of the 500 g supernatant obtained after
centrifugation of the homogenate at 500 g for 10 min is significantly
different (*Figure 8.1,* closed circles). Without Mg^{2+} added to the 500 g

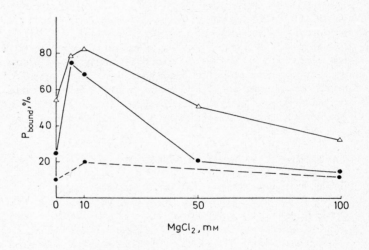

*Figure 8.1   Pelletable zucchini phytochrome as a function of the Mg^{2+} concen-
tration in the extraction medium.   The homogenate was centrifuged at 500 g
for 10 min to remove large debris.   The remaining supernatant (0.5 KS) was
irradiated for 3 min with red light or kept in darkness and centrifuged at 20 000 g
for 30 min.   Pellet, after re-suspension in extraction buffer (without Mg^{2+}), and
supernatant were assayed for phytochrome. P_{bound} is calculated as a percentage
of total phytochrome.   Dark control (●---●); 3 min red light in vivo (△—△);
3 min red light in vitro to 0.5 KS (●——●)*

supernatant, the amount of pelletable phytochrome exceeds the dark control (*Figure 8.1,* broken line) by a factor of two but is still only half that obtained after *in vivo* induction. The maximum for both curves is not significantly different, but at higher Mg^{2+} concentrations more phytochrome is pelletable when red light is applied *in vivo*.

The pH dependence of binding has been investigated both without Mg^{2+} and with 10 mM Mg^{2+} in the extraction buffer. *Figure 8.2a* indicates that in the presence of 10 mM Mg^{2+} more phytochrome is pelletable after irradiation *in vivo* (open triangles) than after red light has been applied to the 500 g supernatant (closed circles), especially at pH 8. The pH values on the abscissa are those of the final homogenates. In the absence of Mg^{2+} (*Figure 8.2b*) the pelletability of phytochrome becomes significantly different from the dark control (broken line) only when red light is applied *in vivo*. It has been shown previously that bound phytochrome obtained after *in vivo* irradiation with red light and extraction without Mg^{2+} at pH 6.6 (in the homogenate) cannot be released to the soluble fraction by increasing the pH of the homogenate after extraction (Marmé, 1975).

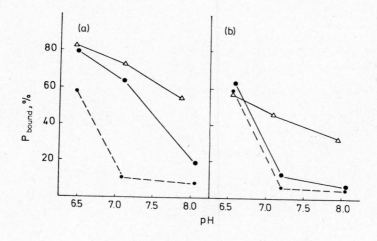

Figure 8.2   Pelletable zucchini phytochrome as a function of pH of the homogenate in the presence of (a) 10 mM Mg^{2+} and (b) in the absence of Mg^{2+}. Experimental protocol is the same as in Figure 8.1. Dark control (●---●); 3 min red light in vivo (△——△); 3 min red light in vitro to 0.5 KS (●——●)

Phytochrome Binding in Corn

Phytochrome binding in corn has been investigated under the same conditions as for zucchini. The Mg^{2+} dependence is shown in *Figure 8.3*. When corn coleoptiles are irradiated *in vivo*, without Mg^{2+} in the extraction buffer, no significant binding is observed. At appropriate Mg^{2+} concentrations, as much phytochrome can be pelleted from crude extracts as for zucchini, about 80 per cent at 10 mM Mg^{2+}. At 100 mM Mg^{2+}, the pelletability decreases only slightly. No considerable induction of binding by red light applied *in vitro* is observed. This result confirms earlier findings from our laboratory that Mg^{2+}-dependent *in vitro* binding as observed with zucchini is absent (Marmé, 1975). Recently published data indicate, in contrast, that for corn there is as much as 37 per cent binding inducible *in vitro* (Quail, 1974). The reason for this discrepancy might be the temperature at which binding was induced: we used 1 °C whereas Quail (1974) irradiated the extracts at room temperature. *In vitro* studies in crude extracts from oats also show no increased phytochrome pelletability induced by red light (Grombein *et al.*, 1975). It has to be noted, however, that even in darkness more than 20 per cent of the pigment can be pelleted in the

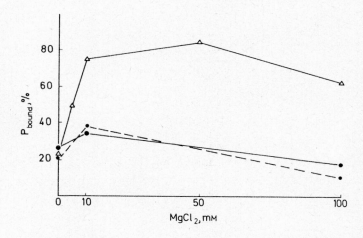

Figure 8.3   Pelletable corn phytochrome as a function of the Mg^{2+} concentration in the extraction medium.   The homogenate was irradiated for 3 min with red light or kept in darkness and centrifuged at 20 000 g for 30 min.   Pellet, after re-suspension in extraction buffer (without Mg^{2+}), and supernatant were assayed for phytochrome.   Dark control (●---●); 3 min red light in vivo (△——△); 3 min red light in vitro to the homogenate (●——●)

absence of Mg^{2+} and nearly 40 per cent in the presence of 10 mM Mg^{2+} (*Figure 8.2,* broken line).

The pH dependence of phytochrome pelletability after irradiation with red light *in vivo* (*Figure 8.4,* open triangles) does not differ qualitatively from zucchini. Binding can be obtained only by *in vivo* irradiation and extraction with Mg^{2+} present in the extraction buffer. Again, *in vitro* induced binding has not been detected (*Figure 8.4,* closed circles).

To summarise the data presented in *Figures 8.1–8.4,* in both corn and zucchini irradiation with red light *in vivo* enhances the pelletability of phytochrome from crude extracts. Zucchini phytochrome becomes bound to particulate material even in the absence of Mg^{2+}. Red light *in vitro* induces considerable binding in crude extracts from zucchini only in the presence of Mg^{2+}, while *in vitro* binding cannot be detected in crude extracts from corn coleoptiles.

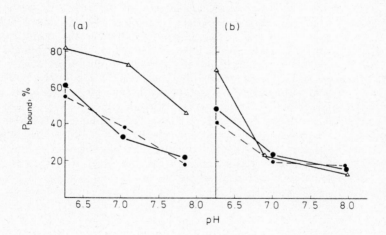

Figure 8.4 Pelletable corn phytochrome as a function of pH of the homogenate in the presence of (a) 10 mM Mg^{2+} and (b) in the absence of Mg^{2+}. Experimental protocol is the same as in Figure 8.3. *Dark control (●---●); 3 min red light* in vivo *(△——△); 3 min red light* in vitro *to the homogenate (●——●)*

Specificity of Binding

Maximum specificity can be defined as the molecular interaction of a molecule A with a molecule B with no substitute for either component. A might be a substrate and B an enzyme or A might be a peptide–hormone and B a membrane receptor. Limited specificity is defined by the possibility of a limited number of substitutes for A and/or B. A lack of specificity means that molecule A interacts with a large number of chemically different molecules. It is obvious that phytochrome binding is specific as far as the ligand is considered, since Pfr is bound in both corn and zucchini to a much higher degree than Pr. But for the entire binding process, which includes the nature of the receptor, the specificity must be proved. For practical reasons, we used the *in vitro* binding system from zucchini to determine whether Pfr binding is specific. To different amounts of a dark extract of zucchini hooks, a constant volume of Pfr from zucchini, as obtained by purification through a brushite column (Pratt, 1973), is added. Bound Pfr is then pelleted at 20 000 *g* for 30 min. In *Figure 8.5,* the bound Pfr is plotted as a function of the amount of dark extract. The broken line represents the amount of Pfr added to the samples. At low concentrations of extract, not all of the Pfr added to the homogenate is bound, which might indicate that the reaction partner or the binding sites are limited. A limited number of binding sites is *more* indicative of specificity of binding. Addition of bovine serum albumin at a concentration 100 times that of the total soluble proteins does not change the amount of bound Pfr. This again is indicative of specific Pfr binding. One should not, however, overstress the importance of the need for specificity of the binding reaction as long as one knows nothing about the function of phytochrome binding. It might be that some function requires Pfr binding which does not fulfil the postulate of specific binding. On the other hand, there might be a limited number of sites where Pfr binds preferentially but which have no physiological significance.

Evidence for Phytochrome in a Plasma Membrane Fraction

Phytochrome binding in corn coleoptiles can be induced only by irradiation with red light *in vivo* (*Figure 8.3*). Increased pelletability of Pfr requires Mg^{2+} in the extraction medium. In the absence of Mg^{2+}, the amount of pelletable phytochrome after irradiation with red light is not different from the dark control. This pelletable phytochrome cannot be washed off with 10 mM EDTA.

Such phytochrome-containing pellets were obtained from corn coleoptiles in the absence of Mg^{2+} and in the presence of 10 mM EDTA. Larger cell debris and most of the mitochondria were removed by differential centrifugation (500 *g* for 10 min and 12 000 *g* for 15 min). The pellets were carefully re-suspended, layered on top of a 20–50 per

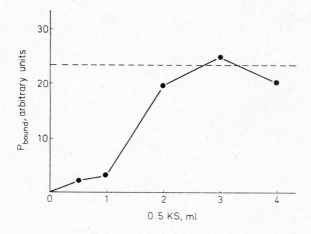

0.5 KS, ml

Figure 8.5 Pelletable phytochrome in arbitrary units as a function of the amount of binding structures. Phytochrome from zucchini hypocotyl hooks was purified on a brushite column. Various amounts of 0.5 KS (containing phytochrome binding material and Pr) were added to buffer of the same pH and Mg^{2+} concentration (10 mM) to give a final volume of 4 ml. A constant amount of phytochrome in the form Pfr was added to each sample (---). The mixture was centrifuged at 20 000 g for 30 min and phytochrome was assayed in the pellets (●)

cent (w/w) linear sucrose gradient and centrifuged for 3 h at 23 000 r.p.m. in a Beckman SW 25/2 rotor. After fractionation of the gradients, the profiles of phytochrome, cytochrome c oxidase, NADH–cytochrome c reductase, specific NPA binding and the sterol/phospholipid-phosphor were measured.

Phytochrome activity is found in a broad band with the greatest activity between 30 and 40 per cent sucrose (*Figure 8.6*). Mitochondria and endoplasmic reticulum, identified by cytochrome c oxidase and NADH-cytochrome c reductase, band sharply at 40 and 25 per cent, respectively (data not shown). Specific NPA binding, which is believed to be a marker for plant plasma membrane (Hertel, Thomson and Russo, 1972), as well as a high sterol content, characteristic for plasma membrane in animals (Colbeau, Nachbaur and Vignais, 1971) and plants (Hodges *et al.*, 1972; Hartmann, Normand and Benveniste, 1975), match the phytochrome activity (*Figure 8.6*).

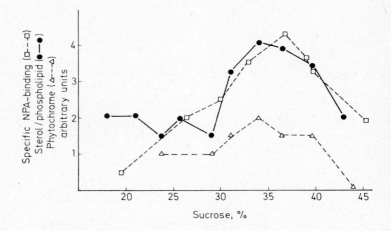

Figure 8.6   Sucrose density gradient profiles for phytochrome (Δ---Δ), specific NPA binding (□ --- □) and the ratio of sterols to phospholipid-phosphorus (●——●). Phytochrome-containing particulate material was obtained from corn coleoptiles by extracting dark-grown or red-irradiated plants in a MOPS–Tris buffer containing no Mg^{2+} but 10 mM EDTA.   Larger cell debris and most of the mitochondria were removed by differential centrifugation (500 g for 10 min and 12 000 g for 15 min).   The resulting supernatant was centrifuged at 50 000 g for 30 min.   The resulting pellet was carefully re-suspended, layered on top of a 20–50 per cent (w/w) linear sucrose gradient and centrifuged for 3 h at 23 000 r.p.m. in a Beckman SW 25/2 rotor.   For NPA binding assay, see Hertel, Thomson and Russo (1972); for sterol and phospholipid assay, see Kohsen (1974)

Evidence of Phytochrome Binding to Rough Endoplasmic Reticulum

Bound phytochrome, obtained by irradiation of crude extracts from zucchini hypocotyl hooks with red light, is pelletable at 20 000 g for 30 min only in the presence of Mg^{2+} (*Figure 8.1*).   In the absence of Mg^{2+}, Pfr becomes pelletable after *in vitro* irradiation only at higher centrifugation speeds.   Centrifugation at 100 000 g for 1 h pellets about 50 per cent of the extractable pigment whereas at 20 000 g for 30 min only 25 per cent is pelleted (*Figure 8.1*).   In previous work, we characterised the particulate fraction to which Pfr is bound in the absence of Mg^{2+} on sucrose density gradients (Marmé, 1975).   *Figure 8.7a* shows gradient profiles for phytochrome.   Phytochrome is plotted as a function of sucrose density.   The soluble zucchini Pr migrates

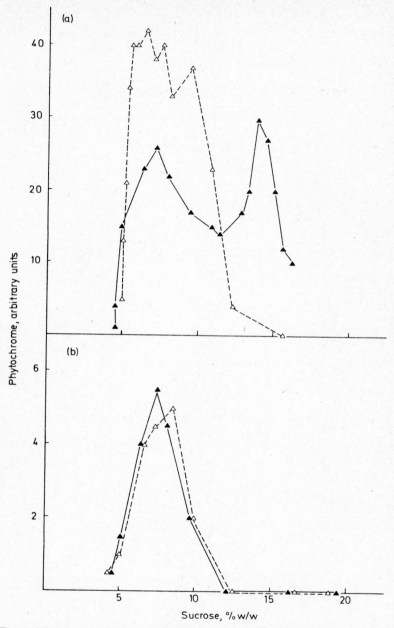

Figure 8.7 (a) Pr (△---△) and Pfr (▲——▲) profiles from linear sucrose density gradients. The crude homogenate obtained from dark-grown zucchini hooks was freed from cell organelles and most of the membrane residues by centrifuging for 1 h at 50 000 g. The resulting supernatant was separated into two parts. One sample was layered under dim green light on top of a gradient. The second sample was irradiated with red light before layering on the gradient. Both gradients were run for 16 h at 23 000 r.p.m. in a Beckman SW 25/2 rotor. (b) Corresponding profiles for Pr (△---△) and Pfr (▲——▲) from corn coleoptiles

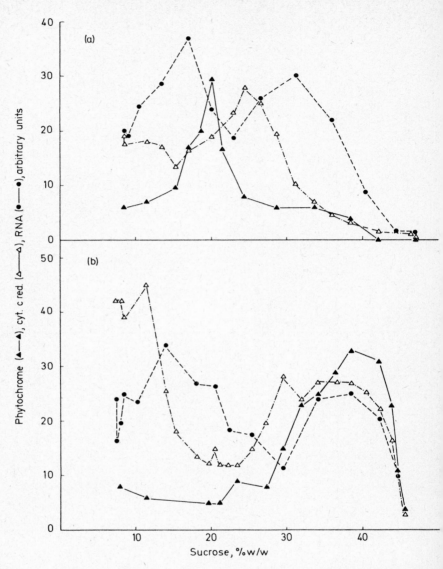

Figure 8.8 (a) Phytochrome (▲——▲), NADH–cytochrome c reductase (△-·-△) and RNA (●---●) profiles from linear sucrose density gradients. Dark-grown hypocotyl hooks from zucchini were extracted in the absence of Mg^{2+}, the homogenate was centrifuged for 10 min at 500 g and subsequently for 15 min at 10 000 g to remove large debris and most of the mitochondria. The resulting supernatant was irradiated for 3 min with red light and layered on top of a sucrose gradient ranging from 10 to 50% (w/w) sucrose. The gradient was centrifuged for 3 h at 23 000 r.p.m. in an SW 25/2 rotor. (b) Same experimental protocol but 5 mM Mg^{2+} was included in the extraction buffer

after 16 h at 23 000 r.p.m. in a Beckman SW 25/2 rotor to 7 per cent sucrose (*Figure 8.7a,* open triangles). When red light is applied to the zucchini homogenate prior to centrifugation, a Pfr-containing particulate fraction migrates further into the gradient to about 15 per cent sucrose (*Figure 8.7a,* closed triangles). This complex is probably identical with the 31S particle, containing bulk ribonucleic acid and proteins, described by Quail and Gressel (Chapter 9). *Figure 8.7b* indicates that Pfr (closed triangles) from corn does not show such behaviour. Pfr remains as soluble as Pr (open triangles). These findings might be related to the fact that in crude extracts from corn no significant binding of Pfr can be induced by red light (*Figures 8.3 and 8.4*).

Either addition of 5 mM Mg^{2+} causes this complex, banding at 15 per cent sucrose (*Figure 8.7a*), to sediment together with· the NADH–cytochrome c reductase, or the presence of Mg^{2+} prevents the 31S particle from being formed. *Figure 8.8* shows the sucrose gradient profiles of phytochrome, RNA and NADH–cytochrome c reductase as a function of sucrose density at two different Mg^{2+} concentrations. In the absence of Mg^{2+} (*Figure 8.8a*), one again obtains a 'light' phytochrome peak, which probably represents the 31S particle observed by Quail and Gressel (Chapter 9), and a heavier shoulder. The shoulder matches the 'heavy' RNA peak but not the NADH–cytochrome c reductase. In the presence of 5 mM Mg^{2+}, all three peaks overlap (*Figure 8.8b*). The endoplasmic reticulum is shifted to higher sucrose densities while the 'heavy' RNA peak does not move markedly. The phytochrome activity profile matches the NADH–cytochrome c reductase and the 'heavy' RNA peak.

Conclusion

The results presented in this chapter clearly indicate that the properties of phytochrome binding depend on the plant used, on whether red light, needed to induce binding, is applied to the intact plant or to the crude extract, and on the composition of the extraction buffer. In zucchini, one is able to obtain pelletable phytochrome after irradiation of the hypocotyl hooks with red light prior to extraction and after irradiation of the crude extracts with red light (*Figure 8.1,* open triangles and closed circles). An increase in pelletable phytochrome from corn coleoptiles occurs only when red light is applied *in vivo;* irradiation of the crude extract with red light does not induce a significant increase in pelletable phytochrome (*Figure 8.3,* open triangles and closed circles). The corn system thus reflects the most pronounced difference between pelletable phytochrome after irradiation *in vivo* and *in vitro.* The same is true for oats: there is no effect of red light *in vitro,* but after irradiation with red light *in vivo* about 70 per cent of the extractable phytochrome becomes pelletable (Grombein *et al.,* 1975). Even in zucchini, where red light applied to crude extracts also has a pronounced effect on the pelletability of phytochrome, differences between the results of red light applied *in vivo* and *in vitro* appear when pelletable phytochrome is studied as a function of Mg^{2+} concentration and pH. The Mg^{2+} dependence is not as pronounced when red light is applied *in vivo* (*Figure 8.1*, open triangles) as it is *in vitro* (*Figure 8.1*, closed circles).   The same is true for the pH dependence in the presence of 10 mM Mg^{2+} (*Figure 8.2a*). The pH dependence is much more pronounced after irradiation *in vitro* (*Figure 8.2a*, closed circles).   Using the correct conditions for zucchini, one is able to separate the binding which takes place *in vivo* from that which takes place *in vitro*: in the absence of Mg^{2+} at pH 8, binding can be induced only by red light *in vivo* (*Figure 8.2b*, open triangles).   These results indicate that there are at least two phytochrome binding systems.   In corn, one is dealing with just one system: binding can be induced at 0 °C by red light only *in vivo*.   In zucchini, both binding systems are present: pelletability of phytochrome can be increased after irradiation with red light *in vivo* or *in vitro*.

In the absence of Mg^{2+}, extracts obtained from corn coleoptiles contain more than 20 per cent of total phytochrome in a pelletable fraction, regardless of whether the pigment is in its Pr or Pfr form (*Figure 8.3*).   Fractionation on sucrose gradients yields most of the phytochrome activity in a broad band between 30 and 40 per cent sucrose. Maximum activity of specific NPA binding and an increased ratio of sterol to phospholipid are found in the same region (*Figure 8.6*).   Both are suggested to be markers of the plant plasma membrane (Hertel, Thomson and Russo, 1972; Hartmann, Normand and Benveniste, 1975). These results are in agreement with the suggestion that phytochrome is a component of the plasma membrane in corn coleoptile cells.   This suggestion arose also from measurements of phytochrome phototransformation in corn coleoptile cells using polarised light (Marmé and Schäfer, 1972), and agreed with the conclusion of Haupt based on physiological experiments (Haupt, Mörtel and Winkelnkemper, 1969).

The particulate material from zucchini capable of binding phytochrome preferentially in its Pfr form *in vitro* has been fractionated on sucrose density gradients.   In the absence of Mg^{2+} and after centrifugation at 50,000 *g* for 1 h, the remaining Pr is mostly soluble and bands between 5 and 12 per cent sucrose (*Figure 8.7a*, open triangles).   In previous work, it was shown that the Pfr activity at about 15 per cent sucrose can be eliminated by adding 100 mM sodium chloride or by adjusting the pH of the gradient to 8 (Marmé, 1975).   Both conditions prevent pelletability of phytochrome (Marmé, 1975; this chapter, *Figure 8.2*). It is probable that in this case Pfr is associated with 31S ribonucleo-proteins, as suggested by Quail and Gressel (Chapter 9).   Corn phytochrome seems to remain soluble after irradiation with red light *in vitro* (*Figure 8.7b*).

Sucrose gradients prepared without Mg^{2+} again show, in the presence of phytochrome binding structures, a 'light' phytochrome peak, apparently identical with the 31S particle (Quail and Gressel, Chapter 9), and a broad shoulder at higher sucrose densities (*Figure 8.8a*, closed triangles).

Activity of NADH–cytochrome *c* reductase (*Figure 8.8a*, open triangles), a common marker for endoplasmic reticulum, does not coincide with this phytochrome activity, but RNA (*Figure 8.8a*, closed circles) does. The RNA activity at a higher sucrose density value is probably due to ribosomes attached to the endoplasmic reticulum (Lord *et al.*, 1973). This observation might indicate that the endoplasmic reticulum fraction in the absence of Mg^{2+} is very heterogeneous. All endoplasmic reticulum vesicles are represented by the NADH–cytochrome *c* reductase, whereas the 'rough' fraction might be represented by the RNA peak between 25 and 35 per cent sucrose. On the basis of the above consideration, the shoulder of phytochrome activity at higher sucrose densities (*Figure 8.8a*) would represent phytochrome associated with the 'rough' fraction of the endoplasmic reticulum. After addition of 5 mM Mg^{2+}, the NADH–cytochrome *c* reductase activity shifts to higher sucrose densities, presumably owing to fixation of ribosomes. Most of the phytochrome as well as the RNA is found in a broad band around 38 per cent sucrose (*Figure 8.8b*). This phytochrome appears to sediment with the heavier fractions of the endoplasmic reticulum, indicating an association with rough endoplasmic reticulum. Using an independent approach, association of phytochrome with rough-surfaced fragments of endoplasmic reticulum has also been suggested, using soybean hypocotyls (Williamson, Morré and Jaffe, 1975).

This tentative identification of phytochrome-containing plasma membrane and phytochrome binding to rough endoplasmic reticulum does not exclude phytochrome in or bound to other organelles or structures as suggested by others (Quail and Gressel, Chapter 9; Wellburn and Wellburn, 1973; Manabe and Furuya, 1974). We are dealing with large amounts of pelletable phytochrome (20–80 per cent of extractable pigment) and it cannot be excluded that other particles contain minor amounts of phytochrome which would not be seen in our assays.

As has been outlined in the section on specificity of Pfr binding, the binding data become significant only when a physiological response can be related to the process of binding. Thus, a search for such a physiological response will be a major aim in our further work.

Acknowledgements

The authors are grateful to Ms Angelika Schäfer for valuable technical assistance. This work was supported by the Deutsche Forschungsgemeinschaft (SFB46).

References

BOISARD, J., MARMÉ, D. and BRIGGS, W.R. (1974). *Plant Physiol.*, **54**, 272

BUTLER, W.L. and NORRIS, K.H. (1960). *Archs Biochem. Biophys.*, **87**, 31

COLBEAU, A., NACHBAUR, J. and VIGNAIS, P.M. (1971). *Biochim. Biophys. Acta*, **249**, 462

COLEMAN, R.A. and PRATT, L.H. (1974). *Planta*, **121**, 119

GORDON, S.A. (1961). In *Progress in Photobiology*. Ed. Christensen, B.C. and Buchman, B. Elsevier, Amsterdam

GROMBEIN, S., PRATT, L.H., RÜDIGER, W. and MARMÉ, D. (1975). *Plant Sci. Lett.*, **5**, 275

HARTMANN, M.A., NORMAND, G. and BENVENISTE, P. (1975). *Plant Sci. Lett.*, **5**, 287

HAUPT, W., MÖRTEL, G. and WINKELNKEMPER, J. (1969). *Planta*, **88**, 183

HERTEL, R., THOMSON, K.-St. and RUSSO, V.E.A. (1972). *Planta*, **107**, 325

HODGES, T.K., LEONARD, R.T., BRACKER, C.E. and KEENAN, T.W. (1972). *Proc. Nat. Acad. Sci. U.S.A.*, **69**, 3307

KOHSEN, M. (1974). *Staatsexamensarbeit*, University of Freiburg, W. Germany

LORD, J.M., KAGAWA, T., MOORE, T.S. and BEEVERS, M. (1973). *J. Cell Biol.*, **57**, 659

MANABE, K. and FURUYA, M. (1974). *Plant Physiol.*, **53**, 343

MARMÉ, D. and SCHÄFER, E. (1972). *Z. Pflanzenphysiol.*, **67**, 192

MARMÉ, D., BOISARD, J. and BRIGGS, W.R. (1973). *Proc. Nat. Acad. Sci. U.S.A.*, **70**, 3861

MARMÉ, D., MACKENZIE, J.M. Jr., BOISARD, J. and BRIGGS, W.R. (1974). *Plant Physiol.*, **54**, 263

MARMÉ, D. (1975). *J. Supramol. Struct.*, **2**, 751

NEWMAN, I.A. and BRIGGS, W.R. (1972). *Plant Physiol.*, **50**, 687

PRATT, L.H. (1973). *Plant Physiol.*, **51**, 203

QUAIL, P.H., MARMÉ, D. and SCHÄFER, E. (1973). *Nature New Biol.*, **245**, 189

QUAIL, P.H. (1974). *Planta*, **118**, 345

ROUX, S.J. and YGUERABIDE, J. (1973). *Proc. Nat. Acad. Sci. U.S.A.*, **70**, 762

RUBINSTEIN, D., DRURY, K.S. and PARK, R.B. (1969). *Plant Physiol.*, **44**, 105

SATTER, R.L., MARINOFF, P. and GALSTON, A.W. (1970). *Am. J. Bot.*, **57**, 912

SCHMIDT, W. (1974). *Anal. Biochem.*, **59**, 91

TANADA, T. (1968a). *Proc. Nat. Acad. Sci. U.S.A.*, **59**, 376

TANADA, T. (1968b). *Plant Physiol.*, **43**, 2070

WELLBURN, F.A.M. and WELLBURN, A.R. (1973). *New Phytol.*, **72**, 55

WILLIAMSON, F.A., MORRÉ, D.J. and JAFFE, M.J. (1975). *Plant Physiol.*, **56**, 738

9

PARTICLE-BOUND PHYTOCHROME: INTERACTION OF THE PIGMENT WITH RIBONUCLEOPROTEIN MATERIAL FROM *CUCURBITA PEPO* L.[1]

PETER H. QUAIL
JONATHAN GRESSEL[2]
Research School of Biological Sciences, Australian National University, Box 475, P.O., Canberra, A.C.T. 2601, Australia

Introduction

There is ample evidence that under the appropriate conditions substantial levels of phytochrome will co-pellet, at relatively low g forces, with particulate material from tissue homogenates (Rubinstein, Drury and Park, 1969; Boisard, Marmé and Briggs, 1974; Marmé, 1975; Marmé, Boisard and Briggs, 1973; Marmé *et al.*, 1974; Quail, 1974a,b, 1975a,b; Quail, Marmé and Schäfer, 1973; Quail and Schäfer, 1974; Schäfer, 1975). The major outstanding problems have been firstly, to identify and isolate the component(s) with which the pigment associates in these pellets, and secondly, to demonstrate a biologically meaningful, phytochrome-induced alteration in the properties of that component(s). An approach to the first of these problems has been reported by Marmé and colleagues. On the basis of data obtained with *Cucurbita* hypocotyl hooks, two major conclusions were reached:

(1) That a phytochrome-containing membrane fraction can be isolated from seedlings irradiated with red light *in vivo* prior to extraction (Marmé *et al.*, 1974).
(2) That red irradiation of homogenates from dark-grown seedlings induces the binding *in vitro* of phytochrome to a membrane fraction (Marmé, Boisard and Briggs, 1973) or some 'solubilised' receptor therefrom (Marmé, 1975).

This chapter advances evidence which suggests an alternative interpretation of these data, namely that the phytochrome in the fractions

[1] Paper presented at the Annual European Photomorphogenesis Symposium and invited by Editor for publication in these Proceedings
[2] On sabbatical leave from the Weizmann Institute of Science, Rehovot, Israel

obtained using these published procedures (Marmé, Boisard and Briggs, 1973; Marmé *et al.*, 1974; Marmé, 1975) is associated electrostatically with a ribonucleoprotein component which is probably degraded ribosomal material. In essence, the approach taken was to subject various fractions from homogenates of *Cucurbita* hooks to sucrose gradient centrifugation and to seek correlations between the distribution profiles of phytochrome and other components under a variety of conditions. Gradients were examined for the distributions of phytochrome, cytochrome *c* oxidase (a mitochondrial marker enzyme), NADPH–cytochrome *c* reductase (an endoplasmic reticulum marker), protein and RNA.

Materials and Methods

Details of the plant material, irradiation procedures, homogenisation, sucrose gradient centrifugation and assays used appear elsewhere (Quail, 1975a,b).

A transition-state analogue of the ribonuclease reaction – a complex of uridine and vanadyl sulphate ($VOSO_4$) – was used as an inhibitor of ribonuclease activity in some experiments (Gray, 1974). $VOSO_4$ and uridine at final concentrations of 2 and 10 mM respectively, were added to the normal (Quail, 1975a) extraction, re-suspension and gradient media before titration to the desired pH. Sterile water and autoclaved glassware and utensils were used in the preparation of the media and through out such experiments in order to minimise extraneous ribonuclease activity.

Ribosomal pellets were prepared according to Leaver and Dyer (1974) with the following modifications. All media contained 2 mM $VOSO_4$ and 10 mM uridine. Ribosomes were pelleted from the Triton X-100 treated 10 000 *g* supernatant through 1.8 M sucrose at 160 000 *g* for 8 h. Pre-fixation of hypocotyl hooks was performed on ice using 2.5 per cent glutaraldehyde for 30 min as described by Yu (1975). Deproteinisation and electrophoresis of RNA on polyacrylamide gels was carried out according to Loening (1969).

Results and Discussion

About 35–40 per cent of the total extractable phytochrome is pelletable at pH 7.0 in the absence of added Mg^{2+} at 20 000 *g* from homogenates of red-irradiated hypocotyl hooks of *Cucurbita*. This amount is little affected by the presence or absence of 3 mM EDTA in the media. The chelator does, however, affect the pattern obtained when 20 000 *g* pellets are subjected to sucrose gradient centrifugation.

Figure 9.1a,b shows the distribution profiles of a pellet from red-irradiated tissue where extraction, re-suspension and centrifugation were performed in the absence of either added Mg^{2+} or EDTA. All of the activities measured are localised more or less in a single broad band in the region of density 1.15–1.20 g cm^{-3}. The inclusion of 3 mM EDTA in the extraction, re-suspension and gradient media has three major effects (*Figure 9.1c,d*):

(1) The phytochrome profile splits into two main bands, one at density 1.16 g cm^{-3} ('heavy') and the other near the top of the gradient ('light').

(2) The main RNA population becomes localised in a sharp peak which is entirely coincident with the light phytochrome band.

(3) The main peak of NADPH–cytochrome *c* reductase activity shifts to a lower density at 1.12 g cm^{-3}.

The apparent association of NADPH–cytochrome *c* reductase and RNA in the absence of EDTA (*Figure 9.1a,b*) suggests the presence of an intact rough ER fraction. Conversely, the separation of these two activities in the presence of 3 mM EDTA (*Figure 9.1c,d*) into a ribonucleoprotein (RNP) component and a major reductase peak of lower isopycnic buoyant density is indicative of an EDTA-induced dissociation of ribosomes from the ER membrane. Such an effect is well documented (Lord *et al.*, 1973; Sabatini, Tashiro and Palade, 1966; Wibo *et al.*, 1971). Further, the density of 1.12 g cm^{-3} seen here for the main reductase peak is identical with that previously reported for plant ER membranes stripped of their ribosomes (Lord *et al.*, 1973). Electron micrographs of the fraction reveal only smooth-surfaced vesicles consistent with this notion (Hughes, Gunning and Quail, unpublished results). The behaviour of at least part of the phytochrome under these conditions is qualitatively parallel to that of the RNP fraction, suggesting that the two are associated.

The predominantly non-pelletable phytochrome extracted from non-irradiated tissue has different sedimentation properties (*Figure 9.1e,f*) to the pigment in the 20 000 *g* pellets from the red-irradiated material (*Figure 9.1a–d*). The major phytochrome peak from the dark controls remains at the interface between the gradient and the loading buffer and is clearly separated from the main RNP band. This implies that Pfr associates more readily with the RNP material than does Pr from non-irradiated tissue.

If the RNP material does derive from dissociated ribosomes, it would not be expected to be at isopycnic equilibrium after the 4 h of centrifugation used in the above experiments. The sedimentation properties of the various components in 20 000 *g* pellets from the red-irradiated tissue were therefore examined using EDTA-containing media (Quail, 1975a) with the following results:

(1) The phytochrome, marker enzymes, RNA and protein in the 'heavy' region of the gradient ($\rho = 1.12$–1.18 g cm^{-3}) come to isopycnic equilibrium within 90 min and remain so for up to 12 h centrifugation.

(2) The lighter phytochrome and RNP peaks, in contrast, continue to

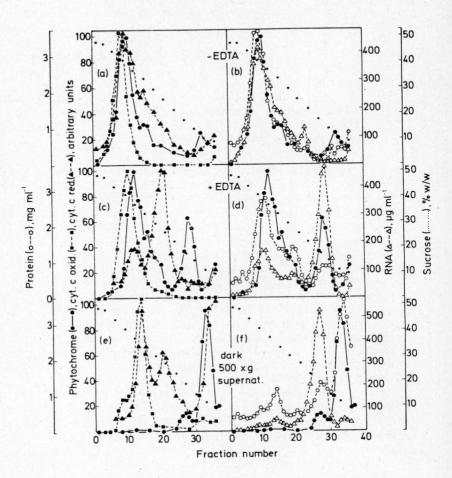

Figure 9.1 Effects of EDTA and in vivo irradiation on sucrose gradient profiles of phytochrome and other cellular components from hypocotyl hooks of Cucurbita seedlings. The distributions of phytochrome, cytochrome c oxidase, cytochrome c reductase, RNA and protein are shown following 4 h centrifugation on 10–50 per cent (w/w) linear sucrose gradients, pH 7.0, at 27 000 r.p.m. in a Beckman SW 27 rotor. Each pair of adjacent panels represents a single gradient. The phytochrome, cytochrome c oxidase and cytochrome c reductase profiles are directly compared in the left-hand panels (a, c, e) and the phytochrome, RNA and protein profiles in the right-hand panels (b, d, f). Gradient a/b: distribution profiles of a 20 000 pellet from red-irradiated hooks using extraction, re-suspension and gradient media containing neither EDTA nor Mg^{2+}. Gradient c/d: distribution profiles of a 20 000 g pellet from red-irradiated hooks using extraction, re-suspension and gradient media free of Mg^{2+} but containing 3 mM EDTA. Gradient e/f: distribution profiles of a 500 g supernatant from non-irradiated hooks extracted and centrifuged in Mg^{2+}-free media containing 3 mM EDTA. (From Quail, 1975a)

co-sediment for up to 12 h of centrifugation. A constant phytochrome to RNA ratio is maintained across the sedimenting peak
throughout the centrifugation period. This is further evidence of
an association between the phytochrome and the RNP material.
This material has a 260:235 nm absorption ratio of 1.36 and a
260:280 nm ratio of 1.87, indicating an RNA content of 35-40
per cent. These values are consistent with those observed for
ribosomal material with substantial levels of extraneously bound
proteins (Petermann, 1964).

   A sedimentation coefficient of 31S has been estimated for the light
phytochrome-RNP band (Quail, 1975a). This indicates that the fraction
does not consist of intact ribosomes or even subunits. Aberrant sedimentation patterns of partially degraded ribosomal material are well
documented, however, particularly where EDTA has been used (Petermann, 1964). The chelator dissociates the subunits and increases
susceptibility to RNase attack (Petermann, 1964; Dyer and Payne, 1974).
Since the present pattern was observed only in the presence of EDTA
(*Figure 9.1*) and no precautions were taken to minimise RNase activity,
the 31S fraction could consist of degraded ribosomal particles. Extraction and analysis of the RNA component supports this view. Both
SDS sucrose gradient analysis (Quail, 1975a) and polyacrylamide gel
electrophoresis of the deproteinised RNA (*Figure 9.3a*) indicate that
the 31S fraction is an aggregate of small RNA fragments with associated
protein.
   Both the origin of the RNP material and the nature of its association
with the phytochrome in the 31S fraction were explored further using
the RNase inhibitor VOSO$_4$-uridine (Gray, 1974). The inhibitor modifies the gradient distribution profiles of phytochrome, RNP and cytochrome *c* reductase (*Figure 9.2*). The phytochrome and RNP, which
co-sediment at 31S in the absence of the inhibitor, both migrate
considerably further into the gradient when VOSO$_4$-uridine is present.
Both the amount of total precipitable RNA recovered from the gradient
and the proportion of the total cytochrome *c* reductase located at
higher buoyant densities are doubled when the inhibitor is used. This
result can be interpreted as indicating that the VOSO$_4$-uridine has

(1)   reduced the degradation of free ribosomes resulting in a higher
      sedimentation coefficient in those particles which would otherwise
      be reduced to 31S
(2)   maintained a higher residual RNP load on rough ER membranes,
      leading to a redistribution of that fraction towards higher buoyant
      densities.

   This view is supported by polyacrylamide gel analyses of the RNA
in the relevant fractions of gradients prepared either with (*Figure 9.2*)
or without (*Figure 9.1c,d*) VOSO$_4$-uridine. Aliquots from the 'heavy'
ER and 31S regions of two such gradients were deproteinised and

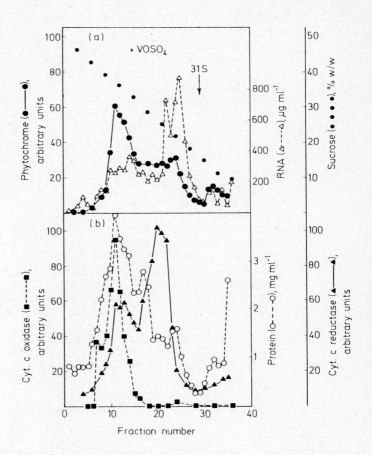

Figure 9.2 Effect of the RNase inhibitor $VOSO_4$-uridine on the sucrose gradient profiles of a 20 000 g pellet from red-irradiated hypocotyl hooks of Cucurbita seedlings. All media were Mg^{2+}-free and contained 3 mM EDTA, 2 mM $VOSO_4$ and 10 mM uridine. The pH of the homogenate and of the re-suspension and gradient media was 7.0. The distributions of (a) phytochrome and RNA and (b) cytochrome c oxidase, cytochrome c reductase and protein are shown following 4 h centrifugation on a 10–50 per cent (w/w) linear sucrose gradient at 27 000 r.p.m. in a Beckman SW 27 rotor

subjected to electrophoresis. When the inhibitor is absent, most of the RNA in all three fractions remains in the supernatant upon ethanol precipitation. This indicates that most of the nucleic acid is present as very small fragments, thus confirming and extending the pattern previously observed for the 31S fraction (Quail, 1975a). Of the small amount of RNA which does precipitate, a large proportion is low-molecular-weight material which migrates with the front, while the remainder appears as a few small discrete peaks of higher molecular weight. *Figure 9.3a,b* shows the gel traces for the 31S and ER fractions. In the presence of the RNase inhibitor, in contrast, most of the RNA is ethanol precipitable and discrete species with molecular weights characteristic of rRNA are recovered from both the major RNP peak (*Figure 9.3c*) and 'heavy' fractions (data not shown). The relative proportions of these RNA species are abnormal, however, and change across the gradient RNP peak, indicating an incomplete inhibition of ribosome nicking (Hsiao, 1968; Payne and Loening, 1970; Leaver, 1973; Dyer and Payne, 1974). Nevertheless, these data provide further strong evidence of the ribosomal origin of the 31S RNP fraction. The recovery of a small proportion of the 31S RNA from the minus inhibitor gradient in the form of discrete species with molecular weights resembling those of rRNA (*Figure 9.3a*) is likewise in accord with this view (Dyer and Payne, 1974).

A final major point arising from these data is that the sedimentation behaviour of the phytochrome in the 'light' fraction is clearly determined by the sedimentation velocity of the RNP material (*Figure 9.2*). This in turn depends on the degree of integrity of the RNA component, as discussed above. This is further evidence for a *direct* association between the phytochrome and RNP material and against the suggestion that the pigment in this fraction might be associated with tiny vesicles or membrane fragments which fortuitously also have a sedimentation coefficient of 31S. The possibility of a lipid or lipoprotein 'bridge' in a phytochrome–membrane fragment–RNP complex is, of course, not excluded by these data. However, preliminary experiments have shown that incubation of 20 000 *g* pellets with pancreatic RNase prior to gradient centrifugation results in:

(1) Abolition of the 31S RNA peak with the release of nucleotides and/or low-molecular-weight fragments to the loading buffer.
(2) Concomitant elimination of the 31S phytochrome peak with the released pigment having apparent sedimentation properties similar to those of the soluble molecule. This tends to argue against the phytochrome–RNP association being mediated via a third component, but confirmation of this point awaits more accurate measurement of the S value of the RNase-treated phytochrome.

What other phenomena are potentially explicable in terms of phytochrome–RNP interaction? Arguments presented below suggest that

(1) the effects of divalent cations (Marmé, Boisard and Briggs, 1973; Marmé *et al.,* 1974; Marmé, 1975),

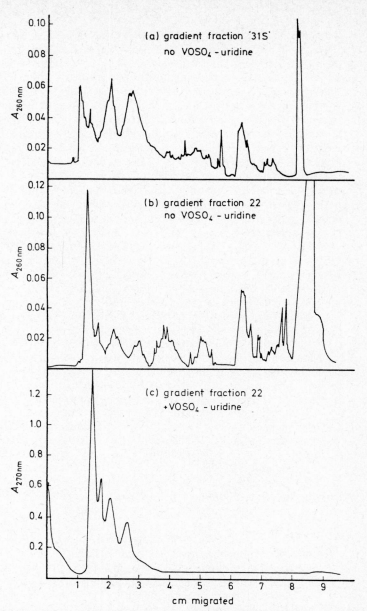

Figure 9.3 Polyacrylamide gel analysis of RNA extracted from sucrose gradient fractions of 20 000 g pellets from red-irradiated hooks of Cucurbita *seedlings prepared in the presence (Figure 9.2) or absence (Figure 9.1c,d) of the RNase inhibitor $VOSO_4$–uridine. (a) The 31S fraction isolated in the absence of $VOSO_4$–uridine (the equivalent of fraction 28, Figure 9.1c,d); (b) fraction 22 of a gradient equivalent to that shown in Figure 9.1c,d where the media contained no $VOSO_4$–uridine; (c) fraction 22 from the major RNA peak of a gradient equivalent to that shown in Figure 9.2 where all media contained $VOSO_4$–uridine. 10 cm long 2.5 per cent polyacrylamide gels prepared according to Loening (1969) were used. These gels were run for 100 min at 50 V in order that low-molecular-weight species would remain in the gels. No additional peaks were resolved by longer electrophoresis. The gels were scanned with a Gilford linear transport. Note the different absorbance scales and wavelengths*

(2) the irradiation-induced association *in vitro* of phytochrome with sedimentable structures (Marmé, Boisard and Briggs, 1973; Marmé, 1975; Quail, 1974b) and

(3) the lack of photoreversibility of this association (Marmé, Boisard and Briggs, 1973; Quail, 1974b) are all manifestations of the same electrostatic adsorption of phytochrome to RNP material.

The behaviour of the 31S fraction towards added Mg^{2+} is identical with that reported by Marmé *et al.* (1974) for their isolated phytochrome-containing 'membrane' fraction. When the cation is added to the isolated 31S material, the samples become turbid, and at 10 mM Mg^{2+} over 90 per cent of the phytochrome and RNA and more than 80 per cent of the protein are pelleted at 20 000 *g* (*Figure 9.4*). Without added Mg^{2+}, less than 5 per cent of any of these components is pelletable. As the Mg^{2+} concentration increases above 10 mM, the amount of pelletable phytochrome decreases dramatically, such that at 100 mM or higher virtually all of the pigment is released to the supernatant. The RNA, in contrast, remains more than 85 per cent pelletable up to 500 mM Mg^{2+}. Protein is released from the pellet with increasing Mg^{2+}, but to a lesser extent than phytochrome. Such effects of Mg^{2+} on ribosomes are well documented (Petermann, 1964). At

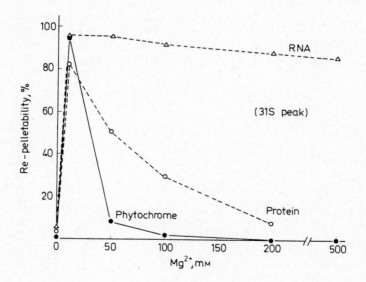

Figure 9.4 Pelletability of the phytochrome, RNA and protein in the 31S ribonucleoprotein fraction in response to added Mg^{2+}. Peak fractions from the 31S band of a 4–h Beckman SW 27 gradient similar to that in Figure 9.1c,d were pooled and divided into aliquots. Mg^{2+} was added to give a range of final concentrations (pH 7.0) and the samples were centrifuged at 20 000 g for 30 min. (From Quail, 1975a)

relatively low concentrations, the cation induces the aggregation and precipitation of ribosomal material, including associated protein. High Mg^{2+} levels also precipitate ribosomes, but in addition, simultaneously displace much of the extraneously bound protein to the supernatant. This procedure is in fact used for the preparation of ribosomes free of bound proteins (Petermann, 1964).

Further confirmation of the electrostatic nature of the interaction between phytochrome and particulate fractions is provided by the effects of high KCl concentrations on the sucrose gradient profiles of 20 000 g pellets (*Figure 9.5*). KCl at 200 mM releases phytochrome from both 'heavy' and 31S fractions. The released pigment accumulates at the loading buffer–gradient interface, clearly separated from the 31S RNA peak, in a manner characteristic of soluble phytochrome (compare with *Figure 9.1e,f*). KCl at 500 mM exaggerates this effect still further, eliminating the minor phytochrome peak associated with the 31S RNA and further decreasing the pigment level in the heavy band.

Mg^{2+} at 10 mM in the extraction medium causes additional amounts of both phytochrome (Marmé *et al.*, 1974) and RNA (*Table 9.1*) to pellet at relatively low g forces. These Mg^{2+}-enhanced levels of both

Table 9.1 Percentage of pelletable phytochrome and RNA from hypocotyl hooks of *Cucurbita* seedlings (20 000 g for 30 min, pH 7.0) as a function of previous *in vivo* irradiation and the presence or absence of 10 mM $MgCl_2$ in extraction medium containing 3 mM EDTA (from Quail, 1975a)

Irradiation[1]	Pelletability (%)[2]			
	Phytochrome		*RNA*	
	$-Mg^{2+}$	$+Mg^{2+}$	$-Mg^{2+}$	$+Mg^{2+}$
Far-red	5	6	30	39
Red	39	54	29	42
Red + far-red	21	24	28	41

[1] Red = 660 nm; far-red = 730 nm; duration = 3 min each wavelength
[2] Percentage pelletability is the amount of a given component in the re-suspended pellet expressed as a percentage of the total amount of that component in the pellet plus the supernatant

components are localised mainly in the 31S fraction (Quail, 1975b). The phytochrome–RNP association is itself not dependent on the added cation, however (e.g. *see Figure 9.1c,d*). This suggests that Mg^{2+}-enhanced phytochrome pelletability results indirectly from the self- or cross-aggregation of phytochrome-associated RNP material into more readily sedimentable complexes.

The phytochrome previously reported to have bound *in vitro* to a 'membrane' fraction following red irradiation of homogenates from

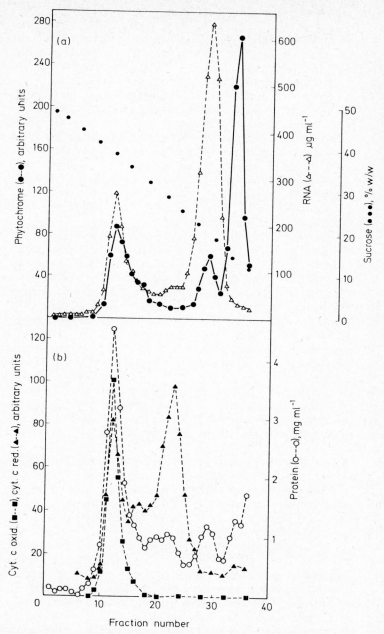

Figure 9.5   Effect of 200 mM KCl on the sucrose gradient profiles of a 20 000 g pellet from red-irradiated hypocotyl hooks of Cucurbita seedlings.   All media were Mg^{2+}-free and contained 3 mM EDTA.   Only the re-suspension buffer contained KCl, pH 7.0.   The distributions of (a) phytochrome and RNA and (b) cytochrome c oxidase, cytochrome c reductase and protein are shown following 4 h centrifugation on a 15–50 per cent (w/w) linear sucrose gradient at 27 000 r.p.m. in a Beckman SW 27 rotor.   (From Quail, 1975b)

non-irradiated tissue (Marmé, Boisard and Briggs, 1973) can also be
shown to be almost exclusively associated instead with the 31S RNP
material.   The phytochrome, RNA and protein pelleted at 50 000 g
from a 20 000 g supernatant adjusted to 10 mM Mg^{2+} are all localised
primarily in a single peak at 31S upon sucrose gradient centrifugation
(*Figure 9.6*).   This is true whether the tissue is irradiated with red
light *in vivo* prior to extraction (*Figure 9.6a*) or the 20 000 g supernatant from non-irradiated, dark-grown material is irradiated with red
light at 25 °C prior to the second centrifugation (*Figure 9.6b*).   The
latter procedure was that used previously to demonstrate the binding
of phytochrome *in vitro* to a putative membrane fraction (Marmé,
Boisard and Briggs, 1973).   Thus the pigment associates readily with
the same 31S RNP fraction whether converted to Pfr *in vivo* prior to
extraction or *in vitro* in the cell-free extract.

Once bound in either case, however, phytochrome is not released by
re-conversion to Pr *in vitro* (Quail, 1975b).   This is again consistent
with the general pattern of electrostatic binding of proteins to ribosomal
material.   It is commonly observed that once binding has occurred during tissue maceration, the proteins remain firmly attached (Bloemendal
and Vennegoor, 1969; Dyer and Payne, 1974; Acton, 1974).   Subsequent dissociation is then considerably more difficult than prevention
of the initial association.   Thus, empirically, the fact that Pfr has a
greater *initial* affinity than Pr for the RNP material does not necessarily
mean that ready dissociation of the pigment should be expected upon
re-conversion of bound Pfr to Pr.   This provides an alternative interpretation of the lack of photoreversibility of binding to the previously
proposed alteration of binding properties of phytochrome 'receptors'
(Marmé, Boisard and Briggs, 1973; Quail, Marmé and Schäfer, 1973;
Boisard, Marmé and Briggs, 1974; Quail and Schäfer, 1974; Schäfer,
1975).

Although the binding of phytochrome to the 31S RNP fraction
resembles the known artifactual adsorption of proteins to ribosomal
material, the question remains as to whether this might not be a
reflection of an *in vivo* phytochrome–ribosome interaction of some
functional significance.   Two types of experiments were performed in
order to examine this problem.

Firstly, attempts were made to determine whether phytochrome
would associate with undegraded ribosomes following irradiation with
red light *in vitro,* with the ultimate aim of testing for phytochrome-
induced functional changes in the particles.   These experiments were
inconclusive.   Ribosomes were isolated from dark-grown *Cucurbita* hooks
as described under *Materials and Methods.*   The ribosomal pellet was
phytochrome-free and the rRNA undegraded, as determined by electrophoresis.   A Pr-containing, ribosome-free supernatant (160 000 g for 2 h)
was also prepared from a homogenate of dark-grown hooks, using the
buffer and extraction procedures normally employed for extracting
pelletable phytochrome (Mg^{2+}-free, no EDTA, final pH 7.0) but containing the $VOSO_4$-uridine complex.   The purified ribosomal pellet was

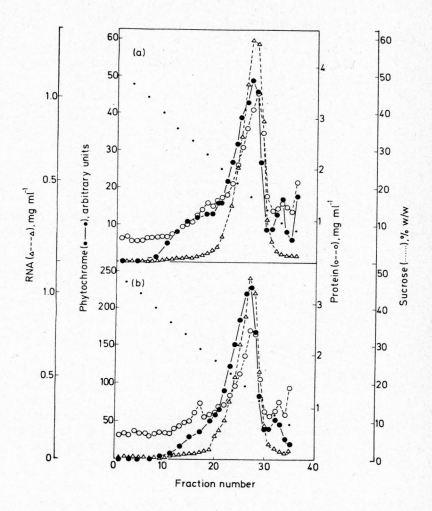

Figure 9.6 Sucrose gradient centrifugation of 20,000–50 000 g (60 min) pellets obtained following addition of 10 mM Mg²⁺ to the 20 000 g (30 min) super-natant fraction from Mg²⁺-free 3 mM EDTA extracts of hypocotyl hooks of Cucurbita *seedlings. The pellets were re-suspended and centrifuged in Mg²⁺-free media containing 3 mM EDTA.   The distributions of phytochrome, RNA and protein are shown following 4 h centrifugation on a 10–50 per cent (w/w) linear sucrose gradient at 27 000 r.p.m. in a Beckman SW 27 rotor.   (a) Pellet obtained by adding 10 mM Mg²⁺ to the 20 000 g supernatant of extracts from hooks irradiated with red light in vivo prior to extraction; (b) pellet obtained by adding 10 mM Mg²⁺ to the 20 000 g supernatant of extracts from non-irradiated hooks, warming the solution to 25 °C and irradiating in vitro with red light prior to the 50 000 g centrifugation.   (From Quail, 1975b)*

Table 9.2 Binding of phytochrome to isolated ribosomes *in vitro*.

A ribosomal pellet was isolated from dark-grown *Cucurbita* hooks as described under *Materials and Methods*. This pellet was added to a ribosome-free, 160 000 g (2 h) supernatant (pH 7.0) from dark-grown hooks extracted in Mg^{2+}- and EDTA-free media containing 2 mM $VOSO_4$ and 10 mM uridine. Samples were irradiated on ice and re-centrifuged at 160 000 g for 2 h.

Irradiation in vitro[1]	Pelletable phytochrome (%)	
	No ribosomal pellet added	Plus ribosomal pellet
Far-red	9	10
Red	16	23

[1] Red = 660 nm; far-red = 730 nm; duration = 3 min each wavelength

then added to the Pr-containing supernatant, the mixture irradiated with red light on ice and the ribosomes re-pelleted at 160 000 g × 2 h. Red light caused about 8 per cent more phytochrome to co-pellet with the ribosomes than in the minus ribosome control (*Table 9.2*). Two factors made this result unsatisfactory. Firstly, it can be calculated that the red light-enhanced binding represents only one phytochrome molecule per 80 ribosomes. Secondly, despite precautions against RNase attack in the supernatant, the integrity of the rRNA was partially lost during the second centrifugation. The original objective of the experiment was thus defeated to a large extent.

In a second type of experiment, ribosomal pellets prepared as described under *Materials and Methods* were tested for phytochrome content as a function of *in vivo* irradiation and pre-fixation in glutaraldehyde (*Table 9.3*). Virtually no phytochrome was associated with ribosomal pellets prepared from red-irradiated tissue with no pre-fixation. The procedures used are designed to prevent adsorption of extraneous proteins to ribosomes (Triton X-100 extraction, pelleting through 1.8 M sucrose, 50 mM KCl, pH 8.5).

Glutaraldehyde fixation of red-irradiated tissue prior to extraction yielded a ribosomal pellet containing about 11 per cent of the total extractable phytochrome, significantly greater than the far-red control. The pre-fixation was designed to preserve any existing intracellular phytochrome–ribosome association during isolation. The 11 per cent phytochrome, however, represents only about one pigment molecule per 60 ribosome equivalents. Further, the pellet thus obtained did not exhibit a sedimentation pattern characteristic of ribosomes upon subsequent sucrose gradient centrifugation. Instead, the material sedimented rapidly to the bottom of the gradient tube, suggesting very large complexes. The composition and state of the material in the pellet were thus in doubt so that conclusions could not be drawn regarding possible phytochrome–ribosome interaction within the cell. The data further suggest that care should be exercised in the interpretation of data obtained with homogenates of glutaraldehyde-treated tissue (Yu, 1975).

Little is known of the nature of the component(s) with which phytochrome is associated in the 'heavy' fraction (*Figure 9.1*). This fraction appears to have been largely discarded by the procedures used in several previous reports of the isolation and analysis of phytochrome-containing components from *Cucurbita* (Marmé, Boisard and Briggs, 1973; Marmé *et al.*, 1974; Marmé, 1975). Therefore, much of the data in these reports would appear to refer primarily to the 31S material.

The rapid arrival at isopycnic equilibrium of the phytochrome in the 'heavy' peak (i.e. within 90 min), and the maintenance of this density over a 12-h centrifugation period (Quail, 1975a), suggest some form of association of the pigment with a membrane fraction(s). The lack of agreement between the phytochrome and the major cytochrome *c* oxidase and reductase peaks (*Figure 9.1b,c*) is evidence against a unique association between the pigment and the mitochondrial or ER membranes *per se*. On the other hand, the apparent coincidence of the phytochrome with the heavy reductase shoulder is of interest as the latter may represent a residual rough ER fraction from which all the ribosomal material has not been dissociated (Sabatini, Tashiro and Palade, 1966; Wibo *et al.*, 1971; Lord *et al.*, 1973). The presence of RNA in that part of the gradient is consistent with this suggestion, although organellar RNA might also account for the observation. One possibility under investigation, therefore, is that the phytochrome in the heavy fraction is also RNP-associated, in this case with residual ER-bound ribosomal material. In addition, attempts are being made to ascertain the distribution of plasma membrane fragments under these conditions for the purpose of seeking correlations with the phytochrome distribution. Such attempts are hampered at present by the unavailability of a reliable plasmalemma marker.

The observation that phytochrome can interact with RNP material presents a serious obstacle to attempts to establish specificity of interaction between pigment and particulate fractions. This problem is

Table 9.3 Percentage of total extractable phytochrome in ribosomal pellets as a function of *in vivo* irradiation and pre-fixation in glutaraldehyde

Hypocotyl hooks from *Cucurbita* were irradiated at 25 °C incubated on ice for 30 min in 2.5 per cent (v/v) glutaraldehyde, washed 6 times in extraction buffer and ribosomal pellets prepared as described under *Materials and Methods*.

Irradiation in vivo[1]	Glutaraldehyde pre-fixation	Pelletable phytochrome (%)
Far-red	+	
Red	+	1.4
Red	−	11.5
		0.6

[1] Red = 660 nm; far-red = 730 nm; duration = 3 min each wavelength

Table 9.4 Estimated ratios of phytochrome to protein, RNA and ribosome equivalents in various fractions

The number of micrograms of phytochrome in each fraction was estimated from the $\Delta(\Delta A)$ readings corrected for $CaCO_3$ scatter, non-saturation of photoconversion by the actinic irradiation and sucrose 'quenching' of the $\Delta(\Delta A)$ signal (Marmé and Gross, 1975). An extinction coefficient of 4×10^4 l mol^{-1} cm^{-1} at 730 nm and a single chromophore per 120 000-dalton monomer were assumed. Ribosome equivalents were calculated by assuming that all of the RNA in the fraction was ribosomal and that a plant ribosome is about 40 per cent RNA and has a molecular weight of 4.3×10^6.

Fraction	Phytochrome per mg of protein (μg)	Phytochrome per mg of RNA (μg)	Phytochrome per mg of ribosome equivalent (μg)	Ribosome equivalents per molecule of phytochrome
500 *g* supernatant[1]	1.0	8.6	−	−
20,000 *g* supernatant[1]	0.8	7.0	−	−
20,000 *g* pellet[1]	1.7	12.2	4.1	6
'Heavy' membrane[1]				
pH 7.0[2]	1.5	34.6	11.5	2
pH 7.6[2]	0.3	6.9	2.3	11
31S RNP[1]				
pH 7.0[2]	2.7	5.4	1.8	14
pH 7.6[2]	8.1	16.2	5.4	5
31S/160KP gradient[1,3]	1.5	2.3	0.9	28
'Ribosomes,' *in vitro* binding[4]	0.5	1.1	0.35	80
'Ribosomes,' glutaraldehyde pre-fixed[5]	0.7	1.1	0.43	58

[1] Red irradiation *in vivo* − from *Table 1* in Quail (1975a) and *Figures 2* and *5* in Quail (1975b)
[2] For discussion of pH effect, *see* Quail (1975b)
[3] For details of 160KP (= 160 000 *g* pellet), *see* Quail (1975b)
[4] From *Table 9.2* (this chapter)
[5] From *Table 9.3* (this chapter)

highlighted in *Table 9.4*, where the numbers of ribosome equivalents per phytochrome molecule in various fractions have been estimated. In no phytochrome-containing fraction so far obtained are there less than the equivalent of two ribosome particles for every phytochrome molecule, and in most cases the number is greater. This large molar excess of RNP over phytochrome, coupled with the well known capacity of ribosomal material to bind large amounts of extraneous proteins, provides adequate potential for the artifactual, electrostatic adsorption of the pigment to the RNP in any of these fractions. An awareness of this problem should hopefully generate caution in the interpretation of studies on the binding of phytochrome to particulate fractions.

Conclusions

Phytochrome can associate electrostatically with a 31S RNP fraction from *Cucurbita* hypocotyl hooks under conditions that have commonly been used in attempts to isolate membrane-bound phytochrome (Marmé, Boisard and Briggs, 1973; Marmé *et al.*, 1974; Marmé, 1975). This RNP fraction is probably degraded ribosomal material with adsorbed proteins, including phytochrome. The phytochrome-containing 'membrane' fraction isolated from red-irradiated *Cucurbita* hooks by Marmé *et al.* (1974) is almost certainly this 31S RNP material. Likewise, the isolated 'membrane' fraction (Marmé, Boisard and Briggs, 1973) and 'solubilised receptor' (Marmé, 1975), to which the pigment could be induced to bind *in vitro*, can both be accounted for by the 31S RNP fraction. The effects of Mg^{2+} on phytochrome pelletability, both in initial homogenates (Marmé *et al.*, 1974) and the isolated 31S fraction (Marmé, Boisard and Briggs, 1973; Quail, 1975a), can be explained by Mg^{2+}-induced aggregation of the RNP material either with itself or other particulate material; as can electron micrographs purporting to document divalent cation-induced 'vesicularisation' of a 'partially solubilised' phytochrome-containing membrane (Marmé *et al.*, 1974). The greater initial affinity of Pfr than of Pr for the RNP material might be the result of a shift in charge distribution in the pigment molecule upon photoconversion (Tobin and Briggs, 1973) as ribosomes are known to adsorb basic proteins preferentially (Petermann, 1964). Likewise, the lack of release of *bound* Pfr upon reconversion to Pr is consistent with the observed differential in the adsorption and release of proteins by ribosomes (Bloemendal and Vennegoor, 1969; Dyer and Payne, 1974). No evidence was obtained in the present study for substantial phytochrome-ribosome interaction within the cell.

It should be emphasised that the present findings are not meant to imply that no potentially meaningful phytochrome–membrane interactions have been recorded (Evans, Chapter 10; Evans and Smith, 1975; Cooke and Saunders, 1975). They do, however, highlight the need for caution in the interpretation of studies on pelletable phytochrome. 'Pelletability' should not be considered synonymous with 'membrane binding' in the absence of further rigorous analysis.

Acknowledgements

The authors thank Professor D.J. Carr for research facilities and Drs T. Kagawa and R. Wettenhall for helpful discussions. Mrs A. Gallagher and Mrs H. Edwards provided valuable technical assistance.

128 *Particle-bound phytochrome*

References

ACTON, G.J. (1974). *Phytochemistry,* **13**, 1303
BLOEMENDAL, H. and VENNEGOOR, C. (1969). In *Techniques in Protein Biosynthesis*, pp.181-207. Ed. Campbell, P.N. and Sargent J.R. Academic Press, London
BOISARD, J., MARMÉ, D. and BRIGGS, W.R. (1974). *Plant Physiol.,* **54**, 272
COOKE, R.J. and SAUNDERS, P.F. (1975). *Planta,* in the press
DYER, T.A. and PAYNE, P.I. (1974). *Planta,* **117**, 259
EVANS, A. and SMITH, H. (1975). *Proc. Nat. Acad. Sci. U.S.A.,* in the press
GRAY, J.C. (1974). *Arch. Biochem. Biophys.,* **163**, 343
HSIAO, T.C. (1968). *Plant Physiol.,* **43**, 1355
LEAVER, C.J. (1973). *Biochem. J.,* **135**, 237
LEAVER, C.J. and DYER, J.A. (1974). *Biochem. J.,* **144**, 165
LOENING, U.E. (1969). *Biochem. J.,* **113**, 131
LORD, J.M., KAGAWA, T., MOORE, T.S. and BEEVERS, H. (1973). *J. Cell Biol.,* **57**, 659
MARMÉ, D. (1975). *J. Supramol. Struct.,* in the press
MARMÉ, D., BOISARD, J. and BRIGGS, W.R. (1973). *Proc. Nat. Acad. Sci. U.S.A.,* **70**, 3861
MARMÉ, D., MACKENZIE, J.M., BOISARD, J. and BRIGGS, W.R. (1974). *Plant Physiol.,* **54**, 263
MARMÉ, D. and GROSS, J. (1975). In *Book of Abstracts, Ann. Eur. Photomorphogenesis Symp.*, p.88. Ed. Smith, H.
PAYNE, P.I. and LOENING, U.E. (1970). *Biochim. Biophys. Acta,* **224**, 128
PETERMANN, M.L. (1964). *The Physical and Chemical Properties of Ribosomes.* Elsevier, Amsterdam
QUAIL, P.H. (1974a). *Planta,* **118**, 345
QUAIL, P.H. (1974b). *Planta,* **118**, 357
QUAIL, P.H. (1975a). *Planta,* **123**, 223
QUAIL, P.H. (1975b). *Planta,* **123**, 235
QUAIL, P.H., MARMÉ, D. and SCHÄFER, E. (1973). *Nature New Biol.,* **245**, 189
QUAIL, P.H. and SCHÄFER, E. (1974). *J. Membr. Biol.,* **15**, 393
RUBINSTEIN, B., DRURY, K.S. and PARK, R.B. (1969). *Plant Physiol.,* **44**, 105
SABATINI, D.D., TASHIRO, Y. and PALADE, G.E. (1966). *J. Mol. Biol.,* **19**, 503
SCHÄFER, E. (1975). *Photochem. Photobiol.,* **21**, 189
TOBIN, E. and BRIGGS, W.R. (1973). *Photochem. Photobiol.,* **18**, 487
WIBO, M., AMAR-COSTESAC, A., BERTHET, J. and BEAUFAY, H. (1971). *J. Cell Biol.,* **51**, 52
YU, R. (1975). *Aust. J. Plant Physiol.,* **2**, 237

10

ETIOPLAST PHYTOCHROME AND ITS *IN VITRO* CONTROL OF GIBBERELLIN EFFLUX

AUDREY EVANS

Department of Physiology and Environmental Studies, University of Nottingham, School of Agriculture, Sutton Bonington, Loughborough, Leics. LE12 5RD, U.K.

Introduction

An association between phytochrome and certain critical membrane structures of the cell is clearly indicated by physiological evidence (Hendricks and Borthwick, 1967; Tanada, 1968; Haupt, 1970) and spectrophotometric measurements of isolated subcellular fragments support this view (Williamson and Morré, 1974; Manabe and Furuya, 1974). However, the study of a cell-free system in which phytochrome controls a specific biochemical process can be of value in the study of the localisation of phytochrome and its primary action at its location. In this paper, evidence is presented that a proportion of total cellular phytochrome is located at the etioplast where it controls the transport of gibberellin across the envelope membranes *in vitro*.

Circumstantial evidence for the presence of phytochrome in etioplasts is found in the control of ultra-structural development of isolated etioplasts where red irradiation enhances development and far-red light reverses this enhancement (Wellburn and Wellburn, 1973). In addition, etioplasts *in situ* exhibit a wide range of photomorphogenic phenomena.

Changes in the level of endogenous gibberellins in response to red irradiation have been found in several plant systems. Some of the most interesting data have come from etiolated cereal leaves whose unrolling is phytochrome mediated (Virgin, 1962). Observations of large transient increases in the level of extractable gibberellin within 10–15 min of the termination of a red irradiation (Reid, Clements and Carr, 1968; Beevers *et al.*, 1970; Loveys and Wareing, 1971) have prompted the view that phytochrome may act, in part, through an effect on endogenous gibberellin status to produce the observed developmental phenomena. Equally rapid increases in gibberellin level can be demonstrated *in vitro* following the irradiation with red light of crude homogenates of barley leaves (Reid *et al.*, 1972).

The chloroplasts of light-grown plants contain significant amounts of gibberellin (Stoddart, 1968) and recent studies of gibberellin metabolism by Railton and Reid (1974a,b) have suggested that the chloroplast may be an important site of gibberellin biosynthesis and regulation. As a major constituent of the homogenate originally used by Reid *et al.* (1972), the etioplast therefore appears as a likely subcellular site for the regulation of endogenous gibberellin status by phytochrome.

Isolation of Intact Etioplasts

Etioplasts were isolated from the first leaf of 6-day-old etiolated barley seedlings by a modification of the technique of Wellburn and Wellburn (1971) which involves the use of a Sephadex column as a final purification step. Segments of the first leaf were cut 3 cm long, 1 cm from the apex and homogenised in extraction buffer containing 25 mM *N*-morpholino-3-propanesulphonic acid (MOPS), 3 mM EDTA (disodium salt), 250 mM sucrose, 0.2 per cent of bovine serum albumin and 14 mM 2-mercaptoethanol, adjusted to a final pH of 7.5. The resulting homogenate was filtered to give a filtrate with a final pH of 7.1, which was maintained throughout the isolation procedure. All procedures were carried out under a dim green safelight and, after harvesting, the temperature was maintained at 5 °C. The homogenate was centrifuged at 6 000 g for 1 min and the pellet washed and re-centrifuged. The final crude plastid pellet was contaminated by nuclei, mitrochondria and broken plastid fragments. The intact etioplasts were separated from these contaminants on a loosely packed Sephadex G-50 (coarse) column, 50 cm long and 1.1 cm in diameter. Fractions 0.5 ml in volume were eluted from the column and 1-drop fractions were taken at intervals for routine measurement of absorbance at 260 nm. Unlike that of Wellburn and Wellburn (1971), the elution profile (*Figure 10.1*) shows only one peak of absorbance before the stationary phase of the column packs and further elution is prevented. When the re-suspended pellet is layered on top of the column, fragments of broken plastids and nuclei remain at the top of the column while mitochondria and intact etioplasts are rapidly separated. Fractions 40-55 were bulked and examined by electron microscopy. These fractions were found to be composed predominantly of intact etioplasts with minimal contamination from other fragments (*Figure 10.2*). It is presumed that the packing of the column after the elution of those fractions containing intact etioplasts prevents the elution of those fractions which would be contaminated by mitochondria.

In an attempt to obtain an estimation of the degree of purity of the etioplast suspension, total nucleic acids were extracted and fractionated on polyacrylamide gels according to the method of Jackson and Ingle (1973). Gel scans of the filtered homogenate (A), the washed, crude plastid pellet (B) and the purified etioplast fractions are shown in *Figure 10.3*. Quantification of the various components is made difficult by the presence of various breakdown products; however, it

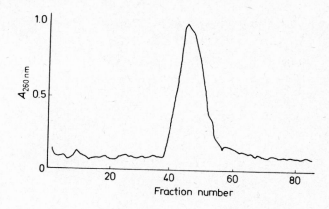

Figure 10.1 Separation of crude etioplast preparation by Sephadex G-50 (coarse)

is evident that the final stage of separation greatly enhances the proportion of plastid RNA as compared with previous stages (*Figure 10.3*). The small amount of cytoplasmic RNA, shown most accurately by the fraction of molecular weight 1.3×10^6, still present in the etioplast fraction may represent the adsorption of cytoplasmic ribosomes to the exterior of the etioplast envelope. The relatively large 0.7×10^6 molecular weight peak is due mainly to breakdown of the chloroplast component of molecular weight 1.1×10^6. It is notable, however, that the extent of breakdown observed after column separation (C) is much less than in the crude plastid pellet (B) and can be taken as an indication of the intactness of the etioplasts obtained after purification.

Detection of Etioplast Phytochrome

The possible presence of phytochrome in fractions eluted from the Sephadex column was determined by direct difference spectrophotometry. Total phytochrome was measured at 25 °C using a Perkin-Elmer dual-wavelength spectrophotometer. All samples were irradiated with sufficient red light to effect the maximum conversion of protochlorophyllide to chlorophyllide prior to determining the difference in absorbance at 660 and 730 nm. Phytochrome is detectable only in those fractions eluted from the column which have been shown to contain etioplasts (*Figure 10.4*) and at no other point along the elution profile. At pH 7.1 in the absence of Mg^{2+} ions, the binding of phytochrome to membranes isolated from non-irradiated tissue is extremely low (Marmé, Boisard and Briggs, 1973). The conclusion can therefore be drawn from these data that a proportion of total cellular phytochrome is located at the etioplast as any phytochrome not specifically associated

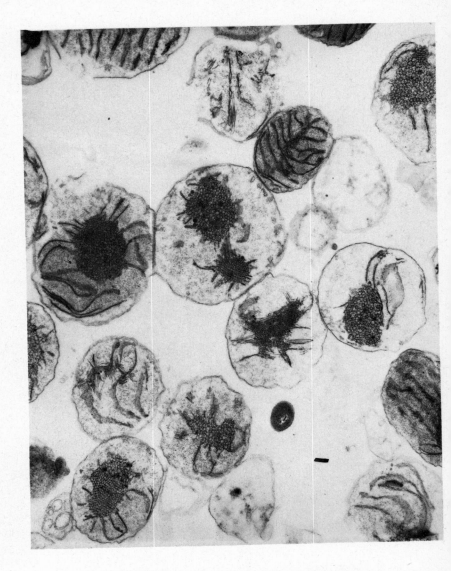

Figure 10.2 Electron micrographs of an etioplast suspension purified on a Sephadex G-50 column (× 10,400)

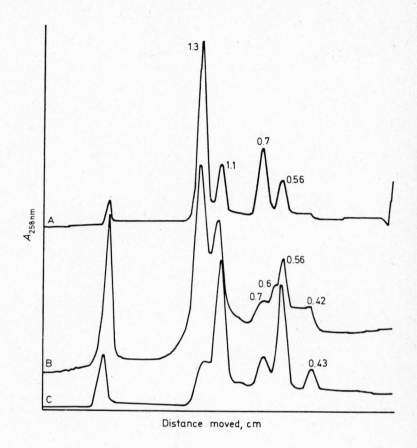

Figure 10.3 Polyacrylamide gel electrophoresis of nucleic acids extracted from A, filtered homogenate; B, washed, crude plastid pellet; C, purified etioplast suspension. The numbers refer to the estimated molecular weights of the respective RNA molecules $\times 10^6$ daltons

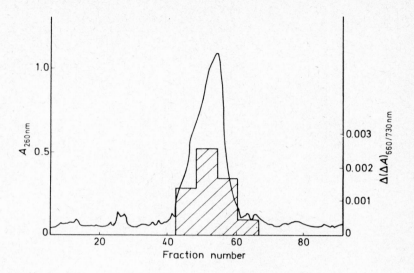

Figure 10.4   Distribution of etioplasts and phytochrome on a Sephadex G-50 (coarse) column.   Solid line, absorbance at 260 nm; cross-hatched, $\Delta(\Delta A)$ at 660 and 730 nm

with the etioplast would remain in the supernatant during the initial stages of isolation.   While the difference spectrum of isolated etioplasts, as seen in *Figure 10.5,* shows some distortion, it is undoubtedly that of phytochrome.   It is possible that the red to far-red ratio of almost two and the shift of the red maximum from 660 to 655 nm may be attributed to interference from high concentrations of protochlorophyllide present in these samples.

The physiological role of etioplast phytochrome was investigated by a study of gibberellin status in suspensions of isolated etioplasts.

Light-mediated Changes in Gibberellin Levels

It was found that etioplasts prepared in this manner contained substances with gibberellin-like activity, the level of which could be enhanced by a short period of irradiation with red light.   Etioplast suspensions were rapidly brought to 24 °C, given a saturating irradiation of red light and placed in darkness or far-red light.   Controls were retained at 24 °C, given a saturating irradiation of red light and placed in darkness.   The treatments were terminated by the addition of ice-cold absolute methanol to a final concentration of 75 per cent.   The

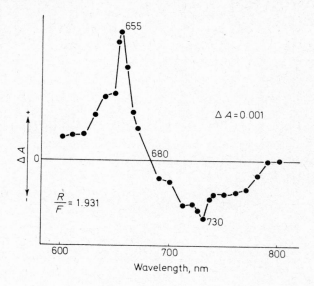

Figure 10.5 Difference spectrum of etioplast phytochrome. ΔA, *absorbance of far-red irradiated sample minus absorbance of red irradiated sample; R/F, ratio of ΔA at 655 nm to −ΔA at 730 nm*

resulting methanolic extract was extracted by conventional solvent partition methods and the acidic ethyl acetate-soluble fraction subjected to thin-layer chromatography. The activity eluted from the plate was assessed in the barley endosperm bioassay (Jones and Varner, 1967). α-Amylase activity induced in the barley half grain was measured by estimating the amount of maltose produced in a hydrolysis reaction with a pure amylose substrate (Bernfield, 1955). The protein content of etioplast pellets washed with bovine serum albumin-free buffer was determined according to the method of Lowry *et al.* (1951). The level of gibberellin-like substances extractable from dark controls is not significantly changed (*Figure 10.6*), nor is there any increase in the level of activity in samples irradiated with far-red light immediately following red light. Red light alone, however, led to a marked subsequent rise in activity.

Thin-layer chromatography of extracts (*Figure 10.7*) indicates a major zone of activity in the dark controls of R_F 0.3 to 0.6. After 5 min of red light followed by 5 min of dark (*Figure 10.7f*), the level of activity in this zone is significantly increased but migrates into two distinct peaks. The level of activity at R_F 0.2 and 0.9 is also enhanced. This pattern is maintained after 10 min of darkness following irradiation with red light. Irradiation with far-red light following

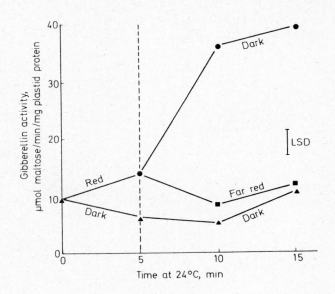

Figure 10.6   The effect of red and far-red light on the level of gibberellin-like activity extractable from suspensions of intact etioplasts

red light prevents the increase in activity seen in those samples placed in the dark after irradiation with red light.   After 5 min of far-red irradiation, the major zone of activity was distinguishable between R_F 0.4 and 0.6.   No distinct pattern could be seen after 10 min of far-red light.

The rapidity of this response to red light suggested that red light caused the production of a freely extractable gibberellin from a previously unextractable form.   As it is known that more gibberellin can be extracted from chloroplasts whose membranes have been ruptured by ultrasonication than from extracts of intact chloroplasts (Stoddart, 1968) it is possible that substances with gibberellin-like activity are confined within the etioplast envelope and are not freely extractable from suspensions of intact etioplasts.   This hypothesis was tested by taking an etioplast suspension prepared in the normal manner and subjecting half to a 1-min ultrasonication given in 20-s bursts, the sample tube being embedded in ice so as to prevent any increase in temperature during the process.   The remaining half was retained untreated.   Both the intact and ultrasonicated samples were brought to 24 °C and extracted for acidic ethyl acetate-soluble gibberellin as previously.   *Figure 10.8* shows that approximately three times the activity extractable from the intact etioplast suspension was extracted following membrane rupture by ultrasonication.   In addition, there is

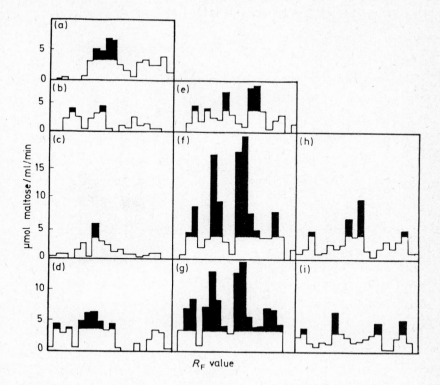

Figure 10.7 Thin-layer chromatography of acidic ethyl acetate fractions of extracts from intact etioplast suspensions. (a), Zero time in dark; (b), 5 min dark; (c), 10 min dark; (d), 15 min dark; (e), 5 min red light; (f), 5 min red light and 5 min dark; (g), 5 min red light and 10 min dark; (h), 5 min red light and 5 min far-red light; (i), 5 min red light and 10 min far-red light. ■, *Significant promotion at 5 per cent level of probability*

a marked increase in the level of non-polar, i.e. rapidly migrating, substances in the ultrasonicated extract. The intact etioplast membrane, therefore, prevents the expression of activity of gibberellin-like substances present within etioplasts held in darkness.

In the light of these data, the change of endogenous gibberellin status in response to red light in isolated etioplasts is most simply interpreted as being due to the transport of pre-existing gibberellin from the etioplast into the ambient medium brought about by the photoconversion of phytochrome from the Pr to the Pfr form. It seems unlikely that phytochrome actually transports gibberellin across the membrane because of the 5 min lag usually seen before the onset of a detectable increase in gibberellin level. It is more likely that

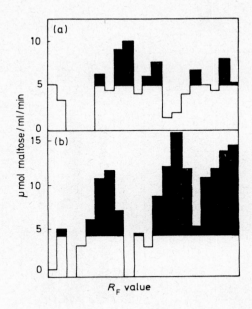

Figure 10.8   Thin-layer chromatography of acidic ethyl acetate fractions of extracts from (a) intact etioplast preparation (total activity = 49 μmol of maltose) and (b) ultrasonicated preparation (total activity = 131.6 μmol of maltose), at time zero in dark. ■ *Significant promotion at 5 per cent level of probability*

phytochrome changes the permeability of the membrane with respect to gibberellins. The far-red reversibility of this response is consistent with this concept. The level of gibberellin-like substances extractable at the onset of far-red irradiation is, in most experiments, not significantly greater than at time zero, so that a change in the permeability of the membrane at this point effected by the photoconversion of Pfr to Pr would prevent any further release of gibberellin from the etioplast. Any release of gibberellin from the etioplast appears to be in some degree selective, as demonstrated by the high proportion of non-polar substances extractable from ultrasonicated preparations which are not detectable in extracts of intact etioplasts.

Table 10.1 shows a comparison of the levels of gibberellin-like activity extractable from suspensions of intact etioplasts and suspensions ultrasonicated immediately prior to extraction. The level of activity in intact preparations increases rapidly following red irradiation. In contrast, the level of activity in ultrasonicated preparations remained constant over the period of irradiation and increased only slightly after

Table 10.1 Comparison of gibberellin-like activity extractable from intact etioplasts and etioplast preparations ultrasonicated immediately prior to extraction

Treatment	Extractable gibberellin-like activity (μmol maltose min^{-1} mg plastid protein^{-1})	
	Intact suspensions	*Sonicated suspensions*
Zero time dark	4.00	17.50
5 min red light	10.80	17.91
5 min red light + 5 min dark	22.03	24.90

a 5-min dark period following red irradiation. It therefore appears that the initial increase in gibberellin level in response to red irradiation is due entirely to release from the envelope but there are rapidly initiated reactions which involve a detectable stimulation of gibberellin production.

Conclusions

Phytochrome can be detected in preparations of isolated etioplasts by dual-wavelength difference spectrophotometry and can be shown to play a distinct physiological role in the regulation of endogenous gibberellin levels. Phytochrome has previously been reported in association with other subcellular membrane structures such as the plasma-lemma (Haupt, 1970), nucleus (Galston, 1968), rough endoplasmic reticulum (Williamson and Morré, 1974) and mitochondria (Manabe and Furuya, 1974) and has a demonstrable affinity for membranes *in vitro* (Roux and Yguerabide, 1973). Taken with this published work, the evidence presented here supports the hypothesis that phytochrome is located in association with certain critical membranes of the cell over which it exerts a regulatory function (Hendricks and Borthwick, 1967; Smith, 1970). The question as to whether phytochrome regulates the passage of gibberellin in a specific manner involving gibberellin 'transport factors' or a non-specific manner such as by the regulation of ATP levels is open to speculation. While the first alternative represents a mechanism specific to gibberellin control, the second is more generally applicable to phytochrome phenomena. The mode of action of phytochrome at the membrane level has yet to be determined.

Investigations of membrane-bound phytochrome indicate that Pfr can bind to specific membrane sites which are not available to Pr. In this way, phytochrome may act as a ligand which allosterically interacts with membrane-binding sites (Boisard, Marmé and Briggs, 1974). Etioplast phytochrome may therefore be envisaged as existing as a free soluble form in the stroma which upon photoconversion to Pfr migrates to the membrane, where it could specifically interact with gibberellin transport proteins or permeases. On the other hand, there is, as yet, little evidence to oppose the view that both Pfr and Pr are normal membrane components and the observed photomorphogenic phenomena are a result of photoconversion *in situ* in the membrane.

Acknowledgements

My thanks are due to Miss J. Elliot for technical assistance, to Dr M. Jackson for help with the nucleic acid analysis and to Mrs A. Tomlinson for the electron microscopy. My special thanks go to Professor H. Smith for his supervision and helpful discussion of this work and to the Ministry of Agriculture, Food and Fisheries for its financial support.

References

BEEVERS, L., LOVEYS, B., PEARSON, J.A. and WAREING, P.F. (1970). *Planta,* **90,** 286
BERNFIELD, B. (1955). *Methods Enzymol.,* **1,** 149
BOISARD, J., MARMÉ, D. and BRIGGS, W.R. (1974). *Plant Physiol.,* **54,** 272
GALSTON, A. (1968). *Proc. Nat. Acad. Sci. U.S.A.,* **61,** 454
HAUPT, W. (1970). *Physiol. Veg.,* **8,** 551
HENDRICKS, S.B. and BORTHWICK, H.A. (1967). *Proc. Nat. Acad. Sci. U.S.A.,* **58,** 2125
JACKSON, M. and INGLE, J. (1973). *Plant Physiol.,* **51,** 412
JONES, R.L. and VARNER, J. (1967). *Planta,* **72,** 155
LOVEYS, B.R. and WAREING, P.F. (1971). *Planta,* **98,** 109
LOWRY, O.H., ROSENBOROUGH, N.J., FARR, A.L. and RANDALL, R.J. (1951). *J. Biol. Chem.,* **193,** 265
MANABE, K. and FURUYA, M. (1974). *Plant Physiol.,* **53,** 343
MARMÉ, D., BOISARD, J. and BRIGGS, W.R. (1973). *Proc. Nat. Acad. Sci. U.S.A.,* **70,** 3861
RAILTON, I.D. and REID, D.M. (1974a). *Plant Sci. Lett.,* **2,** 157
RAILTON, I.D. and REID, D.M. (1974b). *Plant Sci. Lett.,* **3,** 303
REID, D.M., CLEMENTS, J.B. and CARR, D.J. (1968). *Nature, Lond.,* **217,** 580
REID, D.M., TUING, M.S., DURLEY, R.C. and RAILTON, I.D. (1972). *Planta,* **108,** 67
ROUX, S.J. and YGUERABIDE, J. (1973). *Proc. Nat. Acad. Sci. U.S.A.,* **70,** 762
SMITH, H. (1970). *Nature, Lond.,* **227,** 655
STODDART, J.L. (1968). *Planta,* **81,** 106
TANADA, T. (1968). *Proc. Nat. Acad. Sci. U.S.A.,* **59,** 376
VIRGIN, H. (1962). *Physiol. Plant.,* **15,** 380

WELLBURN, A.R. and WELLBURN, F.A.M. (1971). *J. Exp. Bot.,* **23**, 972
WELLBURN, F.A.M. and WELLBURN, A.R. (1972). *New Phytol.,* **72**, 55
WILLIAMSON, F.A. and MORRÉ, D.J. (1974). *Plant Physiol.,* Annu.
  Suppl. 46

11

PHYTOCHROME IN MITOCHONDRIAL AND MICROSOMAL FRACTIONS ISOLATED FROM ETIOLATED PEA SHOOTS

MASAKI FURUYA
KATSUSHI MANABE[1]
Department of Botany, Faculty of Science, University of Tokyo, Hongo, Tokyo 113, Japan

Introduction

The importance of phytochrome-induced changes in membrane properties has been emphasised by many workers (Hendricks and Borthwick, 1967; Haupt, 1970; Smith, 1970; Briggs and Rice, 1972), but little is known about the relationship between red and far-red photoreversible effects and particulate-bound phytochrome. Gordon and Surrey (1960) first suggested that mitochondria contain phytochrome since the rate of oxidative phosphorylation in mitochondria isolated from rat liver was affected by red and far-red irradiations. This result could not be confirmed (Ikuma and Bonner, 1964). Recently, we found that when a 1 000–7 000 g sedimentary fraction prepared from etiolated bean hypocotyls (Manabe and Furuya, 1973) and purified pea mitochondria (Manabe and Furuya, 1974) were briefly exposed to red light, their ability to reduce exogenous NADP was enhanced within a few minutes and this red-light effect was reversed by far-red light, and that photoreversible absorbance changes between 730 and 800 nm were detected spectrophotometrically in the purified mitochondria and mostly in their membrane fraction.

Evidence also has accumulated that a light-dependent binding of phytochrome to particulate cell fractions occurs in various plants *in vivo* (Quail, Marmé and Schäfer, 1973; Quail and Schäfer, 1974; Boisard, Marmé and Briggs, 1974; Marmé *et al.*, 1974; Yu, 1975) and *in vitro* (Marmé, Boisard and Briggs, 1973; Quail, 1974, 1975). The question therefore arises of whether the light-induced binding triggers the red/far-red reversible reaction in mitochondria described above, or the binding itself is one of the very early phytochrome-dependent responses.

[1] Present address: MSU/AEC Plant Research Laboratory, Michigan State University, East Lansing, Michigan 48824, U.S.A.

In this chapter we describe the intracellular distribution of phytochrome in etiolated pea shoots before and after light treatments, some of the characteristics of experimentally induced binding of phytochrome to particulate fractions and the fates of Pfr in each subcellular fraction during dark incubation, and consider the nature of particulate-bound phytochrome-dependent effects.

Fractionation Procedures and Terminology

Seeds of *Pisum sativum* L. cv. Alaska were germinated in a dark room on vermiculite saturated with water at 26 °C. The apical 2-cm portions of 5-day-old etiolated pea shoots were used in all experiments. For routine preparations of particulate fractions (Manabe and Furuya, 1975a), 10 g of pre-chilled tissue were homogenised by grinding with 20 ml of 10 mM Tris–HCl buffer (pH 7.2) containing 0.4 M sucrose and 1 mM dithiothreitol. The extract, adjusted to pH 7.2, was centrifuged for 10 min at 1 000 g and the supernatant was centrifuged for 10 min at 12 000 g. The particle fraction was washed with 20 ml of the extraction buffer and centrifuged for 5 min at 400 g. The supernatant was centrifuged for 10 min at 7 000 g and the precipitate was re-suspended in 1.75 ml of the same extraction buffer (*mitochondrial* fraction). When the fraction was further purified in a Ficoll or sucrose gradient, the term *mitochondria* is used. After preparation of the 12 000 g precipitate, the supernatant was centrifuged for 30 min at 105 000 g and the 12 000–105 000 g sedimentary particle fraction was suspended in 1.5 ml of the extraction buffer (*microsomal* fraction). The 105 000 g supernatant fraction is called *cytosol*.

Phytochrome was measured at 0 °C with a dual-wavelength difference spectrophotometer (Hitachi Ltd., Japan, Model 261), as described previously (Pjon and Furuya, 1968). The amounts of protein in the samples were determined by the method of Lowry *et al.* (1951).

Intracellular Distribution of Phytochrome in Etiolated Pea Shoots

Although the intracellular distribution of phytochrome had long been an open question, recent studies revealed that spectrophotometrically detectable phytochrome was not only located in the cytosol (Butler *et al.*, 1959) but was also found in other subcellular fractions. Amount of particle-bound phytochrome in etiolated tissues of oat (Rubinstein, Drury and Park, 1969), corn (Quail, Marmé and Schäfer, 1973) and squash (Marmé, Boisard and Briggs, 1973) were found to be 4, 6 and 7. per cent, respectively, most of the phytochrome being in the cytosol.

An attempt was then made to determine how phytochrome and protein are distributed in the subcellular fractions obtained by the routine extraction procedure from the apical 2-cm shoots of 5-day-old totally etiolated pea seedlings (*Table 11.1*).

Table 11.1 Intracellular distribution of phytochrome in etiolated pea shoots

| Fraction | Distribution | | | | $\Delta(\Delta A) \times 10^5/$ mg protein |
| | Protein | | Phytochrome | | |
	mg/gfw	%	$\Delta(\Delta A) \times 10^4$/gfw	%	
1 000 *g* supernatant	8.61	100	146	100	170 ± 31
Mitochondrial	0.36	4.1	3.9	2.7	109 ± 12
12 000 *g* super-natant	8.00	92.9	134	91.8	167 ± 27
105 000 *g* precipi-tate	0.88	10.2	6.5	4.5	74 ± 18
105 000 *g* super-natant	7.02	81.5	106	72.6	151 ± 30

Although we did not add any divalent cations to the grinding medium, *ca.* 2.7 per cent of total extractable phytochrome was bound to the mitochondrial fraction and 4.5 per cent to the microsomal fraction, while *ca.* 70 per cent was in the cytosol.

The $\Delta(\Delta A)$ per milligram of protein values were significantly higher for the 12 000 *g* and 105 000 *g* supernatants than for the particulate fractions. The percentages of protein distribution with these subcellular fractions were not always consistent with those of phytochrome distribution.

Effect of Red Light Irradiation on Photoconversion in Subcellular Fractions

In earlier work, many attempts failed to correlate the spectrophoto-metrically detectable status of *in vivo* phytochrome with the physiolo-gical responses to red and far-red light (Hillman, 1967; Furuya, 1968). In this connection, Briggs and Chon (1966) explained the paradox by postulating that a small physiologically active fraction of phytochrome is much more easily converted to Pfr than the bulk phytochrome. Therefore, the relationship between the amount of incident irradiation energy and the degree of photoconversion of phytochrome in each subcellular fraction was determined. Apical 2-cm segments kept on ice in the dark for at least 15 min were irradiated with 0.75 W m^{-2} red light for various periods of time at *ca.* 0 °C. The percentage of Pfr transformed was determined with intact segments and their sub-cellular fractions. The results (Manabe and Furuya, 1975b) showed that the bulk phytochrome in 12 000 *g* and 105 000 *g* supernatants and the membrane-bound phytochrome in mitochondrial fraction were equally photoconverted, depending simply on the total incident energy irrespective of phytochrome concentration, as in intact tissues (Furuya and Hillman, 1964). Thus the result did not support the explanation of the Briggs and Chon (1966) paradox. When the shoot tissues were irradiated with red light of 16 J m^{-2} or more, the irradiation always caused 100 per cent or greater Pfr in the microsomal fraction. These results still do not support the interpretation of Briggs and Chon (1966), but suggests that there is a cooperativity between cytosolic phytochrome and microsomal membranes.

Red Light-induced Rapid Binding of Phytochrome to Mitochondrial and Microsomal Fractions at Low Temperature

In the above study with pea segments, we noticed that phytochrome binding to mitochondrial and microsomal fractions occurred very rapidly after brief exposure to red light when the segments were kept on ice. Pre-chilled segments of pea shoots were therefore irradiated at 0 °C with various dosages of red light just before extraction, and the phytochrome contents and the percentage of Pfr formed were determined in mitochondrial (Manabe and Furuya, 1975a) and microsomal fractions and the 12 000 g supernatant.

The total amount of phytochrome in the 12 000 g supernatant was not affected by the irradiation, while the amount in the mitochondrial fraction increased up to *ca.* 170 per cent of the dark control, depending upon the percentage of Pfr formed. The degree of rapid binding of Pr and Pfr in the mitochondrial fraction was consistent with the ratio of Pr to Pfr in the cytosol at every dosage of red light. The increase in phytochrome bound to the mitochondrial fraction occurred without a concomitant decrease in the 12 000 g supernatant, because less than 5 per cent of total extractable phytochrome from the tissue was found in the mitochondrial fraction.

When 70 J m^{-2} or higher dosages of red light was applied to the pea segments, the percentage of Pfr observed with the microsomal fraction became more than 100 per cent of the photochemical transformation (*Figure 11.1*). This result may suggest that the affinity of microsomal membrane to Pfr was stronger than that to Pr.

Hill coefficients of 1.4 with the mitochondrial fraction and of 3.7 with the microsomal fraction (*Figure 11.2*) were calculated from the data described above. The dotted line in *Figure 11.2* shows the theoretical curve for no cooperativity between ligand and membrane, and it is clearly different from the plots of the particle fractions.

Apical 2–cm shoot segments kept on ice for at least 15 min were exposed to alternating red and far-red light, and the phytochrome contents in the mitochondrial and microsomal fractions were determined in order to find any changes in phytochrome distribution at *ca.* 0 °C (*Tables 11.2* and *11.3*).

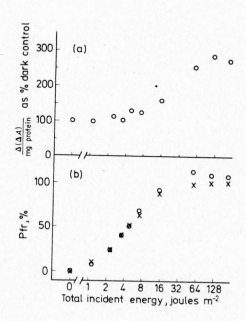

Figure 11.1 Effect of the energy of red light on photoconversion (b) and the phytochrome contents (a) in the microsomal fraction (○) and the 12 000 g supernatant (x) from pea shoots kept at 0 °C during the light treatment and analysed immediately afterwards

Table 11.2 Phytochrome in mitochondrial fraction isolated from pea shoot segments

Irradiation at $0\,^{\circ}C^1$	Phytochrome $(\Delta(\Delta A) \times 10^5/\text{gfw})$	Protein (mg/gfw)	$\Delta(\Delta A) \times 10^5/\text{mg protein}$		
			Pr	Pfr	Total
Dark	48	0.34	142	0	142
FR	49	0.33	144	2	146
R	71	0.33	43	174	217
R–FR	72	0.30	228	8	236
R–FR–R	87	0.35	45	205	250

[1] R, 0.75 W m^{-2} red light for 3 min; FR, 15 W m^{-2} far-red light for 2 min

Table 11.3 Phytochrome in microsomal fraction isolated from pea shoot segments

Irradiation at $0\,^{\circ}C^1$	Phytochrome $(\Delta(\Delta A) \times 10^5/\text{gfw})$	Protein (mg/gfw)	$\Delta(\Delta A) \times 10^5/\text{mg protein}$		
			Pr	Pfr	Total
Dark	57	0.83	68	0	68
FR	72	0.96	72	3	75
R	189	0.97	27	169	196
R–FR	192	0.99	190	3	193
R–FR–R	200	0.99	28	173	201

[1] As in *Table 11.2*

The first brief irradiation of the tissue with red light caused a significant increase in the phytochrome content in both mitochondrial and microsomal fractions, while the 12 000 *g* supernatant was not signi ficantly affected by the light treatments. Irradiation with far-red light did not significantly affect the phytochrome contents in either particle fraction. The red-light effect in the mitochondrial fraction was not far-red reversible, but somewhat cumulative in successive irradiations with red light. This very rapid increase in phytochrome contents in the particulate fractions at 0 °C is termed 'rapid phytochrome binding'. The degree of rapid binding was not changed for at least 1 h after irradiation with red light if the tissues were kept on ice.

Photoreversible Binding of Pfr to the Mitochondrial Fraction at 30 °C

The intracellular distribution of phytochrome in etiolated pea shoots has been demonstrated to be greatly influenced not only by a brief irradiation with red light at 0 °C but also by dark incubation at 30 °C after the irradiation (Manabe and Furuya, 1975b). To observe such changes in the intracellular distribution of phytochrome, pea shoot segments kept at 0 °C were irradiated with red and/or far-red light for 3 min on ice. The segments were returned to darkness at 30 °C

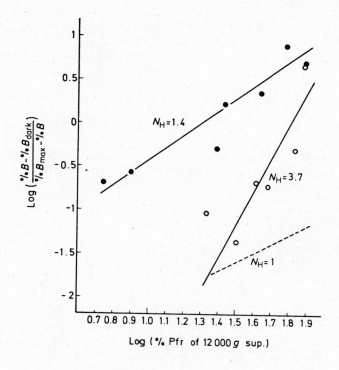

Figure 11.2 Hill plot of data with mitochondrial fraction (Manabe and Furuya, 1975a) and microsomal fraction from Figure 11.1. •, *Mitochondrial fraction;* o, *microsomal fraction;* ---, *theoretical.* N_H = *Hill coefficient*

immediately after the treatment. The distribution of *in vivo* phytochrome in the mitochondrial fraction of the segments was determined 5 min after irradiation (*Table 11.4*).

If the data in *Tables 11.2* and *11.4* are compared, the phytochrome contents are found to decrease during the dark incubation for all light-treated samples, especially far-red treated samples, although the protein contents were not significantly altered. Therefore, the total amount of phytochrome in the mitochondrial fraction apparently became higher after red-light exposure than after far-red irradiation, and the effect was red/far-red reversible. This photoreversible binding to the mitochondrial fraction can be interpreted as follows: Pr rapidly bound to the mitochondrial fraction at 0 °C was mostly dissociated from the membrane at 30 °C so that the difference in dissociation rates between Pr and Pfr may result in apparent photoreversibility with the amount of bound phytochrome in the mitochondrial fraction at the physiological temperature.

Table 11.4　Phytochrome in mitochondrial fraction isolated from pea shoot segments after dark incubation for 5 min at 30 °C in Tris–HCl buffer solution, pH 7.2

Irradiation at 0 °C[1]	Phytochrome ($\Delta(\Delta A)$ × 10^5/ gfw)	Protein (mg/gfw)	$\Delta(\Delta A)$ × 10^5/mg protein		
			Pr	Pfr	Total
Dark	47	0.38	121	3	124
FR	39	0.38	101	0	101
R	55	0.33	37	129	166
R–FR	46	0.34	139	−1	138
R–FR–R	57	0.34	35	134	169

[1] As in *Table 11.2*

Table 11.5　Phytochrome in microsomal fraction isolated from pea shoot segments after dark incubation for 5 min at 30 °C in Tris–HCl buffer solution, pH 7.2

Irradiation at 0 °C[1]	Phytochrome ($\Delta(\Delta A)$ × 10^5/gfw)	Protein (mg/gfw)	$\Delta(\Delta A)$ × 10^5/mg protein		
			Pr	Pfr	Total
Dark	59	0.79	72	3	75
FR	55	0.78	67	3	70
R	80	0.94	18	67	85
R–FR	102	0.95	104	3	107
R–FR–R	111	0.90	29	95	124

[1] As in *Table 11.2*

However, the effect of red and far-red light in the microsomal fraction was not reversible and appeared to be cumulative in terms of both phytochrome and protein contents (*Table 11.5*).

Non-photochemical Transformation of Phytochrome in Subcellular Fractions

Since the early work of Furuya and Hillman (1964), dark transform-ations of phytochrome in pea tissues have been extensively studied by several workers, but the fate of phytochrome in each subcellular fraction after initial red-light irradiation has so far been obscure. Manabe and Furuya (1975b) found that, after brief irradiation of intact shoots with red light at 26 °C each subcellular fraction showed different patterns of dark transformation *in vivo* at the physiological temperature.　The pattern of dark phytochrome changes in the 12 000 *g* and 105 000 *g* supernatant fractions obtained from red-light treated shoots reflects well that in intact pea tissues (Furuya and Hillman, 1964), whereas the dark reaction curves of the particulate fractions are different from those of the supernatants and the intact tissue, in that the amounts of the particulate-bound phytochrome increased immediately after the irradiation, and a reversion of Pfr to Pr in the dark was indicated in the cytosol for the first 2 h, but not at all in the mitochondrial and microsomal fractions. The amount of Pr in the mitochondrial and microsomal fractions remains constant during the dark period, keeping a level of Pr which had been induced by an initial red irradiation, while those in the cytosol increased significantly with time.

Very similar results were obtained with apical shoot segments after exposure to red light at 0 °C and subsequent dark incubation in 10 mM Tris-HCl buffer at 26 °C (*Figure 11.3*). It is evident that the contents of particulate-bound phytochrome at zero time in segments kept on ice during the light treatment were significantly higher than in intact shoots kept at 26 °C, whereas the rates of dark transformation of Pfr with the 12 000 *g* supernatant fraction was the same in both cases.

The pattern of non-photochemical transformation of phytochrome in pea subcells seems to be different from that in squash seedlings (Boisard, Marmé and Briggs, 1974) in which phytochrome destruction is related exclusively to the fraction which becomes membrane-bound, while the soluble fraction of phytochrome remains constant. Pratt, Kidd and Coleman (1974) demonstrated immunochemically that in oat

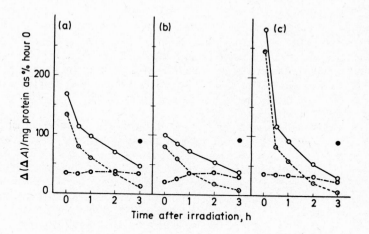

Figure 11.3   Dark changes of phytochrome at 26 °C in fractions of (a) mito-chondrial, (b) 12 000 g supernatant and (c) microsomal pea shoot segments after brief irradiation with red light.   Multiplying measured values by a factor of 1.25 gives corrected values for total phytochrome (○), Pfr measured (⊙), Pr calculated (◐), total phytochrome of dark control (●)

shoots *in vivo* phytochrome destruction involves the loss of extractable phytochrome protein.

It is interesting to note that the pattern of dark phytochrome changes in the particulate fractions of peas is the same as that in intact coleoptiles of *Zea* (Hopkins and Hillman, 1965; Pratt and Briggs, 1966), *Hordeum* (Hopkins and Hillman, 1965), *Oryza* (Pjon and Furuya, 1968) and *Pharbitis* (Miyoshi, Furuya and Takimoto, 1974), while the pattern for the 12 000 *g* supernatant appears to be comparable with that in root tips of *Pisum* (Furuya and Hillman, 1964) and hypocotyls of *Phaseolus, Raphanus* and *Glucine* (Hopkins and Hillman, 1965).

Particulate-bound Phytochrome Induced by Pfr Decay Inhibition

As Quail, Marmé and Schäfer (1973) found that the pelletability of phytochrome decreased with time in the dark, following first-order kinetics with a half-time of *ca.* 50 min, phytochrome bound to particulate fractions in red-light treated pea segments also disappears within *ca.* 3 h at 26 °C. However, if the Pfr destruction in the segments was prevented by exogenous inhibitors (Furuya, Hopkins and Hillman, 1965), the phytochrome content in the mitochondrial and microsomal fractions continued to increase in the dark for several hours (Manabe and Furuya, 1975a). After brief exposure of the pea shoot segments to red light at 0 °C and subsequent dark incubation at 30 °C in Tris–HCl buffer containing dithiothreitol or EDTA, which inhibits Pfr destruction in the dark, the increase in phytochrome contents in the particulate fractions reached a level several times higher than that of a dark control by 2 h. The effect of red light was reversed in a typical manner by far-red light (*Figure 11.4*). We call this the *long-term* effect.

The results with EDTA and dithiothreitol were essentially the same in terms of changes of intracellular phytochrome distribution, so that the photoreversible increase of Pfr content in the mitochondrial and microsomal fractions can be concluded to be due to the inhibition of Pfr destruction rather than an unexpected effect of dithiothreitol or EDTA itself. In the case in which the red-light treated segments were immersed in a buffer containing 5 mM EDTA and 1 mM dithiothreitol, the long-term effect was so enhanced that phytochrome contents in the mitochondrial and microsomal fractions increased to levels 8.4 and 6.2 times higher than that of the dark control, respectively.

It is not yet certain whether the 'long-term' effect is due to the binding of cytosolic Pfr to the particulate fractions or to *de novo* synthesis of phytochrome in the particles. The experimental evidence that phytochrome is synthesised as Pr and remains as Pr in the dark, while the newly bound phytochrome in the particulate fractions is mainly Pfr, may support the former explanation.

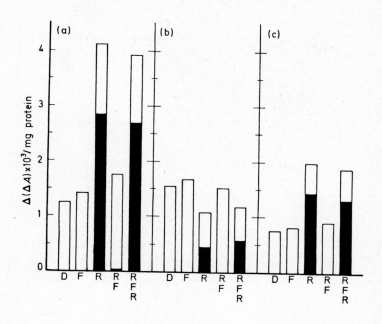

Figure 11.4 Photoreversible changes of phytochrome distribution in particle fractions: (a) mitochondrial, (b) 12 000 g supernatant and (c) microsomal isolated from segments of pea shoots incubated in 10 mM Tris–HCl buffer containing 5 mM dithiothreitol solution, pH 7.2, for 2 h at 30 °C. R, 0.75 W m⁻² red light for 3 min; F, 15 W m⁻² far-red light for 2 min; D, dark

Red/Far-red Reversible Reactions in Particulate Fractions

In the past few years, we have attempted to find some red/far-red reversible reactions which can be ascribed to a definite subcellular fraction. However, we have so far failed to discover them, except the phytochrome-dependent $NADP^+$ reduction in mitochondria (Manabe and Furuya, 1974). In fact, no red/far-red reversible effect on respiratory control values and ADP:O ratios was observed polarographically at 25 °C in the mitochondrial fraction (Miyoshi, Yamamoto and Furuya, 1974). Although many trials have been made in this laboratory, we have not succeeded in detecting any correlation between the known red/far-red reversible reactions and the experimentally induced binding of phytochrome to particulate fractions in pea tissues.

It is well known that physiological responses do not correlate well with the state and content of *in vivo* phytochrome as measured by spectrophotometry in most red/far-red reversible effects (Hillman, 1967), with the exception of a few phenomena (Pjon and Furuya, 1968). However, as we now know that each subcellular fraction obtained from a tissue of one plant shows not only different patterns of dark transformations of Pfr (Manabe and Furuya, 1975b) but also somewhat different responses in the above-mentioned photoinduced binding of phytochrome to particulate fractions, the conventional spectrophotometry with *in vivo* phytochrome, which mainly detects phytochrome in cytosol (*Table 11.1*), would certainly not be expected to reflect upon the state of phytochrome in a particular active site. Further, even spectrophotometric determinations with one of the subcellular fractions may again show a paradox, since the present fractionation procedures still appear to be too crude to identify a physiologically active site. However, the so-called 'small active fraction' of *in vivo* phytochrome controlling physiological responses may eventually be obtained from a particulate fraction with certain phytochrome-dependent phenomena.

References

BOISARD, J., MARMÉ, D. and BRIGGS, W.R. (1974). *Plant Physiol.*, **54**, 272

BRIGGS, W.R. and CHON, H.P. (1966). *Plant Physiol.*, **41**, 1159

BRIGGS, W.R. and RICE, H.V. (1972). *Annu. Rev. Plant Physiol.*, **23**, 293

BUTLER, W.L., NORRIS, K.H., SIEGELMAN, H.W. and HENDRICKS, S.B. (1959). *Proc. Nat. Acad. Sci. U.S.A.*, **45**, 1703

FURUYA, M. (1968). *Prog. Phytochem.*, **1**, 347

FURUYA, M. and HILLMAN, W.S. (1964). *Planta*, **63**, 31

FURUYA, M., HOPKINS, W.G. and HILLMAN, W.S. (1965). *Biophys. J.*, **112**, 180

GORDON, S.A. and SURREY, K. (1960). *Radiat. Res.*, **12**, 325

HAUPT, W. (1970). *Z. Pflanzenphysiol.*, **62**, 287

HENDRICKS, S.B. and BORTHWICK, H.A. (1967). *Proc. Nat. Acad. Sci. U.S.A.*, **58**, 2125

HILLMAN, W.S. (1967). *Annu. Rev. Plant Physiol.*, **18**, 301

HOPKINS, W.G. and HILLMAN, W.S. (1965). *Am. J. Bot.*, **52**, 427

IKUMA, H. and BONNER, W.D. (1964). *Proc. Int. Congr. Biochem.*, **6**, 780

LOWRY, O.H., ROSEBROUGH, N.J., FARR, A.L. and RANDALL, R.J. (1951). *J. Biol. Chem.*, **193**, 265

MANABE, K. and FURUYA, M. (1973). *Plant Physiol.*, **51**, 982

MANABE, K. and FURUYA, M. (1974). *Plant Physiol.*, **53**, 343

MANABE, K. and FURUYA, M. (1975a). *Planta*, **123**, 207

MANABE, K. and FURUYA, M. (1975b). *Plant Physiol.*, **56**, 772

MARMÉ, D., BOISARD, J. and BRIGGS, W.R. (1973). *Proc. Nat. Acad. Sci. U.S.A.*, **70**, 3861

MARMÉ, D., MACKENZIE, J.M. Jr., BOISARD, J. and BRIGGS, W.R. (1974). *Plant Physiol.,* **54**, 263

MIYOSHI, Y., FURUYA, M. and TAKIMOTO, A. (1974). *Plant Cell Physiol.,* **15**, 1115

MIYOSHI, Y., YAMAMOTO, K. and FURUYA, M. (1975). *Proc. 39th Ann. Mtg Bot. Soc. Japan,* p.105

PJON, C.J. and FURUYA, M. (1968). *Planta,* **81**, 303

PRATT, L.H., and BRIGGS, W.R. (1966). *Plant Physiol.,* **41**, 467

PRATT, L.H., KIDD, G.H. and COLEMAN, R.A. (1974). *Biochim. Biophys. Acta,* **365**, 93

QUAIL, P.H. (1974). *Planta,* **118**, 357

QUAIL, P.H. (1975). *Planta,* in the press

QUAIL, P.H., MARMÉ, D. and SCHÄFER, E. (1973). *Nature New Biol.,* **245**, 189

QUAIL, P.H. and SCHÄFER, E. (1974). *J. Membrane Biol.,* **15**, 393

RUBINSTEIN, B., DRURY, K.S. and PARK, R.B. (1969). *Plant Physiol.,* **44**, 105

SMITH, H. (1970). *Nature, Lond.,* **227**, 665

YU, R. (1975). *Aust. J. Plant Physiol.,* **2**, 273

III

CELLULAR ASPECTS OF PHYTOCHROME ACTION

12

LIGHT, CLOCKS AND ION FLUX: AN ANALYSIS OF LEAF MOVEMENT

ARTHUR W. GALSTON
RUTH L. SATTER
Department of Biology, Yale University, New Haven, Connecticut 06520, U.S.A.

Introduction

The consequences of phytochrome transformation are many, and their mode of display throughout the plant kingdom are varied (Hendricks and Borthwick, 1967). The selection of a system to investigate the mode of action of phytochrome depends on the kinds of questions the investigator wishes to answer. If the aim is to delineate the most rapid possible reaction, then investigation of the 'Tanada (1967) effect' or chloroplast movements (Haupt, 1965) could be a wise choice. If, on the other hand, one insists on studying phytochrome in relation to morphogenetic changes, then de-etiolation is a more likely area of exploration (Mohr, 1969). The nature of induced biochemical changes can probably best be studied at the level of enzyme induction or activation (Attridge and Smith, 1967), while the relation of phytochrome-mediated changes to those controlled by rhythms can best be approached through photoperiodic analysis (Hillman, 1973).

We have chosen to work on phytochrome-controlled leaf movements, which seemed to us to maximise several advantages within a single biological system. Leaf movements in *Albizzia* and *Samanea* are reasonably rapid and easy to monitor, and are an accurate indicator of K^+ flux and thus of membrane properties. They also have measurable electrical components, are amenable to some types of biochemical analysis and show interaction with rhythmic phenomena, as required in any explanation of photoperiodism. Since most of the conclusions of our earlier investigations have recently been gathered together (Galston and Satter, 1972; Satter and Galston, 1973), we will summarise them only very briefly before going on to the newer experimental material which forms the major part of this contribution.

159

Materials and Methods

PLANTS

We utilised the pinnule and pinna movements of *Albizzia julibrissin* and *Samanea saman*, respectively, in our experiments. Nyctinastic leaf movements in these two leguminous species had been observed for many years, and had formed the basis for experimental approaches to the control of rhythmic behaviour and light action in plants (Palmer and Asprey, 1958a,b; Hillman and Koukkari, 1967; Jaffe and Galston, 1967). The demonstration that phytochrome controls nyctinastic movement in *Mimosa pudica* (Fondéville, Borthwick and Hendricks, 1966) was quickly extended to *Albizzia* (Hillman and Koukkari, 1967; Jaffe and Galston, 1967) and *Samanea* (Sweet and Hillman, 1969), whose lack of thigmonastic sensitivity made them more convenient subjects for investigation of light and rhythmically mediated phenomena.

LEAF MOVEMENT STUDIES

As the methods employed in growing and habituating these plants have already been described in great detail (Satter and Galston, 1971b), we need not repeat these facts here, but we should reiterate details of the *in vitro* treatments which form the basis for so many of the rhythmic experiments. These rest on the fact that excised pinnule pairs of *Albizzia*, attached to a portion of rachilla, continue to open and close for several days when floated on appropriate aqueous solutions in a petri dish. Hundreds of these butterfly-like pinnule pairs, carefully selected from a single leaf (*Figure 12.1*), show great uniformity of response to environmental stimuli such as light, temperature and chemicals included in the bathing medium. We usually placed 5–10 such excised pinnule pairs in a 10–cm petri dish on *ca.* 15 ml of medium. Rhythmic movements can be conveniently visualised by angular measurements every 60 or 120 min, while rapid responses to single stimuli, whether chemical or physical, can be noted within 5–10 min (Satter, Marinoff and Galston, 1970). Angle measurements were usually made by comparison with drawn figures; some attempts were also made to automate data gathering through mechanical strain-gauge transduction and time-lapse cinematography with green safelight illumination.

K^+ FLUX AS THE BASIS FOR LEAFLET MOVEMENT

In *Albizzia*, the movement of the tertiary pulvini is due to changing turgor pressure patterns of ventral and dorsal motor cells (Satter, Sabnis and Galston, 1970). When ventral cells are turgid and dorsal cells are somewhat flaccid, pinnules are open. When ventral cells

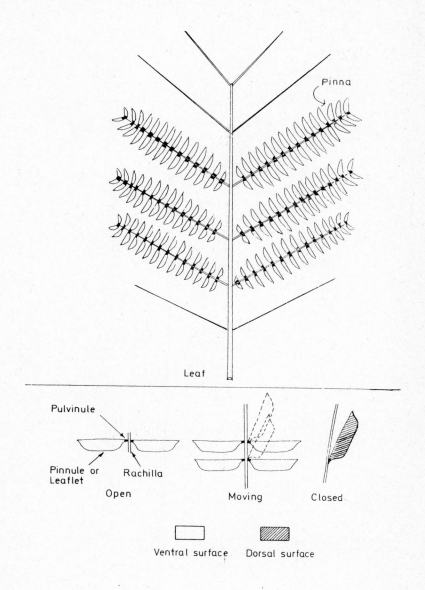

Figure 12.1 Mature Albizzia julibrissin *leaf with several pairs of pinnae, each divided into* ca. *15 pairs of pinnules.* (From Satter, Sabnis and Galston, 1970; reprinted with permission of Am. J. Bot.)

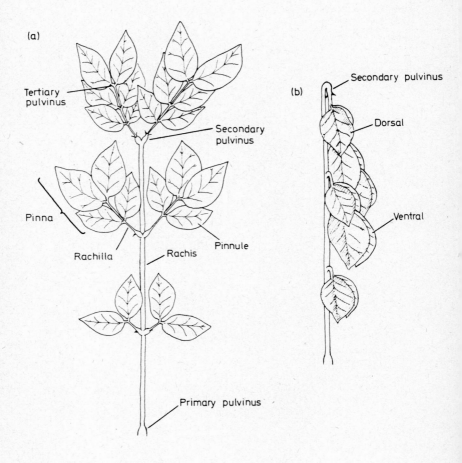

Figure 12.2 Leaf from Samanea saman *in the open (a) and closed (b) positions. (From Satter* et al., *1974; reprinted with permission of* J. Gen. Physiol.*)*

become less turgid and dorsal cells more turgid, upward closure of pinnule pairs ensues (*Figure 12.1*). In *Samanea,* the large secondary pulvini control *downward* closure of entire pinnae (*Figure 12.2*). In this instance, high turgor of dorsal pulvinal motor cells makes for opening, and high turgor of ventral motor cells makes for closure. In both plants, the change in angular orientation depends on movement of water, which in turn is based on movement of an osmotically active agent, which we have shown to be the K^+ ion (Satter and Galston, 1971a; Satter *et al.,* 1974).

POTASSIUM ANALYSES

Initially, analyses were made on pulvinal homogenates by flame atomic-absorption spectrophotometry with a Perkin-Elmer instrument. K^+ concentrations of 0.4–0.5 N were shown. Later, as it became clear that *in situ* analyses were required because most of the K^+ movements were intrapulvinal, we adopted the electron microprobe technique with an Acton instrument, examining the K^+ content of similar motor tissue exposed to different environmental conditions. Tissue preparation required extreme care so that no post-harvest leakage or transport of K^+ from the motor cells would occur. The method employed involved rapid freezing of pulvini in modified Tissue-Tek over dry-ice, followed by sectioning at 24–40 μm thickness. Sections were then lyophilised at 10 μmHg pressure, flattened between glass slides, mounted on graphite rods with an epoxy resin as adhesive, sketched with the aid of a *camera lucida,* coated with a thin layer of carbon in a vacuum evaporator and stored in a desiccator until required for microprobe analysis. Details of the operation of the instrument can be found in Satter, Marinoff and Galston (1970). In our early papers, K^+ scintillations were expressed per unit Ca^{2+}, the latter acting as an internal standard to correct for changes in the number of cells in the path of the electron beam. We soon found that this correction was not necessary (Satter *et al.,* 1974) and our more recent data express K^+ scintillations directly.

Results

LIGHT AND RHYTHMICALLY CONTROLLED LEAF MOVEMENTS IN *ALBIZZIA* AND *SAMANEA*

When darkened, closed leaflets of *Albizzia julibrissin* are exposed to white light, they open rapidly. Blue light, the effective region of the spectrum (Jaffe and Galston, 1967; Evans and Allaway, 1972), sets in motion two distinct processes: (a) the rapid change in turgor of pulvinal motor cells and (b) the phasing of a circadian rhythm. The former process provides the mechanical basis for immediate leaflet open-ing   while the latter determines that closure will commence at the latest in about 12 h, whether the leaves are darkened or illuminated. If the leaves are darkened, the rate and extent of closure are deter-mined by the state of phytochrome (Hillman and Koukkari, 1967; Jaffe and Galston, 1967).

Whether *Albizzia* leaflet closure is nyctinastic and under the control of phytochrome or rhythmic, the main chemical changes correlated with movement are the loss of potassium from the ventral pulvinal motor cells and gain in potassium by the dorsal motor cells (Satter and Galston, 1971a).   During the 90 min or so of nyctinastic closure, the state of phytochrome continuously governs the rate of the move-ment; thus, at any time, the photoconversion of Pfr to Pr slows the reaction, and at any time Pr → Pfr conversion speeds it up (Satter, Marinoff and Galston, 1970).   In this sense, the action of light on closure is not inductive.

Nyctinastic closure is sensitive to anaerobiosis and respiratory inhibitors such as NaN_3 and DNP (Satter, Marinoff and Galston, 1970; Satter and Galston, 1971b).   There are also sharp temperature limitations for the reaction.   Thus, at 29 °C, closure is slow and incomplete, but phyto-chrome control is pronounced.   At 18 °C, the extent of closure is increased, but there is less phytochrome control. K^+ analyses obtained by microprobe scan of longitudinal pulvinal sections show a loss of K^+ from ventral pulvinal cells and a gain in the dorsal motor cells (*Figure 12.3*).   This, together with the fact that exogenously supplied K^+ salts were most effective in inhibiting closure, led to the belief that relative K^+ concentrations in dorsal and ventral motor cells actually control the leaf angle.

When the light is extended beyond *ca.* 12 h, leaflets close in response to a rhythmic control (Satter and Galston, 1971a).   Similarly, leaflets kept in the dark for more than *ca.* 12 h open spontaneously, then close spontaneously *ca.* 12 h later.   These rhythmic oscillations in leaflet angle persist for several days under appropriate conditions. Although Pfr reduces the angle of opening during the first day, it is promotive on subsequent days, owing to rhythmic damping under Pr conditions (*Figure 12.4*).   Rhythmic movements also involve movements of K^+ into and out of key pulvinal motor cells, but these tend to be less rapid than light-initiated movements.   It is clear, however, that leaflet movement in *Albizzia* involves joint control by rhythm and phytochrome-mediated light absorption.   Because rhythmic closure shows a lesser sensitivity to NaN_3 than does opening, it appeared that rhythmic oscillations in K^+ might result from alternation of periods in which active K^+ absorption and passive K^+ leakage predominate.   Our most recent studies have indicated the desirability of a slight modification of this view, as discussed below.

Previous workers (Yin, 1941; Williams and Raghavan, 1966) had reported control of rhythmic leaf movements by auxin.   Auxin (10^{-5} M

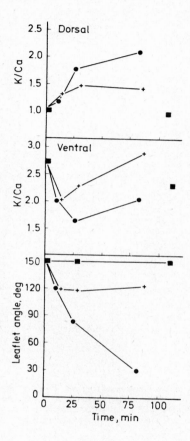

Figure 12.3 *Leaflet movement and* K^+ *flux in pulvinal motor cells of* Albizzia *leaflets transferred from white light to darkness (D) following red (R) or far-red (FR) pre-irradiation.* K/Ca *represents the ratio of* K^+ *to* Ca^{2+} *scintillations in dorsal and ventral motor tissue, measured with an Acton electron microprobe, 20–μm beam diameter.* ●——● *R,D;* +——+, *FR,D;* ■——■, *light. (After Satter and Galston, 1971b)*

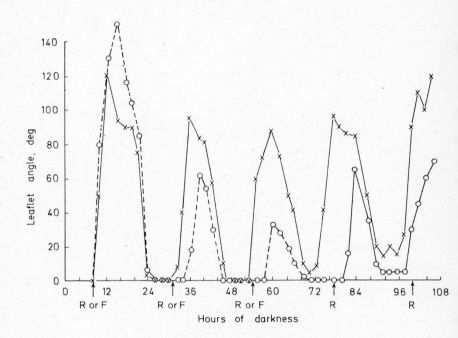

Figure 12.4 *Rhythmic movement of darkened* Albizzia *leaflet pairs floating on 50 mM sucrose. Leaflets were exposed to R (3 min exposure, 1.2 kerg cm⁻² s⁻¹ at* ca. *600–690 nm or FR (1.5 min exposure, 9 kerg cm⁻² s⁻¹ at* ca. *710–760 nm) every 23 h.* x, R *each rhythmic cycle (23 h);* o, F *first 3 cycles, R next 2 cycles*

and higher) does inhibit nyctinastic leaflet closure in *Albizzia* and also promotes rhythmic opening in the dark (Satter, Marinoff and Galston, 1972), but oscillations in auxin level alone cannot explain the behaviour of *Albizzia* leaflets, since they continue to open and close rhythmically even in the presence of optimal IAA in the bathing solution.   Auxin effects are correlated with appropriate control of K^+ level in pulvinal motor cells and are usually reversible on transfer of leaflets to water. The effect of auxin is not due to ethylene production, nor are changes in acetylcholine titre of pulvini correlated with movement (Satter, Applewhite and Galston, 1972).

Additional information on rhythmic oscillation was obtained through a study of timed applications of electrolytes, inhibitors and temperature changes (Applewhite, Satter and Galston, 1973; Satter *et al.*, 1973). For example, rhythmic opening is promoted by a temperature increase in the physiological range, and has a Q_{10} of about 3, while rhythmic closure is *retarded* by a similar temperature change (*Figure 12.5*). Similarly, salts such as $Ca(NO_3)_2$ or $Mg(NO_3)_2$ supplied during the opening phase greatly increased leaflet angle, but were ineffective during the closing phase (*Figure 12.6*). Other salts such as acetates and sulphates inhibited opening, and these effects correlated well with the rank of the anion in the Hofmeister series. These results add weight to the hypothesis of two alternating rhythmic phases, one characterised by high membrane integrity and the other by 'leakiness'.

More precise information linking inhibitory treatments to K^+ fluxes in pulvinal motor cells was obtained by timed pulses of inhibitors during various phases of the opening and closing cycle (Satter, Applewhite and Galston, 1974). Thus, low temperature and NaN_3 inhibit the K^+ flux that leads to opening, but promote net K^+ flux during closure. Cycloheximide and sodium acetate act optimally prior to opening, probably through control of the synthesis of proteins involved in regulation of K^+ movement. $Mg(NO_3)_2$ and similar salts promote best when applied after opening has begun. Aminophylline, known to inhibit activity of a cyclic nucleotide-splitting phosphodiesterase (Butcher and Sutherland, 1962) inhibits white light-promoted K^+ secretion from dorsal motor cells and concomitant opening.

Electron microprobe studies of cross-sections of the larger, fleshier secondary pulvini of *Samanea* permitted answers to be obtained to several questions that were not easily resolved with *Albizzia*. That these answers are probably also relevant to *Albizzia* is indicated by the similarity of phytochrome and rhythmic control in the two systems. The two systems can be encompassed within a single framework by the use of the terms *extensor* to denote cells which enlarge their volume during opening, and *flexor* to denote cells which expand somewhat, but mainly change shape during closure (Satter *et al.*, 1974). Since the leaflets move in two planes (towards and away from each other as well as dorsiventrally), the extensor and flexor regions are not exactly coincident with the dorsal and ventral regions. We asked that if potassium is high in extensor and low in flexor when the leaflet is open, what sort of a circumferential transition is there between the two regions? Is it gradual or abrupt? Studies with *Samanea* opened in the light indicate the latter; there is a high stable level of K^+ throughout the extensor, a low level of K^+ in the flexor, and a sharp transition between the two, extending over about two cell diameters (*Figures 12.7* and *12.8*). This indicates a remarkable differential action in contiguous motor cells. The large differences in salt content also indicated the possibility of sizeable differences in electrical potential between the two regions; this expectation was realised in later experiments (see below).

When leaves are allowed to close in the dark, the K^+ pattern shifts dramatically to a single maximum in the mid-flexor region, and gradual tapering on both sides of the peak to a minimum in the mid-extensor

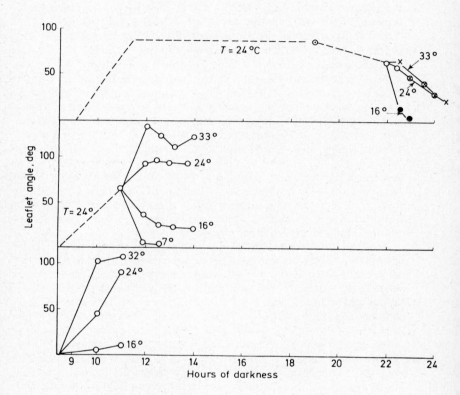

Figure 12.5 Effect of temperature alteration on the rhythmic movement of darkened Albizzia leaflets. Excised leaflet pairs floating on water were transferred from room temperature (24 °C) to incubators at the indicated temperatures for 2.5 h. (From Satter and Galston, 1973; reprinted with permission of BioScience)

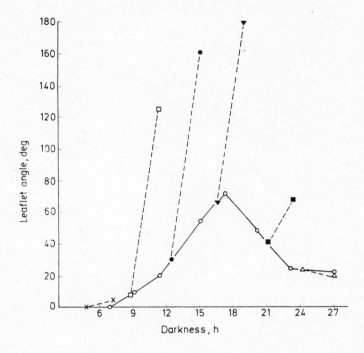

Figure 12.6 Effect on Albizzia *leaflet angle of differently timed 2.5–h pulse incubations in 0.1 N Ca(NO₃)₂. Solid line, control without added salt; dotted line, course of change after addition of salt. (From Satter* et al., *1973; reprinted with permission of* Plant Physiol.*)*

region. Changes produced during rhythmic opening include the formation of two additional peaks in the extensor region. Because temperature changes have their greatest effect on phytochrome-mediated processes during the 'active' period (Satter *et al.*, 1974), it appears that phytochrome interacts with that part of the cycle in which ion pumping predominates.

Phytochrome effects are also shown in *Samanea* during nyctinastic closure, with Pfr leading to lower K^+ in extensor cells, more K^+ in flexor cells, and accordingly smaller leaf angles (Satter, Geballe and Galston, 1974). In *Samanea*, as in *Albizzia*, phytochrome-controlled differences in leaflet closure increase with increasing temperature, providing further support for the concept of a Pfr-regulated active process.

In all cases investigated in *Samanea*, there is a decreasing gradient of K^+ from the boundary of the central vasculature towards the epidermis,

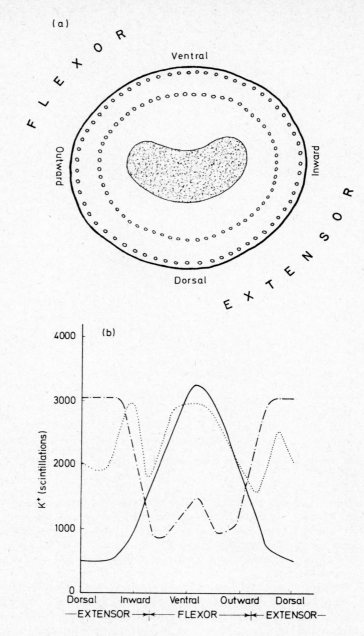

Figure 12.7 (a) Transverse section of a Samanea pulvinus. The central region is the vasculature. (b) Electron microprobe analysis of K^+ (scintillations per 15 s) in the motor tissue, 30 μm from the epidermis (outer circled regions of upper figure). The abscissa indicates distance from mid-dorsal reference point. —·—·—, Open (light); ······, partially open (dark); ———, closed (dark). (From Satter et al., 1974; reprinted with permission of J. Gen. Physiol.)

whether dorsal or ventral. From this result, it is reasonable to conclude that K^+ entering or leaving pulvinal motor cells does so by way of the inner cortex.

In both *Albizzia* and *Samanea,* excision of the leaflet lamina does not prevent the operation of the pulvinar mechanism (Koukkari and Hillman, 1968; Satter and Galston, 1971b). It appears that the pulvinus contains the clock and the pigments necessary for timing and light reception and, at least during short periods following excision, the potassium and the energy source required for movement. The partial excision experiments of Palmer and Asprey (1958b), together with our data, support the view that extensor cell turgor is most sensitive to rhythmic control, while flexor cell turgor responds most to transitions between white light and darkness.

LIGHT, RHYTHMS AND TRANSMEMBRANE POTENTIAL IN *SAMANEA* PULVINI

Sharp differences of K^+ in motor cells on opposite sides of the pulvinus suggested that large differences in electrical potential should exist between them. *Samanea* pulvini are large enough to permit the insertion of fine glass microelectrodes used for recording bioelectric potentials (Racusen and Satter, 1975). An electrode with a tip diameter of less than 0.5 μm and a tip resistance of 15-45 MΩ was connected by means of Ag/AgCl wire to a WPI Model M4-A electrometer and the reference electrode was earthed. Secondary pulvini from the third to seventh uppermost leaves were cut and mounted in a lucite chamber and perfused with a solution containing 1 mM KCl, 1 mM $Ca(NO_3)_2$, 0.25 mM $MgSO_4$ and a 1 mM sodium phosphate buffer, pH 5.6. Pulvini were excised every 2-3 h during a 48-h dark period. Cells were impaled immediately after pulvinal excision, and potentials were recorded during the subsequent 15-120 s. Leaf angles were also estimated periodically on similarly darkened intact plants, in order to provide a physiological base-line for the electrical observations. Potentials ranging from -30 to -90 mV were recorded

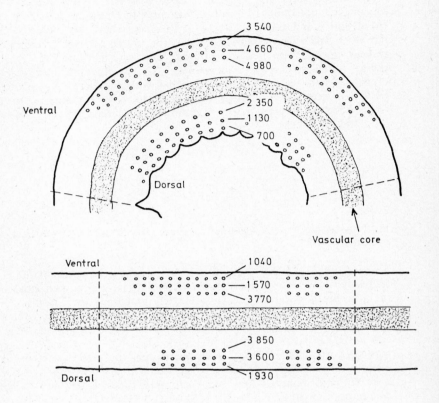

Figure 12.8 K^+ distribution in longitudinal, dorsi-ventral sections of a Samanea pulvinus closed in the dark (top) and open in white light (bottom). K^+ was analysed in the circled regions in the motor tissue (outer cortex, 30 μm from the epidermis) and midcortex (100 and 225 μm from the epidermis). Measured values are scintillations during 15 s. (From Satter et al., 1974; reprinted with permission of J. Gen. Physiol.)

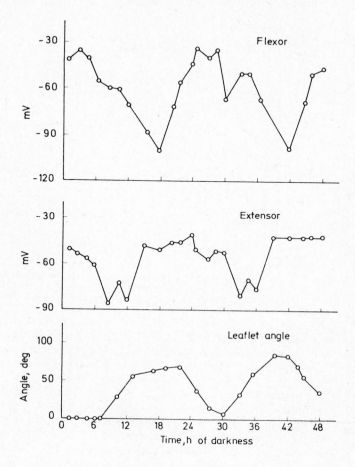

Figure 12.9 Rhythmic changes in leaflet angle and transmembrane potential of extensor and flexor cells of Samanea. *The reference electrode was in a sodium phosphate buffer (1 mM, pH 5.6) containing 1 mM KCl, 1 mM Ca(NO₃)₂ and 0.25 mM MgSO₄. (From Racusen and Satter, 1975; reprinted with permission of* Nature, Lond.*)*

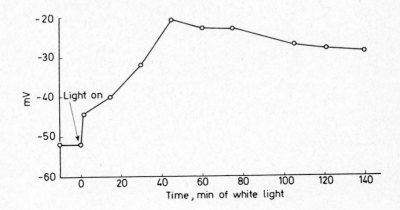

Figure 12.10   Transmembrane potential of flexor motor cells of Samanea *leaflets in cool white fluorescent light (ca. 1000 ft candles).   See legend to Figure 12.9 for experimental conditions.   (From Racusen and Satter, 1975; reprinted with permission of Nature, Lond.)*

from both flexor and extensor; the flexor is in phase with leaflet movement, showing maximum polarisation (interior negative) during the 'active' part of the cycle, when leaf angle is maximal.   The extensor is about 8 h out of phase with the others (*Figure 12.9*).

The potential of the flexor cells was also measured during light–dark transitions.   In these experiments, cells were impaled prior to the light treatments so as to permit the determination of rapid responses to light.   When darkened leaves are exposed to white light, there is a rapid (*ca.* 15–20 min) opening response, presumably due to excretion of K^+ from flexor cells.   There is also a much more rapid depolarisation response in the flexor region (*Figure 12.10*), detectable within 5 s. The depolarisation continues for about 50 min (from *ca.* -50 to -20 mV), then slowly reverses to about -30 mV in continuous bright white light.   When the effect of cool white fluorescent light was compared with that of blue light, the latter obtained by interposing a Cinemoid 20 filter (peak transmission = 40 per cent, at 450 nm) between the tissue and the cool white fluorescent source, blue-irradiated pulvini depolarised even more than the white-irradiated pulvini (Racusen, unpublished data), leading to the conclusion that other wavelengths antagonise the blue light-mediated response.

Open leaflets transferred from bright white light to low-intensity red or far-red light close more rapidly when phytochrome is in the Pfr form.   The flexor cells of red light-treated leaflets hyperpolarise 30 mV

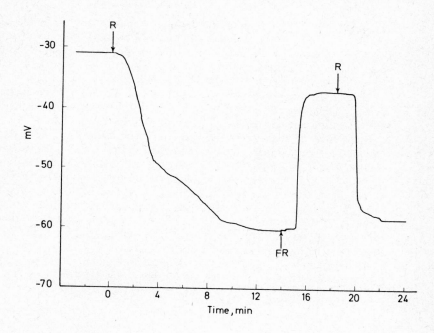

Figure 12.11 Phytochrome-controlled changes in the transmembrane potential of flexor motor cells of Samanea. *See legend to* Figure 12.9 *for experimental conditions. (From Racusen and Satter, 1975; reprinted with permission of* Nature, Lond.*)*

in about 10 min, while leaflets exposed to far-red light maintain a fairly constant potential during this interval. Far-red after red light reverses the red-light effect within 90 s, and red light again reverses the far-red effect fully within an additional 90 s (*Figure 12.11*). Because these electrical changes precede appreciable K^+ movement, it follows that mass K^+ movement cannot be responsible for the observed changes in potential. Rather, it would appear that some other event, such as an anion or proton flux, possibly electrogenic, precedes and causes the K^+ flux that is important in leaflet movement.

NEW EXPERIMENTS ON THE INTERACTION BETWEEN LIGHT AND
RHYTHMS IN *ALBIZZIA*

To determine the interaction between light signals and rhythmic oscillation in leaf movement, excised *Albizzia* leaflet pairs were exposed to
long dark periods of 96–130 h. Periodic exposures to red or far-red
light and periodic changes of solution from water to sucrose permitted
many interesting conclusions to be reached.

The rhythm is phased by the onset of the previous white light
period, mediated, we presume, by the blue-absorbing pigment that promotes leaflet opening and depolarisation of the flexor cells.

Persistence of the rhythm depends on the presence of Pfr and a
carbon or energy source such as sucrose (*Figure 12.12*). Thus, if the
rhythm has damped because of treatment with far-red light, it can be
made to resume, in step with undamped controls, simply by exposure
to red light (*Figure 12.4*). In addition, repeated daily exposures to
red light are far more effective than a single red irradiation for maintenance of the rhythm (*Figures 12.12* and *12.13a*). Repeated daily
exposures to far-red light, in contrast, are only slightly more effective
than a single far-red treatment (*Figure 12.13b*).

Red light-treated leaflets deprived of sucrose damp at some angle
intermediate between open and closed; they can also be made to resume

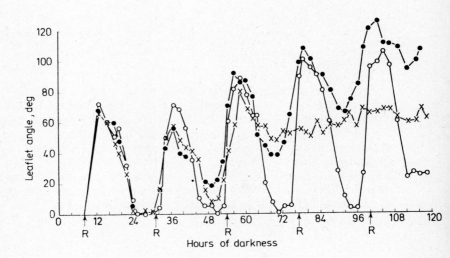

Figure 12.12 Rhythmic leaf movements in vitro. *Darkened* Albizzia *leaflet pairs
floating on water or sucrose were exposed briefly to red light every 23 h.*
x, H_2O; ○, *sucrose (50 mM)*; ●, *sucrose (5 mM)*

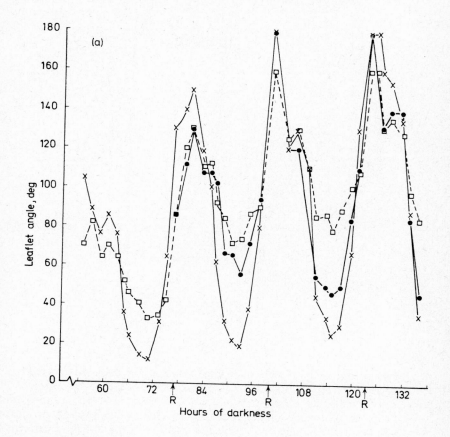

Figure 12.13 Leaflets for (a) and (b) were taken from the same Albizzia leaf, and are physiological replicates. Darkened leaflet pairs floating on 50 mM sucrose were exposed briefly to red or far-red light at hour 8, or at regular 23-h intervals. In (a), an additional group of leaflets was exposed to red light at hour 8, and also at 23-h intervals after the rhythm started to damp (hour 77).
□ , red light at hour 8; x, red light at hour 8 and every 23 h thereafter;
●, red light at hours 8, 77, 100 and 123; Δ, far-red light once at hour 8;
○, far-red light at hour 8 and every 23 h thereafter

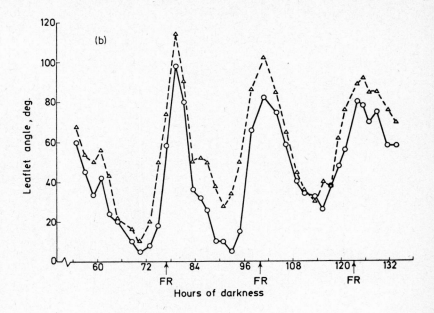

Figure 12.13 (cont.)

oscillation, in step with undamped controls, if given 50 mM sucrose
(*Figure 12.14a*). Sucrose need not be given continuously in such
experiments; administration for any ¼ of the 23-h cycle, during photophile or skotophile phase, is almost as effective as continuous treatment
(*Figure 12.15*). Leaflets damped because of far-red light treatment *and*
sucrose deprivation damp in the open state; when supplied continuously
with sucrose, they close and damp in that position (*Figure 12.14b*).
It is clear that both Pfr and sucrose are required for continued manifestation of the overt leaf movement rhythm. Whether they are
required for operation of the clock itself remains to be determined.

 Similar experiments with *Samanea* have not yet been performed, but
in intact *Samanea* plants maintained at four different temperatures, the
frequency was independent of temperature between *ca.* 20 and 30 °C,
while at 16 °C the rhythm damped quickly (*Figure 12.16*). In excised
Albizzia leaflets, damping did not occur at 16 °C, but did occur when
the temperature was decreased to 11 °C. When such low-temperature
damped leaflets were returned to higher temperatures, oscillations
resumed at a phase indicating that the clock had been stopped by low-
temperature treatments. This contrasts with the effects of sucrose and
phytochrome, described above.

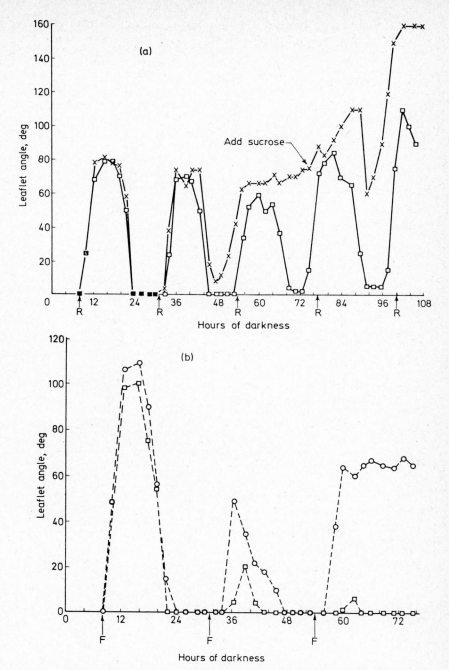

Figure 12.14 Leaflets for (a) and (b) were taken from the same Albizzia leaf, and are physiological replicates. Darkened leaflet pairs, exposed briefly to red or far-red light at regular 23-h intervals, were floated on water or 50 mM sucrose. In (a) leaflets on water were transferred to sucrose after the rhythm had damped. x, H_2O until hour 75, then sucrose; □, sucrose continuously; ○, H_2O

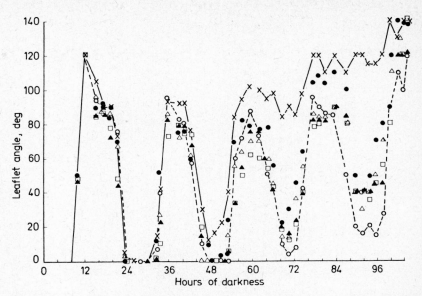

Figure 12.15   Darkened Albizzia leaflets exposed briefly to red light every 23 h were supplied with 50 mM sucrose for a 5½–h period of each rhythmic cycle. x, H₂O; ○, sucrose continuously; △, sucrose first half photophile; ●, sucrose second half photophile; ▲, sucrose first half skotophile; □, sucrose second half skotophile

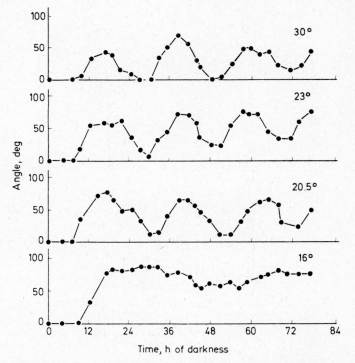

Figure 12.16   Effect of temperature on rhythmic leaf movements in intact Samanea plants.   Note especially damping at 16 °C

Discussion

The combined physiological, chemical and electrical data on leaf movements in *Albizzia* and *Samanea* reinforce the view (Fondéville, Borthwick and Hendricks, 1966) that some membrane-localised process such as ion transport is an early consequence of phytochrome phototransformation. The fact that rhythmic and blue light-mediated processes also regulate ion movements and changes in electrical potential suggests the possibility that all of these phenomena and their interactions are the result of membrane-based changes. Despite recent advances (Singer and Nicholson, 1972), our knowledge of membrane structure is still fragmentary and the possibility of a meaningful analysis in biochemical detail is severely limited, yet some attempt should be made.

It is probably easiest to suggest an interpretation of the temperature data. The damping of the oscillation at such moderate temperatures as 16 °C (*Samanea*) and 11 °C (*Albizzia*) can possibly be understood in terms of solidification points of membrane lipids (Njus, Sulzman and Hastings, 1974). This hypothesis should be directly testable in *Samanea*, where groups of motor cells large enough for lipid analysis can be excised.

The role of phytochrome is more difficult to delineate, but it presumably regulates some energetically driven ion movements, since Pr *vs* Pfr differences are (a) greatest during the active phase of the rhythm; (b) decreased by respiratory inhibitors such as NaN_3 and DNP and (c) decreased by low temperature. The Tanada (1967) effect, the localisation experiments of Haupt (1965) and experiments on model lipid membranes (Roux and Yguerabide, 1973) all support the notion that Pfr is active in a membrane and that it can affect ion flux. Whatever its mode of action, phytochrome seems remarkably stable in these pulvini, since the differential effects of a single irradiation with red *versus* far-red light continue to be manifested even after 124 h (compare *Figures 12.13a* and *12.13b*).

As for the rhythm, we have previously hypothesised that it involves both ion pumping and leakage, with alternating predominance of active transport during leaf opening, and leakage through ion channels during closure. In view of the demonstrated sucrose requirement for closure, this theory requires some modification. Several possibilities suggest themselves. For example, Roux (1971) has found that carbohydrates are an integral part of the phytochrome molecule; about 36 μg of sugar per milligram of purified phytochrome were indicated by the phenol–sulphuric acid method. He has also shown that even *in vitro*, exposure of phytochrome to sucrose results in spectral anomalies in the amino acids released from phytochrome by acid hydrolysis; these include a shoulder on the lysine peak and a new peak between cysteic acid and aspartic acid. Thus, one possibility is that a carbohydrate part of the phytochrome molecule is functionally involved in its action, and that sucrose deprivation results in loss of this moiety, leading to inactivation of phytochrome.

A second possibility is that phytochrome attaches to membranes at a glycoprotein locus, and that sucrose deprivation results in a loss of such loci through normal turnover of membrane constituents. The phytochrome carbohydrate found by Roux (1971) might even represent a portion or all of this receptor locus removed during the extraction procedure. In this connection, it should be noted that all *in vitro* studies of Pfr attachments to membrane fragments (Marmé, Boisard and Briggs, 1973; Quail, Marmé and Schäfer, 1973) have involved membranes harvested after density-gradient centrifugation in sucrose.

A third possibility would envisage activation of phytochrome through a sucrose-mediated conformational change. Such effects have been noted by others (Clement-Métral and Yon, 1968) for β-lactoglobulin *a*. According to this interpretation, Pfr attached to the membrane would not become active until conformationally altered by sucrose. Finally, there is the possibility that sucrose acts as respiratory substrate that provides energy or carbon skeletons for phytochrome-mediated processes such as ion transport. If this view is correct, then sucrose is required only to manifest the changes which flow automatically from fully active, membrane-attached Pfr.

The blue light reaction presumably involves photoreception by a flavoprotein and reduction of a *b*-type cytochrome (Poff and Butler, 1974; Muñoz and Butler, 1975), leading in some way to membrane depolarisation and ion secretion from flexor cells, and setting of the clock for subsequent oscillations. Potential changes, especially depolarisation in the flexor, are detectable within 5 s after white-light irradiation and continue as long as the light is on. The rate of potential change is dependent on the intensity (Racusen, unpublished data).

Since blue light works better than white light, it is possible that some other region of the spectrum, perhaps the green light absorbed by reduced cytochrome *b*, antagonises the effect of blue light. The oxidation of reduced cytochromes following the administration of KCl to salt-starved wheat roots (Lundegårdh, 1960) provides an example of how the photo-activated flavoprotein might lead to salt transfer.

The understanding of the detailed interactions of these control systems will require much additional work. The effort will be worthwhile, since the possible rewards include no less than an understanding of plant photoperiodism.

Acknowledgements

This work was aided by a grant from the National Science Foundation. We are indebted to numerous helpful colleagues for aid and valuable discussions over the years. These include Drs P.B. Applewhite and D.D. Sabnis; graduate students G.F. Gardner, G.T. Geballe, R. Racusen and S. Long; and undergraduate assistants J. Chaudhri, D. Chernoff, D. Kavon, D.J. Kreis, Jr., and P. Marinoff.

References

APPLEWHITE, P.B., SATTER, R.L. and GALSTON, A.W. (1973). *J. Gen. Physiol.*, **62**, 707

ATTRIDGE, T.H. and SMITH, H. (1967). *Biochim. Biophys. Acta*, **148**, 805

BUTCHER, R.W. and SUTHERLAND, E.W. (1962). *J. Biol. Chem.*, **237**, 1244

CLEMENT-MÉTRAL, J. and YON, J. (1968). *Biochim. Biophys. Acta*, **160**, 340

EVANS, L.T. and ALLAWAY, W.G. (1972). *Aust. J. Biol. Sci.*, **25**, 885

FONDÉVILLE, J.C., BORTHWICK, H.A. and HENDRICKS, S.B. (1966). *Planta*, **69**, 357

GALSTON, A.W. and SATTER, R.L. (1972). In *Structural and Functional Aspects of Phytochemistry*, pp.51-79. Eds. Runeckles, V.C. and Ts'o, T.C. Academic Press, New York

HAUPT, W. (1965). *Annu. Rev. Plant Physiol.*, **16**, 267

HENDRICKS, S.B. and BORTHWICK, H.A. (1967). *Proc. Nat. Acad. Sci. U.S.A.*, **58**, 2125

HILLMAN, W.S. (1973). *BioScience*, **23**, 81

HILLMAN, W.S. and KOUKKARI, W.L. (1967). *Plant Physiol.*, **42**, 1413

JAFFE, M.J. and GALSTON, A.W. (1967). *Planta*, **77**, 135

KOUKKARI, W.L. and HILLMAN, W.S. (1968). *Plant Physiol.*, **43**, 698

LUNDEGÅRDH, H. (1960). *Plant Physiology*, p.266. American Elsevier Publishing Co., New York

MARMÉ, D., BOISARD, J. and BRIGGS, W.R. (1973). *Proc. Nat. Acad. Sci. U.S.A.*, **70**, 3861

MOHR, H. (1969). In *An Introduction to Photobiology*, pp.99-141. Ed. Swanson, C.P. Prentice-Hall, Englewood Cliffs, N.J.

MUÑOZ, V. and BUTLER, W.L. (1975). *Plant Physiol.*, **55**, 421

NJUS, D., SULZMAN, F.M. and HASTINGS, J.W. (1974). *Nature, Lond.*, **248**, 116

PALMER, J.H. and ASPREY, G.F. (1958a). *Planta*, **51**, 757

PALMER, J.H. and ASPREY, G.F. (1958b). *Planta*, **51**, 770

POFF, K.L. and BUTLER, W.L. (1974). *Nature, Lond.*, **248**, 799

QUAIL, P.H., MARMÉ, D. and SCHÄFER, E. (1973). *Nature New Biol.*, **245**, 189

RACUSEN, R.H. and SATTER, R.L. (1975). *Nature, Lond.*, **255**, 408

ROUX, S.J. (1971). *Ph.D. Thesis*, Yale University

ROUX, S.J. and YGUERABIDE, J. (1973). *Proc. Nat. Acad. Sci. U.S.A.*, **70**, 762

SATTER, R.L., APPLEWHITE, P.B. and GALSTON, A.W. (1974). *Plant Physiol.*, **54**, 280

SATTER, R.L., APPLEWHITE, P.B. and GALSTON, A.W. (1972). *Plant Physiol.*, **50**, 523

SATTER, R.L., APPLEWHITE, P.B., KREIS, D.J. Jr., and GALSTON, A.W. (1973). *Plant Physiol.,* **52,** 202

SATTER, R.L. and GALSTON, A.W. (1971a). *Science, N.Y.,* **174,** 518

SATTER, R.L. and GALSTON, A.W. (1971b). *Plant Physiol.,* **48,** 740

SATTER, R.L. and GALSTON, A.W. (1973). *BioScience,* **23,** 407

SATTER, R.L., GEBALLE, G.T., APPLEWHITE, P.B. and GALSTON, A.W. (1974). *J. Gen. Physiol.,* **64,** 413

SATTER, R.L., GEBALLE, G.T. and GALSTON, A.W. (1974). *J. Gen. Physiol.,* **64,** 431

SATTER, R.L., MARINOFF, P. and GALSTON, A.W. (1970). *Am. J. Bot.,* **57,** 916

SATTER, R.L., MARINOFF, P. and GALSTON, A.W. (1972). *Plant Physiol.,* **50,** 235

SATTER, R.L., SABNIS, D.D. and GALSTON, A.W. (1970). *Am. J. Bot.,* **57,** 374

SINGER, S.J. and NICHOLSON, G. (1972). *Science, N.Y.,* **175,** 720

SWEET, H. and HILLMAN, W.S. (1969). *Physiol. Plant.,* **22,** 776

TANADA, T. (1967). *Proc. Nat. Acad. Sci. U.S.A.,* **58,** 376

WILLIAMS, C.M. and RAGHAVAN, V. (1966). *J. Exp. Bot.,* **17,** 742

YIN, H.C. (1941). *Am. J. Bot.,* **28,** 250

13

PHOTOCONTROL OF ENZYME LEVELS

T.H. ATTRIDGE
Department of Biological Science, North East London Polytechnic, Romford Road, London E14 4LZ, U.K.

C.B. JOHNSON
Department of Physiology and Environmental Studies, Nottingham University, School of Agriculture, Sutton Bonington, Loughborough, Leics. LE12 5RD, U.K.

During the process of de-etiolation, when dark-grown seedlings first perceive light, extensive developmental and metabolic changes occur. The levels of various enzymes increase and some of these have been demonstrated to be under the control of phytochrome. Phenylalanine ammonia-lyase (PAL) and ascorbic acid oxidase are two such enzymes which have received particular attention from research workers, their intentions being to ascertain the mechanism by which phytochrome is able to increase the level of these enzymes, and whether or not the increases involve a common mechanism.

When dark-grown gherkins are irradiated with blue light, a transient increase in the level of PAL occurs, reaching a maximum between 3 and 4 h (Attridge and Smith, 1973). When mustard seedlings are irradiated with far-red light, a similar increase occurs but a maximum is not reached until about 24 h (Durst and Mohr, 1966). For this reason, gherkin seedlings were chosen as the more favourable tissue for the investigation of this system. For some years it was thought, as a result of experiments using protein synthesis inhibitors, that this blue light-mediated increase in PAL activity was a result of light-stimulated synthesis of PAL (Engelsma, 1967). However, more recently evidence has been put forward to suggest that this is not the case (Attridge and Smith, 1973). When dark-grown gherkin seedlings are chilled (5 °C) for 24 h and then their temperature is increased to 25 °C, an increase in the level of PAL is observed. This increase is not prevented by cycloheximide. Indeed, if cycloheximide is applied at the time of change from 5 to 25 °C, a large increase in the level of PAL, over and above that of the control, is found. Similarly, if dark-grown gherkin seedlings are treated with cycloheximide, the level of PAL increases with a shorter lag-phase than that of the increase promoted by blue light.

The blue light-initiated increase has been extensively investigated using the technique of density labelling (Attridge and Smith, 1974). The results of one such experiment are shown in *Table 13.1*. When dark-grown gherkin seedlings are transferred from H_2O to 100 per cent ^2H_2O for 4 h, an increase in the buoyant density of PAL results. In contrast, when dark-grown gherkin seedlings are transferred simultaneously to 100 per cent ^2H_2O and blue light, no increase in the buoyant

Table 13.1   Effect of blue light on the incorporation of density label into PAL (gherkin hypocotyls)

Treatment[1]	Buoyant density of extractable PAL (kg 1^{-1})
96 h H_2O: 4 h H_2O dark control	1.294
96 h H_2O: 4 h H_2O + blue light	1.293
96 h H_2O: 4 h ^2H_2O dark control	1.300
96 h H_2O: 4 h ^2H_2O + blue light	1.293
72 h H_2O: 28 h H_2O dark control	1.305
72 h H_2O: 24 h ^2H_2O: 4 h ^2H_2O + blue light	1.303

[1] After these treatments, PAL was extracted from the hypocotyls and centrifugation carried out as described by Attridge and Smith (1974)

density of PAL is detected. Pre-incubation with ^2H_2O prior to illumination leads to incorporation of ^2H_2O into PAL but the density of the enzyme from blue light-treated plants is always lower than in its appropriate dark control. The density labelling results can be interpreted in a number of ways. Firstly, one may accept the *prima facie* result that less synthesis of PAL occurs in blue light than in dark-grown material. This interpretation does not explain the increase of PAL activity in blue light-irradiated tissue. Secondly, it may be suggested that blue light increases the stability of PAL, which would allow native PAL molecules to survive longer and thus give a mean value in buoyant density to the deuterated blue light treatment which is lower than its appropriate dark control. An increase in the half-life of PAL would lead to an increase in activity if it occurred as a result of blue light irradiation. No evidence exists either to support or to refute this hypothesis, but when the cycloheximide-initiated increases in PAL are borne in mind (Attridge and Smith, 1973), a third alternative presents itself. These results indicate that there exists, in dark-grown gherkin tissue, a pool of inactive PAL from which increases in the level of this enzyme might originate.

The results of the density labelling experiments can be interpreted in terms of this hypothesis. If PAL synthesis proceeds at the same rate in the dark and the light and these newly formed molecules are subsequently inactivated, they would become part of an inactive pool which was partially formed during the H_2O regime and thus would contain a proportion of native PAL molecules. If the blue light-mediated response originates from this inactive pool, then it would be expected that the buoyant density of the enzyme would be lower than that of the appropriate dark control, which would consist of newly synthesised (and therefore deuterated) PAL molecules which had not yet been inactivated. This suggestion is strongly supported by the fact that when the density labelling technique is applied to investigate the cycloheximide response, the results are identical with those obtained with the blue light treatment; the cycloheximide-treated plants contain PAL lower in buoyant density than in the appropriate dark control, even though there has been a large increase in enzyme activity (Attridge and Smith, 1974). All the evidence points to the fact that the blue light response and the cycloheximide response originate via activation from a pool of inactive enzyme, since in both instances substantial increases in enzyme activity occur with negligible concomitant synthesis.

An entirely different picture emerges when the enzyme ascorbic acid oxidase is studied in mustard cotyledons. The activity of this enzyme increases over a much longer period when dark-grown mustard seedlings are irradiated with far-red light. The results of density labelling experiments first carried out at Sutton Bonington (Attridge, 1974) and repeated at Freiburg (Acton, Drumm and Mohr, 1974) are shown in *Table 13.2*. Although there was a significant difference in the experimental procedures used by these two groups [Attridge (1974) transferred 48–h-old dark-grown mustard seedlings to 100 per cent ^2H_2O, while Acton, Drumm and Mohr (1974) transferred 48–h-old seedlings to 80 per cent ^2H_2O], a close correlation between the results was nevertheless found. It may be noted that synthesis of ascorbic acid oxidase occurs both in the far-red and dark treatments, but in all experiments ascorbic acid oxidase has a higher buoyant density when extracted from far-red light-treated plants than when extracted from dark-grown material. Complicated hypotheses could be proposed to explain this

Table 13.2 Effect of far-red light on the incorporation of density label into ascorbic acid oxidase in mustard cotyledons

Treatment[1]	Buoyant density of ascorbic acid oxidase (kg l^{-1})
48 h H_2O: 12 h H_2O dark control[2]	1.304
48 h H_2O: 12 h H_2O + far-red light[2]	1.303
48 h H_2O: 24 h H_2O dark control[3]	1.303
48 h H_2O: 24 h H_2O + far-red light[3]	1.302
48 h H_2O: 6 h ^2H_2O (100%) dark control[3]	1.306
48 h H_2O: 6 h ^2H_2O (100%) + far-red light[3]	1.315
48 h H_2O: 12 h ^2H_2O (100%) dark control[3]	1.310
48 h H_2O: 12 h ^2H_2O (100%) + far-red light[3]	1.315
48 h H_2O: 12 h ^2H_2O (80%) dark control[2]	1.310
48 h H_2O: 12 h ^2H_2O (80%) + far-red light[2]	1.313
48 h H_2O: 24 h ^2H_2O (100%) dark control[3]	1.312
48 h H_2O: 24 h ^2H_2O (100%) + far-red light[3]	1.316

[1] After these treatments, ascorbic acid oxidase was extracted, centrifuged and assayed as described in the original references
[2] Data from Acton, Drumm and Mohr (1974)
[3] Data from Attridge (1974)

result in terms of activation, but in the absence of evidence the simplest explanation is that far-red light, acting via phytochrome, increases the rate of synthesis of this enzyme.

When the evidence for the light-mediated activation of PAL in gherkin seedlings was first propounded, both the workers at Sutton Bonington and those at Freiburg decided to investigate the response of PAL to far-red light treatment in mustard seedlings. Once again experimental differences existed. In the work of Acton and Schopfer (1975), 36-h-old dark-grown seedlings were transferred to 80 per cent ^2H_2O, whereas in that of Attridge, Johnson and Smith (1974) 48-h-old dark-grown seedlings were transferred to 100 per cent ^2H_2O. These are only the most obvious of the differences in the experimental procedures. Whereas in the experiments with ascorbic acid oxidase similar results were achieved by both groups despite the differences in method, with PAL widely differing results were obtained. A comparison of these results is shown in *Table 13.3*. The results of Acton and Schopfer (1975) clearly demonstrate that, under the conditions used, the buoyant density of PAL extracted from far-red light-treated seedlings is the same as that extracted from dark-grown seedlings. In contrast, the results of Attridge,

Table 13.3 Effect of far-red light on the incorporation of density label into PAL in mustard cotyledons

A: Data from Acton and Schopfer (1975)

Treatment[1]	Buoyant density of PAL (kg 1^{-1})
36 h H_2O: 6 h H_2O dark control	1.2863
36 h H_2O: 6 h H_2O + far-red light	1.2854
36 h H_2O: 6 h ^2H_2O (80%) dark control	1.2946
36 h H_2O: 6 h ^2H_2O (80%) + far-red light	1.2957
36 h H_2O: 12 h ^2H_2O (80%) dark control	1.2962
36 h H_2O: 12 h ^2H_2O (80%) + far-red light	1.2968
36 h H_2O: 24 h ^2H_2O (80%) dark control	1.2984
36 h H_2O: 24 h ^2H_2O (80%) + far-red light	1.2995

B: Data from Attridge, Johnson and Smith (1974)

Treatment[1]	Buoyant density of PAL (kg 1^{-1})
48 h H_2O: 24 h H_2O dark control	1.295
48 h H_2O: 24 h H_2O + far-red light	1.297
48 h H_2O: 24 h ^2H_2O (100%) dark control	1.308
48 h H_2O: 24 h ^2H_2O (100%) + far-red light	1.308
48 h H_2O: 40 h ^2H_2O (100%) dark control	1.315
48 h H_2O: 40 h ^2H_2O (100%) + far-red light	1.309
48 h H_2O: 48 h ^2H_2O (100%) dark control	1.321
48 h H_2O: 48 h ^2H_2O (100%) + far-red light	1.314

[1] After these treatments, PAL was extracted, centrifuged and assayed as described in the original references

Johnson and Smith (1974) demonstrate equally clearly that the buoyant density of the enzyme extracted from far-red light-treated seedlings after 40 and 48 h of far-red light and ^2H_2O treatment is much lower than that of the appropriate dark control. These results reflect those obtained with gherkin tissue under blue-light treatment.

Acton and Schopfer (1975) made a study of the bandwidths of PAL in CsCl gradients and as a result postulated that the half-life of the enzyme is too short to demonstrate an increased rate of synthesis mediated by phytochrome. By the time the extractions are made, both light and dark enzymes are equally labelled and under these conditions it is impossible to test for newly synthesised PAL. Although we have reservations of a detailed nature about the use of bandwidth measurements in investigations of this kind (Smith, Attridge and Johnson, 1975), we accept that these data suggest that it may not be possible to measure a light-stimulated increase in the rate of PAL synthesis in mustard. In our first publication on this subject (Attridge, Johnson and Smith, 1974), we established the protocol of the density labelling technique as applied to a comparison of rates of enzyme synthesis. We pointed out that in situations where the response time was long compared with the half-life of the enzyme under investigation, the density labelling technique would be useless for demonstrating an increased rate of synthesis. We also showed this situation to apply to PAL in mustard, but not in gherkin where the response time is so much shorter. It does not, however, render the technique useless for demonstrating the *activation* of PAL from an existing pool of inactive enzyme and, with the proviso that a far-red light-stimulated PAL stabiliser could produce the same density labelling results, this is our proposal.

A further difference in the results between the two groups of workers is that the magnitude of the PAL response to far-red light found by the workers at Freiburg is reduced by more than a third by treatment with 80 per cent ^2H_2O, whereas in the dark it is increased by about the same percentage. On the other hand, the Sutton Bonington group found that by transferring the plants to 100 per cent ^2H_2O, the lag-phase of the response is increased to 24 h. It may be that when the effect of ^2H_2O *per se* on this system is more fully understood, the reason for the differences between the results from the two groups will be explicable. At present, vague criticisms, e.g. that one procedure or the other causes toxification of the plant material, or that the subsequent response is an artifact, are purely subjective.

Acton, Drumm and Mohr (1974) have also suggested that it is possible, on the basis of bandwidth determinations, to distinguish between mechanisms that involve control of synthesis and activation. This suggestion is based on the assumption that inactive enzyme of lower density diluting out the label in the detectable enzyme pool following stimulation would cause an increase in the heterogeneity of the (now) active enzyme pool. This would contrast with a decreased bandwidth caused by an increased rate of synthesis. We see no reason, however, why this difference need be found. Where there is a significant increase in enzyme activity derived from conversion of inactive enzyme to an active form, the distribution of the activity in a density gradient will simply be a function of the turnover rate of the inactive pool. Thus, while measurement of bandwidth can give, under steady-state conditions, an indication of the half-life of an enzyme, it does not yield any information which enables one to distinguish between control of synthesis and activation.

Since the data obtained from density labelling experiments with PAL extracted from mustard cotyledons have yielded conflicting results, attempts were made to see whether there was any evidence for the existence of inactive PAL in this tissue. Temperature transfer experiments which had previously proved fruitful with gherkin were carried out on mustard seedlings. *Figure 13.1* shows the last phase of such an experiment. Dark-grown mustard seedlings, 48 h old, were chilled (5 °C) for 24 h and then returned to 25 °C conditions. A subsequent increase in the level of PAL was detected. *Figure 13.2* illustrates the result of an experiment which was essentially similar except that at the time of transfer from 5 to 25 °C conditions one treatment was sprayed with cycloheximide at a concentration of 10 μg ml^{-1}. These results

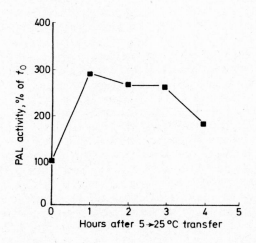

Figure 13.1   Effect of a transfer from 25 to 5 to 25 °C on the extractable activity of PAL from mustard cotyledons. Seedlings grown for 48 h in darkness at 25 °C were chilled at 5 °C for 24 h and then returned to 25 °C, all in darkness. The curve shows the changes in PAL activity with time after the return to 25 °C (zero time)

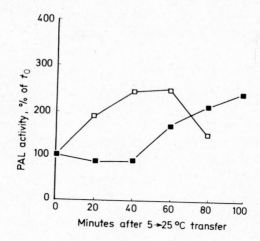

Figure 13.2 Effect of cycloheximide on the temperature transfer-induced increase in PAL activity in mustard cotyledons. Seedlings treated as in Figure 13.1 were either left as controls (■) or sprayed with 10 μg ml⁻¹ cycloheximide solution at zero time (□)

can be interpreted as indicating that the increase in the level of PAL in dark-grown mustard seedlings after temperature transfer treatment is not dependent on protein synthesis, and is evidence for the existence of a pool of inactive PAL in dark-grown mustard seedlings. The mechanisms of these responses are currently being investigated by density labelling.

The evidence to date can be summarised as follows:

(1) There is an inactive pool of PAL in both gherkin and mustard.
(2) The blue light-mediated increases in PAL activity in gherkin hypocotyl tissue do not involve increased rates of synthesis or stabilisation of the enzyme but rather activation of existing PAL molecules.
(3) Evidence obtained in this laboratory concerning the far-red light-mediated increases in PAL activity in mustard cotyledons (which disagrees, however, with evidence obtained by the Freiburg group) suggests that increased synthesis is not involved in the response. This is further supported by the recent evidence for inactive PAL in this tissue. It thus seems likely that phytochrome causes increases in PAL and ascorbic acid oxidase activity through different mechanisms in the same tissue.

Acknowledgements

The authors are grateful to Professor H. Smith for his help and advice throughout this work, which was supported by Research Grants from the Science Research Council.

References

ACTON, G.J., DRUMM, H. and MOHR, H. (1974). *Planta*, **121**, 39

ACTON, G.J. and SCHOPFER, P. (1975). *Biochim. Biophys. Acta*, in press

ATTRIDGE, T.H. (1974). *Biochim. Biophys. Acta*, **362**, 258

ATTRIDGE, T.H., JOHNSON, C.B. and SMITH, H. (1974). *Biochim. Biophys. Acta*, **343**, 440

ATTRIDGE, T.H. and SMITH, H. (1973). *Phytochemistry*, **12**, 1569

ATTRIDGE, T.H. and SMITH, H. (1974). *Biochim. Biophys. Acta*, **343**, 452

DURST, F. and MOHR, H. (1966). *Naturwissenschaften*, **53**, 531

ENGELSMA, G. (1967). *Planta*, **75**, 207

SMITH, H., ATTRIDGE, T.H. and JOHNSON, C.B. (1975). In *Perspectives in Experimental Biology*, pp.325-336. Ed. Sunderland, N. Pergamon Press, Oxford

14

PHOTOCONTROL OF MICROBODY AND MITOCHONDRION DEVELOPMENT: THE INVOLVEMENT OF PHYTOCHROME

P. SCHOPFER
D. BAJRACHARYA[1]
H. FALK
Institut für Biologie II, Universität Freiburg, D-78 Freiburg i. Br., Schänzlestrasse 9-11, West Germany

Introduction

The growth and differentiation of higher plant organs and tissues can be controlled by light acting via the photomorphogenic sensor and effector phytochrome. Since the physiological function of an organ or a tissue is largely determined by the biochemical potential of its cells, this control principally implies a control over the development of cell metabolism and the subcellular compartments (organelles) by which this metabolism is accomplished.

This 'cytophotomorphogenesis' (Schopfer *et al.*, 1975) is most obvious and has been intensively studied in the light-mediated genesis of chloroplasts, which will be dealt with elsewhere in this volume (*see* Chapters 15 and 16). In sharp contrast, information regarding the involvement of phytochrome in the development of other subcellular systems is rather sparse. In this chapter we present some recent results from our laboratory obtained in part of a programme to investigate the molecular and fine structural basis of phytochrome-mediated cytophotomorphogenesis in mustard (*Sinapis alba* L.) seedlings.

The cotyledons of the dormant mustard embryo contain large amounts of fat serving as a source of energy and carbon during the early period of seedling development, which is characterised by a purely dissimilatory energy production. The first step of this 'embryonic' metabolism is the mobilisation of the reserves, i.e. a breakdown of the biochemically inert fat molecules into manageable pieces and their transformation into translocatable carbohydrates (sugars). The pathway of fat mobilisation involves fat hydrolysis by lipases, β-oxidation of fatty acids, formation of succinate from acetate by the glyoxylate cycle, transformation of succinate into oxaloacetate by the citrate cycle and finally the gluconeogenic sequence 'reversing' the direction of glycolysis. On the left-hand side of *Figure 14.1* the main stations of this

[1]Present address: Department of Botany, Tribhuvan University, Kathmandu, Nepal

complex metabolic route are illustrated, indicating their localisation in different cellular compartments.   This part of the scheme is largely based on investigations with the fatty endosperm tissue of germinating *Ricinus* seeds, but there are good reasons to apply it also to the functionally homologous situation in fatty cotyledons (for reviews, *see* Tolbert, 1971, and Vigil, 1973).   A dominant role in this pathway is played by the glyoxysome, a species of microbody that is typical of fat-mobilising plant tissues.   This   single-membrane-bound organelle, which can become intimately associated with oleosomes (lipid bodies, *cf.* Yatsu, Jacks and Hensarling, 1971) containing the stored lipid, are responsible for the degradation of fatty acids to acetate units and their transformation into succinate via the glyoxylate cycle.   The key enzymes of this cycle [isocitrate lyase (E.C. 4.1.3.1) and malate synthase (E.C. 4.1.3.2)] as well as catalase (E.C. 1.11.1.6) are convenient and reliable marker enzymes for the glyoxysomes.

When the cotyledons of the developing mustard seedling reach the light, they develop chloroplasts and accordingly acquire properties of assimilatory leaves.   Hence, photosynthesis can take over the supply of energy and carbon for the growing seedling after the fat reserves have been depleted.   The establishment of the photosynthetic apparatus in the plastids is accompanied by the appearance of another type of microbody, the leaf peroxisome.   This particle carries the metabolic machinery for the removal of glycollate, a by-product of photosynthetic carbon assimilation (oxygenase function of ribulosebisphosphate carboxylase; *cf.* Andrews, Lorimer and Tolbert, 1973) in the presence of oxygen. The right-hand side of *Figure 14.1* shows the pathway of partial oxidative dissimilation of glycollate to amino acids which can eventually be converted back to carbohydrates.   This process, which drains up to 50 per cent of the carbon fixed by the Calvin cycle from the main route, makes a major contribution to light-dependent oxygen uptake and carbon dioxide production ('photorespiration', *cf.* Goldsworthy, 1970) of green plants (for a more detailed account of peroxisome metabolism, *see* reviews by Tolbert, 1971, and Vigil, 1973).   In the photosynthesising leaf, the peroxisomal microbodies are found in close spatial association with chloroplasts (Frederick and Newcomb, 1969), indicating their functional relationship to the photosynthetic metabolism. Marker enzymes of peroxisomes are glycollate oxidase (E.C. 1.1.3.1), glyoxylate (hydroxypyruvate) reductase(NAD) (E.C. 1.1.1.26) and again catalase.

It is obvious from *Figure 14.1* that both glyoxysomes and peroxisomes have some functional connections to the mitochondria.   On the structural level, these connections are not as easily detectable as in the case of oleosomes and glyoxysomes, or chloroplasts and peroxisomes.   Nevertheless, it is clear that the mitochondria also participate in the metabolic co-operation of cellular compartments and cannot be ignored in investigating the development of organelles involved in energy metabolism during the transition from the lipid-degrading to the photosynthetic stage of cotyledons.

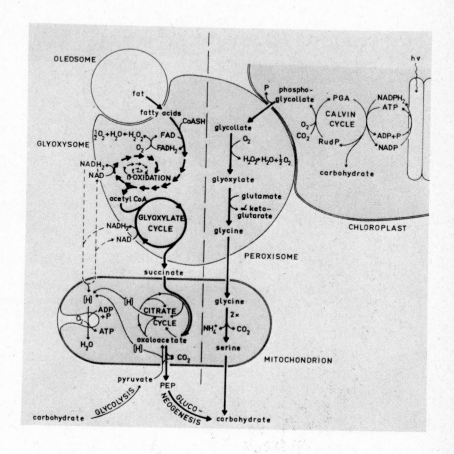

Figure 14.1   Generalised scheme of energy and carbon metabolism in plants, showing the subcellular localisation of major pathways present in fat-containing cotyledons of dicotyledonous seedlings.   The scheme emphasises the dual role of microbodies being involved in lipid mobilisation in the form of glyoxysomes (left-hand side) and in photorespiratory glycollate degradation in the form of peroxisomes (right-hand side).   PEP, phosphoenol pyruvate; PGA, phosphoglycerate; [H], high-energy reduction equivalents; P, inorganic phosphate

Phytochrome-controlled Microbody Development in Mustard Cotyledons

The fat-storing, potentially photosynthetic, cells of the cotyledons of many dicots such as mustard possess the potential to produce both glyoxysomes and leaf peroxisomes during their ontogeny (Gerhardt, 1973; Trelease *et al.*, 1971a,b). Although glyoxysomal and peroxisomal functions may not actually coincide in time, this type of cotyledon can, in some way, be envisaged as a functional combination of fatty endosperm tissue and photosynthetic leaf tissue and thus provides a unique experimental system to study the developmental relationships between two types of functionally well defined microbodies housed in the same cell. So far, all attempts to separate physically glyoxysomes from peroxisomes *in vitro* have been unsuccessful. Therefore, there seems to be no direct biochemical approach to the problem of how glyoxysomal and peroxisomal functions are established in the microbody population. However, there is a very useful physiological tool for discriminating between the two metabolically different microbody species. In greening cotyledons as well as in ordinary leaf tissue, the appearance of peroxisomal enzymes is dependent on light (Feierabend and Beevers, 1972a,b; Gerhardt, 1973; Gruber *et al.*, 1970; Gruber, Becker and Newcomb, 1972, 1973; Kagawa, McGregor and Beevers, 1973; Schnarrenberger, Oeser and Tolbert, 1971; Trelease *et al.*, 1971a). The photoreceptor of this light-mediated enzyme induction is phytochrome (Feierabend, 1975; Klein, 1969; Van Poucke and Barthe, 1970; Van Poucke *et al.*, 1970). On the other hand, glyoxysomal enzymes of cotyledons are produced in darkness as well as in light (Hock, 1969; Karow and Mohr, 1967; Schnarrenberger, Oeser and Tolbert, 1971; Trelease *et al.*, 1971a). Thus, the experimental evidence suggests that formation of glyoxysomes is independent, but that of peroxisomes is strongly influenced by light acting through phytochrome.

In this type of photomorphogenic experiment, it is advisable to stimulate the phytochrome system via the 'high-irradiance reaction', i.e. continuous irradiation with far-red light around 720 nm (Hartmann, 1967; Schäfer, 1975), which produces large responses in the absence of photosynthesis. However, it is a general experience with the mustard seedling that all responses mediated by continuous far-red light are in principle also inducible by brief red light pulses, the effect of which is reversible by brief far-red light pulses. The repeated claims (Schneider and Stimson, 1971, 1972) of a causal relationship between the 'high-irradiance reaction' of phytochrome and photosynthesis have recently been shown to be unjustified (Drumm, Wildermann and Mohr, 1975; Mancinelli *et al.*, 1975).

With regard to the ontogenetic origin of peroxisomes in fatty cotyledon tissue, two opposing hypotheses are currently under debate. Beevers and associates have expressed the opinion that glyoxysomes and peroxisomes, with their characteristic enzyme complement, are generated as independent microbody populations (Beevers, 1971; Kagawa,

McGregor and Beevers, 1973; Kagawa and Beevers, 1975). This view implies a degradation of glyoxysomes and simultaneous formation of peroxisomes when the cotyledon proceeds from the fat-mobilising to the photosynthetic state. On the other hand, Trelease and associates could not detect any fine structural evidence for such an exchange of microbody populations in greening cucumber cotyledons and consequently favour the hypothesis that peroxisomes originate from existing microbodies (Trelease *et al.*, 1971a,b).

The cotyledons of mustard seedlings produce glyoxysomes during the first few days of germination. The rise and fall of the glyoxysomal marker enzyme isocitrate lyase, which is not influenced by phytochrome, is correlated with the depletion of storage fat in the tissue (*Figure 14.2a*). It can be assumed that the time course of this key enzyme of the glyoxylate cycle represents the time course of glyoxysomal activity. An electron microscopic investigation of etiolated cotyledons reveals that the microbodies (glyoxysomes) are subject to a characteristic structural development during this period. Before entering the phase of fat metabolisation, the microbodies appear dense and more or less spherical. At the culmination of fat digestion (60–70 h after sowing), these organelles appear amoeboid and are intimately associated with oleosomes (*Figure 14.3*). Encircling and invagination of oleosomes produce finger-like intrusions in the microbodies which persist and become filled with cytoplasm (containing ribosomes, strands of endoplasmic reticulum and sometimes even mitochondria) after the enclosed fat bodies have vanished. Towards the end of the fat mobilisation period, these microbodies attain a highly perforated appearance in cross-section (*Figure 14.3a*). Similar, but less pronounced, fine structural changes of glyoxysomes have been observed in developing *Ricinus* endosperm (Vigil, 1970) and *Cucumis* cotyledons (Gruber *et al.*, 1970; Trelease *et al.*, 1971a,b). It is reasonable to interpret these cellular events as the structural aspect of the dynamics of glyoxysomal function during the purely dissimilatory phase of seedling development. In this period, the plastids of etiolated mustard cotyledons develop from the proplastid to the etioplast state (Kasemir, Bergfeld and Mohr, 1975). There is no detectable spatial relationship between plastids and microbodies in darkness.

In the cotyledons of light-grown mustard seedlings, the formation of metabolically active glyoxysomes takes place in much the same way as just described for etiolated tissue. However, after this stage (i.e. after about 60 h after sowing), an increasing number of microbodies appear appressed to plastids and thus display a characteristic quality of leaf peroxisomes (*Figures 14.3b* and *14.4*). The effect of white light in producing this phenomenon can be achieved also with far-red light (standard far-red source, cf. Mohr, 1966) which is unable appreciably to support chlorophyll synthesis (Masoner, Unser and Mohr, 1972) in mustard. We have to conclude, therefore, that this spatial arrangement of microbodies in the cell is mediated by phytochrome and that it takes place independently of the building up of an active photosynthetic apparatus.

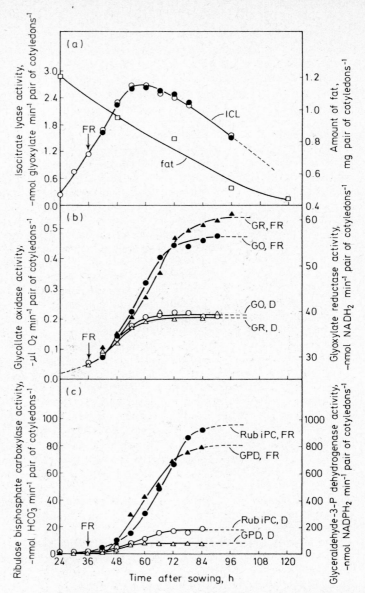

Figure 14.2 Influence of phytochrome [i.e. of continuous standard far-red light (3.5 W m⁻², λ_max = 740 nm, bandwidth 100 nm, less than 0.1 per cent emission at λ < 680 nm), with respect to phytochrome equivalent to 718–nm monochromatic light; cf. Mohr, 1966; Oelze-Karow and Mohr, 1973] on the development of some glyoxysome (a), peroxisome (b) and plastid (c) marker enzymes in the cotyledons of mustard seedlings (25 °C). a, Isocitrate lyase (ICL) activity and decrease of fat content in darkness. [After Karow and Mohr (1967) and Hock, Kühnert and Mohr (1965), respectively]. b, Glycollate oxidase (GO) and glyoxylate (hydroxypyruvate) reductase (GR) activity (after Van Poucke et al., 1970). c, Ribulosebisphosphate carboxylase (RubiPC) and glyceraldehyde phosphate dehydrogenase(NADP) (GPD) activity (after Brüning, Drumm and Mohr, 1975). Onset of irradiation: 36 h after sowing (arrows); D, darkness; FR, far-red light

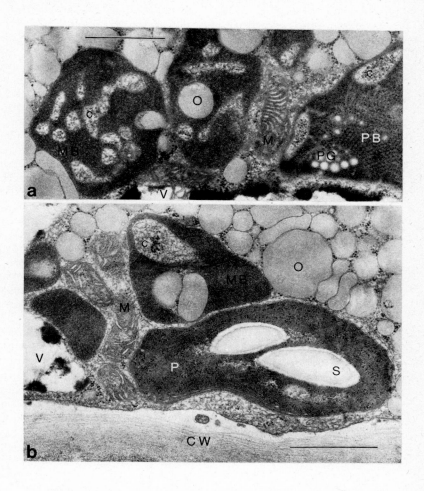

Figure 14.3 Sections of mesophyll cells (spongy parenchyma) from mustard cotyledons showing typically developed and arranged organelles. a, Seedlings grown for 84 h in darkness. Note the large microbodies (glyoxysomes) with invaginated pockets containing cytoplasm with ribosomes and oleosomes (lipid bodies) of different size. The mitochondrion shows a parallel arrangement of cristae-like internal membrane structures. b, Seedlings grown for 36 h in darkness and then irradiated with standard far-red light (cf. Figure 14.2) for 48 h. Note the microbody (peroxisome) with lipid inclusions which appears appressed to a plastid containing some primary thylakoids and starch grains. The mitochondria show an irregular arrangement of sacculi-like internal membrane structures. Experimental: Fixation under green safelight in 3 per cent glutaraldehyde buffered with 50 mM sodium cacodylate/HCl (pH 7.2) containing 10 mM $MgCl_2$ and 200 mM KCl. Post-fixation in 2 per cent OsO_4 in the same buffer for 1 h at 4 °C. Blocks stained with 2 per cent uranyl acetate. After sectioning, double staining with uranyl acetate and lead citrate. The bars represent 1 μm. Abbreviations: C, cytoplasm; CW, cell wall; M, mitochondrion; MB, microbody; O, oleosome (lipid body), P, plastid; PB, prolamellar body; PG, plastoglobuli; S, starch; V, vacuole

Concomitantly to the structural changes in the microbody population, a steep rise of the peroxisomal marker enzymes glycollate oxidase and glyoxylate reductase is induced by phytochrome (*Figure 14.2b*). In the etioplasts, which are also profoundly altered in structure and enzyme content by far-red light (cf. Kasemir, Bergfeld and Mohr, 1975; Schopfer *et al.*, 1975), ribulosebisphosphate carboxylase (E.C. 4.1.1.39) and glyceraldehyde phosphate dehydrogenase(NADP) (E.C. 1.2.1.13) activity also rise in a loosely correlated manner (*Figure 14.2c*; Brüning, Drumm and Mohr, 1975). Obviously, the enzymes of photosynthetic CO_2 fixation and photorespiratory glycollate pathway are prepared in parallel under the influence of phytochrome.

A remarkable feature of the data in *Figure 14.2* is that the far-red light-mediated increase of peroxisomal enzymes does not take place before the period of decreasing glyoxysomal activity. This indicates that phytochrome can mediate the formation of peroxisomal microbodies only at a defined stage of seedling development and that there is a considerable transient period (about 48 h) during which both types of microbodies can be expected to exist. However, with respect to the developmental origin of peroxisomes, these results do not permit any conclusive decision between the two competitive concepts mentioned above. In principle, the data presented so far can be reconciled equally well with a phytochrome-controlled *de novo* appearance of peroxisomes, which is independent of the dynamics in the glyoxysomal population, or a phytochrome-controlled transformation of existing glyoxysomes into peroxisomes.

A closer examination of the microbodies which become associated to plastids during the transition period of glyoxysomal to (potential) peroxisomal activity in far-red light (cf. *Figure 14.2*) reveals that these 'peroxisomes' almost invariably contain invaginated pockets with cytoplasm and lipid bodies of variable size, which are obviously being subjected to degradation (*Figure 14.4*). The discovery of an oleosome–microbody–plastid association during the transient period of changing microbody function in the mustard cotyledons proves the close relationship between glyoxysomes and peroxisomes and provides strong evidence for the hypothesis favoured by Trelease *et al.* (1971a,b), postulating a transformation of existing glyoxysomes into peroxisomes. We are led to the conclusion that these microbodies, which lack their own protein synthesis, have the ability specifically to lose and acquire certain enzyme activities without loss of their compartmental integrity. At the moment there is only vague information available on how this task may be managed by the cell and in which way phytochrome is involved in regulating this transformation at the molecular level. Several reports in the literature (*see* e.g. Vigil, 1973) point to an involvement of the endoplasmic reticulum in feeding the microbodies with enzymes. In rat liver peroxisomes, Lazarow and De Duve (1973) have found direct biochemical evidence for a rapid import of catalase apoprotein subunits, which are probably synthesised by the rough endoplasmic reticulum, into the existing microbodies, where the addition of haeme and the assembly of the enzymatically active tetramers takes place.

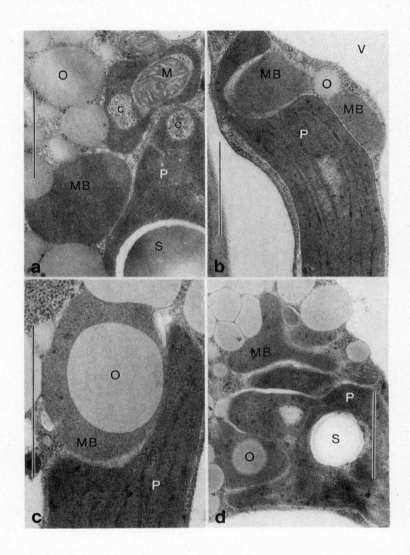

Figure 14.4   Typical microbodies (potential peroxisomes) from mesophyll cells of mustard cotyledons, demonstrating the effect of phytochrome on the structure and localisation of these organelles.   The seedlings were grown in darkness for 36 h and then irradiated with far-red light (cf. Figure 14.2) for a further 48 h.   Note the intimate association of these microbodies with plastids (ranging from mere appression to closely interdigitated configurations) and with oleosomes (demonstrating the glyoxysomal function of these microbodies).   a, Spongy parenchyma; b–d, palisade parenchyma (for experimental details and abbreviations, see Figure 14.3)

Further support for the partial identity of glyoxysomes and peroxisomes in the cotyledons of mustard seedlings comes from an investigation of the control of catalase activity by phytochrome in these organs (Drumm and Schopfer, 1974). This common marker enzyme of glyoxysomes and peroxisomes shows a perfectly intermediary behaviour with respect to its development in dark-grown and far-red irradiated seedlings (*Figure 14.5*). In darkness, the catalase activity follows strictly the dynamics of iso-citrate lyase development (*Figure 14.5b*). Irradiation with far-red light leads to a significant, apparently stable, increment of catalase activity (*Figure 14.5a*) which can be quantitatively correlated with the simultan-eous rise of the peroxisomal marker enzymes glycollate oxidase and glyoxylate reductase (*Figure 14.5c*). Thus, the phytochrome-mediated transition of glyoxysomes to peroxisomes can also be demonstrated with the common marker enzyme catalase.

Since the existence of catalase isoenzymes in plants is well established (e.g. Scandalios, 1968), one can ask the question whether 'glyoxysomal' and 'peroxisomal' catalases can be discriminated on the isoenzyme level. *Figure 14.6* shows some isoenzyme patterns obtained by starch gel electrophoresis of catalase extracts prepared from cotyledons of dark-grown and irradiated seedlings and from upper leaves of white light-grown plants of mustard.

In the etiolated cotyledons only three strong bands are always clearly detectable. Far-red or white light induces the appearance of at least seven additional, slower moving, bands of various strengths, most of which are below the limit of detectability in the case of dark-grown cotyledons. The most important feature of these data is that the action of phytochrome is not to replace one set of isoenzymes by another, but rather to supplement the 'glyoxysomal' catalases by addit-ional molecular species of this enzyme produced in multiple forms. Comparison between the catalase patterns obtained with units of equal activity from cotyledons of far-red treated seedlings and those from cotyledons of older plants grown under white light demonstrate the survival of 'glyoxysomal' isoenzymes in the aged cotyledons in which the period of glyoxysomal activity has been surpassed. Further, it becomes evident that the microbodies of true leaves, which are peroxi-somes by definition, also contain the complete set of catalase isoenzymes characteristic of light-grown cotyledon. This result is, of course, to be expected if peroxisomes can be made from glyoxysomes in this organ.

Phytochrome-controlled Mitochondrion Development in Mustard Cotyledons

In contrast to plastids and microbodies, the mitochondria are usually considered to represent relatively conservative structures which function in essentially the same way in all cells of a differentiated plant. Indeed, there is little electron microscopic and biochemical evidence for a differ-entiation of mitochondria during development of higher plants (Clowes,

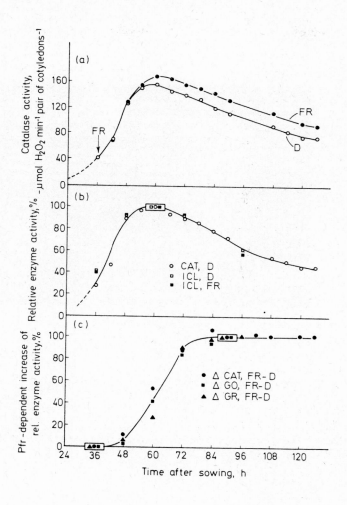

Figure 14.5   Influence of phytochrome (i.e. of continuous standard far-red light, cf. Figure 14.2) on the development of catalase (CAT) activity in the cotyledons of mustard seedlings.   a, Time course of catalase activity per pair of cotyledons in darkness (D) and far-red light (FR; onset at 36 h after sowing, arrow). b, Time course of relative catalase activity in darkness and of isocitrate lyase (ICL) activity in darkness and in far-red light.   Curves are normalised to the maxima at 60 h after sowing (= 100 per cent).   c, Time course of the far-red light-induced increment (light minus dark) of catalase, glycollate oxidase (GO) and glyoxylate reductase (GR) activity.   Curves are normalised to the maxima at 90 h after sowing (= 100 per cent).   (After Drumm and Schopfer, 1974)

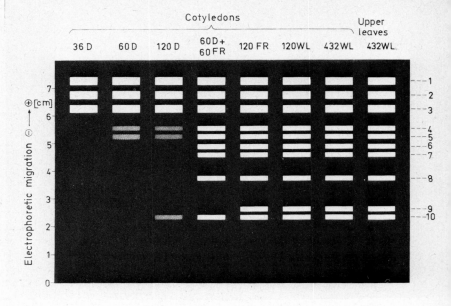

Figure 14.6 Influence of phytochrome (i.e. continuous standard far-red light, cf. Figure 14.2, or white fluorescent light of 7000 lx) on the catalase isoenzyme pattern in the cotyledons of mustard seedlings. Equal amounts of catalase activity units were electrophoresed on horizontal starch gel slabs and catalase bands visualised with $FeCl_3/K_3[Fe(CN)_6]$ staining after incubation with H_2O_2. Extracts were prepared from cotyledons of seedlings which were grown either in darkness (D) for 36, 60 or 120 h or under far-red (FR) or white light (WL) for 60 or 120 h, respectively. In addition, cotyledons and upper leaves of 3-week (432-h)-old plants which were grown in the greenhouse under natural light conditions were used. (After Drumm and Schopfer, 1974).

1971; Öpik, 1968).   This situation is different in animals, where tissue
specificity of mitochondrion size and structure is well established.   This
is true also in certain fungi (e.g. facultatively anaerobic yeasts) which
are able to adapt their dissimilatory metabolism to changing environmental conditions (especially to oxygen supply; cf. Schatz, 1969; Criddle
and Schatz, 1969; Watson, Haslam and Linnane, 1970; Luzikov, Zubatov
and Rainina, 1973).   A similar case of respiratory adaptation has been
discovered in coleoptiles of rice, a plant which is able to germinate and
produce well developed seedlings in the complete absence of oxygen.
In the mitochondria of anaerobically grown rice coleoptiles, the growth
of which can be supported exclusively by fermentation for a considerable period of time, cytochromes, including cytochrome oxidase, are
repressed by the absence of oxygen (Öpik, 1973; Vartapetian, Maslov
and Andreeva, 1975).   The regulatory effect on the level of the electron
transport chain was reported to be accompanied by changes of the inner
mitochondrial membrane system (Vartapetian, Andreeva and Kursanov,
1974).   This example demonstrates that mitochondria also are principally
able to differentiate with respect to their function and that this differentiation is detectable on the biochemical and fine structural level (for
further examples, *see* Schopfer *et al.*, 1975).

Towards the end of seed ripening (desiccation phase) in dicots, the
mitochondria of the embryo undergo a controlled disorganisation which
is characterised by the disappearance of enzymes and membrane structures.   This process can be described as a developmental regression of
metabolically active mitochondria to poorly organised 'promitochondria'
in analogy to the similar situation in anaerobically grown yeast.   During
germination of the seed, these promitochondria are transformed again
into well organised, respiratorily active mitochondria within a few days
(Bain and Mercer, 1966; Kollöffel and Sluys, 1970; Nawa and Asahi,
1971; Malhotra and Spencer, 1973; Öpik, 1973; Solomos *et al.*, 1972;
Wilson and Bonner, 1971).   This developmental process, which is
usually marked by a steep rise in respiratory oxygen consumption, is
dependent on *de novo* protein synthesis (Malhotra and Spencer, 1973).

Does light (and this means possibly phytochrome) have any influence
on the promitochondrion to mitochondrion transformation in germinating
seed embryos?   In recent investigations, we have been able to find an
answer to this question.   As an experimental system we used again the
cotyledons of mustard seedlings, the O_2 uptake (Hock and Mohr, 1964;
Hock, Kühnert and Mohr, 1965) and CO_2 production (Weischet, 1971)
of which have been previously shown to be controllable with continuous
far-red light.   The molecular basis of this phytochrome-dependent respiratory gas exchange, however, is still unclear.

In the cotyledons of the mustard seedling, the extractable activity of
at least three mitochondrial marker enzymes, namely cytochrome *c* oxidase (E.C. 1.9.3.1), succinate dehydrogenase (E.C. 1.3.99.1) and fumarase
(fumarate hydratase, E.C. 4.2.1.2) can be promoted by continuous far-
red light (*Figure 14.7*).   Although the overall kinetics of the three
enzymes appear to be similar, there are characteristic differences in the

extent of the light effect, which is strongest in cytochrome oxidase and least in succinate dehydrogenase. We can conclude from these data (which are supported by red/far-red induction/reversion experiments) that phytochrome does induce the appearance of soluble as well as membrane-bound mitochondrial marker enzymes in mustard cotyledons and that this effect does not merely represent the general growth or proliferation of existing mitochondria. Inhibitors of protein synthesis suppress the increase of these enzyme activities.

The phytochrome-mediated changes in the enzymatic capacity shown in *Figure 14.7* are accompanied by drastic qualitative differences in the physical properties of the mitochondria. *Figure 14.8* illustrates the equilibrium distribution in an isopycnic sucrose gradient of mitochondria which have been isolated from cotyledons of dark-grown and far-red irradiated mustard seedlings. It is evident that the 'far-red mitochondria' display a higher buoyant density than the 'dark mitochondria'. The difference in density was found to increase with seedling age (i.e. with the length of the irradiation). This density difference observed in an artificial medium *in vitro* does not, of course, prove that the mitochondria are of different density also *in vivo*. However, irrespective of the precise cause, this phenomenon indicates that phytochrome alters something in the molecular organisation of the mitochondria, possibly the properties of the mitochondrial membranes. This idea is supported by the finding that the mitochondria of dark-grown cotyledons appear to be more fragile during isolation than those of light-grown material. There remains, for instance, a larger fraction of fumarase activity on top of the sucrose gradient during isopycnic centrifugation of 'dark mitochondria', indicating increased leakage of this enzyme from the particles. Further, solubilisation of cytochrome oxidase activity from the inner mitochondrial membrane requires a considerably lower concentration of detergent (Triton X-100) in 'dark mitochondria' than in 'far-red mitochondria'. In 36-h-old mustard cotyledons, the mitochondria are so fragile that we have been unable to isolate these particles by sucrose density gradient centrifugation.

Finally, the mitochondria of cotyledons (mesophyll cells) from dark- and light-grown mustard seedlings older than about 60 h are clearly distinguishable under the electron microscope. *Figure 14.9* shows some typical pictures to illustrate the fine structural differences. The 'dark mitochondria' appear as more or less elongated cylindrical particles with an electron-dense matrix and parallel orientated sheets of cristae. Favourable sections (cf. *Figure 14.9a*) show large stacks of regularly spaced septum-like double lamellae which lack a detectable connection to the inner membrane of the mitochondrion envelope. The architecture of these organelles appears strikingly similar to that of the mitochondria of some animal tissues, for instance to the mitochondria of mammalian kidney cells (cf. Sjöstrand, 1956). To our knowledge, this kind of mitochondrion has not been observed so far in higher plants. If the seedlings are kept in light subsequent to 36 h after sowing, the mitochondria of the mesophyll present the aspect expected of ordinary

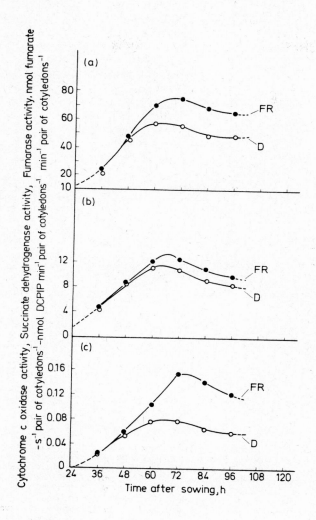

Figure 14.7 Influence of phytochrome (i.e. continuous standard far-red light, cf. Figure 14.2) on the development of some mitochondrial marker enzymes in the cotyledons of mustard seedlings. a, Fumarase (FUM); b, succinate dehydrogenase (SDH); c, cytochrome c oxidase (CO). The seedlings were either kept in darkness (D) or irradiated with far-red light (FR) from sowing (DCPIP = 2,6-dichlorophenolindophenol)

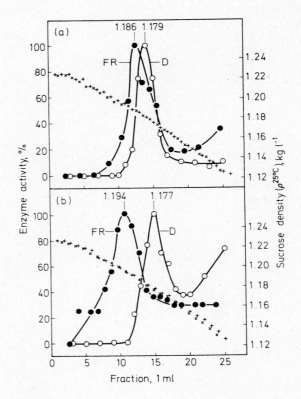

Figure 14.8 Influence of phytochrome (i.e. continuous irradiation with far-red light, cf. Figure 14.2*) on the apparent buoyant density of mitochondria isolated from the cotyledons of (a) 72- and (b) 120-h-old mustard seedlings which were grown either in total darkness or under far-red light following 36 h after sowing. A crude particulate fraction was obtained after homogenisation of the tissue in a mannitol–Ficoll–dextran medium (Bourque and Naylor, 1972; medium D) containing dithioerythritol and polyvinylpyrrolidone 10 000 and layered on top of a linear sucrose gradient.   After centrifugation (5.5 h, 60 000 g) the gradients were fractionated and assayed for cytochrome oxidase activity (measurement of fumarase and succinate dehydrogenase resulted in very similar activity profiles).   The profiles of 'dark mitochondria' and 'far-red mitochondria' (obtained from two tubes of the same run) are superimposed on the basis of the sucrose gradients.   The numbers at the maxima indicate the midpoint density (kg l^{-1}) of the bands. +, D; x, FR*

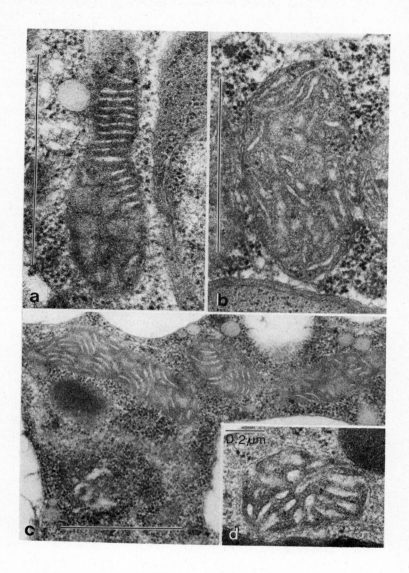

Figure 14.9 Typical mitochondria from mesophyll cells (palisade parenchyma) of mustard cotyledons, demonstrating the effect of phytochrome on the structure of these organelles. The seedlings were either grown in darkness for 72 h (a) and 84 h (c) or irradiated with standard far-red light (cf. Figure 14.2) from 36 until 60 h (b) or from 36 until 120 h (d) after sowing. Note the parallel orientation of the septum-like internal membrane structures (cristae) in the mitochondria of dark-grown cotyledons and the irregular orientation of tubular structures (sacculi) in the mitochondria of far-red light-grown cotyledons. This tubular modification, which represents the 'typical' plant mitochondrion, can also be produced by irradiating the seedlings with white light (for experimental details and abbreviations, see Figure 14.3)

plant mitochondria with an extended system of irregularly arranged, curved tubules (sacculi; cf. *Figure 14.9b,d*). This type of mitochondrion architecture is produced by far-red as well as by white light, indicating the involvement of phytochrome also in this structural feature of mitochondrion development.

In summary, we can present at the moment three independent lines of evidence (cf. *Figures 14.7, 14.8* and *14.9*) indicating that phytochrome has a drastic effect on the development of mitochondria in mustard cotyledons. We have to conclude that in addition to plastids and microbodies (and most likely other subcellular structures also), the mitochondria are also involved in the phytochrome-mediated transition of cotyledonary storage cells from the purely dissimilatory to the mainly assimilatory function. We are now in a position to ask questions regarding the precise role of these organisational changes brought about by phytochrome which can be observed at the subcellular level. At present, the metabolic significance of the control by light of mitochondrion development remains a matter of speculation. However, it is now becoming clear that the plastids are by no means the only cellular structures of which the development is regulated by light via phytochrome. There is little doubt that this photoreceptor, in addition to its many other functions, plays a central role as a regulator of intracellular morphogenesis. The involvement of phytochrome in the coordinate differentiation of metabolically co-operating cellular compartments may be a major factor in the adaptation of an etiolated plant cell to the requirements of life in light.

Acknowledgements

Research supported by the Deutsche Forschungsgemeinschaft (SFB 46) and by a doctoral fellowship of the Deutscher Akademischer Austauschdienst held by D.B. We thank Mrs. K. Ronai for help with the ultramicrotomy.

References

ANDREWS, T.J., LORIMER, G.H. and TOLBERT, N.E. (1973). *Biochemistry*, **12**, 11

BAIN, J.M. and MERCER, F.V. (1966). *Aust. J. Biol. Sci.*, **19**, 69

BEEVERS, H. (1971). In *Photosynthesis and Photorespiration*, pp.483–493. Ed. Hatch, M.D., Osmond, C.B. and Slatyer, R.O. Wiley-Interscience, New York, London, Sydney, Toronto

BOURQUE, D.P. and NAYLOR, A.W. (1972). *Plant Physiol.*, **49**, 826

BRÜNING, K., DRUMM, H. and MOHR, H. (1975). *Biochem. Physiol. Pflanz.*, **168**, 141

CLOWES, F.A.L. (1971). In *Results and Problems in Cell Differentiation*, Vol.2, pp.323–342. Ed. Beermann, W. and Reinert, J. Springer, Berlin, Heidelberg, New York

CRIDDLE, R.S. and SCHATZ, G. (1969). *Biochemistry*, **8**, 322
DRUMM, H. and SCHOPFER, P. (1974). *Planta*, **120**, 13
DRUMM, H., WILDERMANN, H. and MOHR, H. (1975). *Photochem. Photobiol.*, **21**, 269
FEIERABEND, J. (1975). *Planta*, **123**, 63
FEIERABEND, J. and BEEVERS, H. (1972a). *Plant Physiol.*, **49**, 28
FEIERABEND, J. and BEEVERS, H. (1972b). *Plant Physiol.*, **49**, 33
FREDERICK, S.E. and NEWCOMB, E.H. (1969). *Science, N.Y.*, **163**, 1353
GERHARDT, B. (1973). *Planta*, **110**, 15
GOLDSWORTHY, A. (1970). *Bot. Rev.*, **36**, 321
GRUBER, P.J., BECKER, W.M. and NEWCOMB, E.H. (1972). *Planta*, **105**, 114
GRUBER, P.J., BECKER, W.M. and NEWCOMB, E.H. (1973). *J. Cell Biol.*, **56**, 500
GRUBER, P.J., TRELEASE, R.N., BECKER, W.M. and NEWCOMB, E.H. (1970). *Planta*, **93**, 269
HARTMANN, K.M. (1967). *Z. Naturforsch.*, **226**, 1172
HOCK, B. (1969). *Planta*, **85**, 340
HOCK, B. and MOHR, H. (1964). *Planta*, **61**, 209
HOCK, B., KÜHNERT, E. and MOHR, H. (1965). *Planta*, **65**, 129
KAGAWA, T., McGREGOR, D.I. and BEEVERS, H. (1973). *Plant Physiol.*, **51**, 66
KAGAWA, T. and BEEVERS, H. (1975). *Plant Physiol.*, **55**, 258
KAROW, H. and MOHR, H. (1967). *Planta*, **72**, 170
KASEMIR, H., BERGFELD, R. and MOHR, H. (1975). *Photochem. Photobiol.*, **21**, 111
KLEIN, A.O. (1969). *Plant Physiol.*, **44**, 897
KOLLÖFFEL, C. and SLUYS, J.V. (1970). *Acta Bot. Neerl.*, **19**, 503
LAZAROW, P.B. and DE DUVE, C. (1973). *J. Cell Biol.*, **59**, 507
LUZIKOV, V.N., ZUBATOV, A.S. and RAININA, E.I. (1973). *Bioenergetics*, **5**, 129
MALHOTRA, S.S. and SPENCER, M. (1973). *Plant Physiol.*, **52**, 575
MANCINELLI, A.L., YANG, C.-P.H., LINDQUIST, P., ANDERSON, O.R. and RABINO, I. (1975). *Plant Physiol.*, **55**, 251
MASONER, M., UNSER, G. and MOHR, H. (1972). *Planta*, **105**, 267
MOHR, H. (1966). *Z. Pflanzenphys.*, **54**, 63
NAWA, Y. and ASAHI, T. (1971). *Plant Physiol.*, **48**, 671
OELZE-KAROW, H. and MOHR, H. (1973). *Photochem. Photobiol.*, **18**, 319
ÖPIK, H. (1968). In *Plant Cell Organelles*, pp.47–88. Ed. Pridham, B. Academic Press, London
ÖPIK, H. (1973). *J. Cell Sci.*, **12**, 725
SCANDALIOS, J.G. (1968). *Ann. N.Y. Acad. Sci.*, **151**, 274
SCHÄFER, E. (1975). *J. Math. Biol.*, **2**, 41
SCHATZ, G. (1969). *Umschau*, **69**, 11
SCHNARRENBERGER, C., OESER, A. and TOLBERT, N.E. (1971). *Plant Physiol.*, **48**, 566
SCHNEIDER, M.J. and STIMSON, W.R. (1971). *Plant Physiol.*, **48**, 312

212   *Photocontrol of microbody and mitochondrion development*

SCHNEIDER, M.J. and STIMSON, W.R. (1972). *Proc. Nat. Acad. Sci. U.S.A.,* **69**, 2150
SCHOPFER, P., BAJRACHARYA, D., FALK, H. and THIEN, W. (1975). *Ber. Dtsch. Bot. Ges.,* **88**, 245
SJÖSTRAND, F.S. (1956). *Int. Rev. Cytol.,* **5**, 455
SOLOMOS, T., MALHOTRA, S.S., PRASAD, S., MALHOTRA, S.K. and SPENCER, M. (1972). *Can. J. Biochem.,* **50**, 725
TOLBERT, N.E. (1971). *Annu. Rev. Plant Physiol.,* **22**, 45
TRELEASE, R.N., BECKER, W.M., GRUBER, P.J. and NEWCOMB, E.H. (1971a). *Plant Physiol.,* **48**, 461
TRELEASE, R.N., BECKER, W.M., GRUBER, P.J. and NEWCOMB, E.H. (1971b). In *Photosynthesis and Photorespiration,* pp.523-533. Ed. Hatch, M.D., Osmond, C.B. and Slatyer, R.O.   Wiley-Interscience New York, London, Sidney, Toronto
VAN POUCKE, M. and BARTHE, F. (1970). *Planta,* **94**, 308
VAN POUCKE, M., CERFF, R., BARTHE, F. and MOHR, H. (1970). *Naturwissenschaften,* **56**, 132
VARTAPETIAN, B.B., ANDREEVA, I.N. and KURSANOV, A.L. (1974). *Nature, Lond.,* **248**, 258
VARTAPETIAN, B.B., MASLOV, A.I. and ANDREEVA, I.N. (1975). *Plant Sci. Lett.,* **4**, 1
VIGIL, E.L. (1970). *J. Cell Biol.,* **46**, 435
VIGIL, E.L. (1973). *Sub-Cell. Biochem.,* **2**, 237
WATSON, K., HASLAM, J.M. and LINNANE, A.W.I. (1970). *J. Cell Biol.,* **46**, 88
WEISCHET, W.O. (1971). *Diplom Thesis,* University of Freiburg i. Br.
WILSON, S.B. and BONNER, W.D. (1971). *Plant Physiol.,* **48**, 340
YATSU, L.Y., JACKS, T.J. and HENSARLING, T.P. (1971). *Plant Physiol.,* **48**, 675

15

THE PHOTOCONTROL OF CHLOROPLAST DEVELOP-MENT – ULTRASTRUCTURAL ASPECTS AND PHOTOSYNTHETIC ACTIVITY

J.W. BRADBEER
G. MONTES

Department of Plant Sciences, King's College, University of London, 68 Half Moon Lane, London SE24 9JF, U.K.

Introduction

Light exerts a decisive effect on plastid development. In meristematic cells, the plastids are present in the form of proplastids which at their simplest are organelles of diameter less than 0.5 μm with an envelope consisting of two distinct membranes and an absence of any other well defined fine-structural characteristics. Only when the proplastids develop some internal membrane or accumulate starch (Bradbeer *et al.*, 1974a) can they be distinguished from promitochondria on the basis of their fine structure. When seedlings are grown in continuous darkness, the proplastids develop into etioplasts, a process which was described in quantitative terms by Bradbeer *et al.* (1974a). At their maximum development, etioplasts contain a large amount of membrane (80–100 μm^2 of membrane for etioplasts of *Phaseolus vulgaris*) of which, again in the case of bean, about half is in the prolamellar body in the form of a regularly arranged lattice of tubules with the remainder in porous lamellar sheets that may be thought of as flattened sacs of membrane (thylakoids) which possess pores penetrating both membranes. *Figure 15.1a* and *15.1b* shows electron micrographs of etioplasts of the mesophyll and bundle sheath of maize.

The effect of light on chloroplast development is dependent on the quality and amount of light and on the timing of the exposure. We shall discuss in particular the effects of the quality and amount of light on chloroplast development in this chapter.

Figure 15.1   Etioplasts and chloroplasts of the first leaf of Zea mays.   The scale lines represent 0.5 μm.   BSP, bundle sheath plastid; G, grana; M, mito-chondrion; MP, mesophyll plastid; OG, osmiophilic globule; PB, prolamellar body; PR, peripheral reticulum; T, thylakoid.   A, mesophyll etioplasts of 14-day-old dark-grown plants, X 10 000; B, bundle sheath etioplast of 14-day-old dark-grown plant, X 20 000; C, chloroplasts of bundle sheath and mesophyll of light-grown plant, X 17 500

Control of Chloroplast Development by Light Quality

When seedlings are grown under continuous far-red irradiation, they resemble seedlings grown in continuous white light in most of their visible characteristics except that they remain etiolated and accumulate little chlorophyll (Mohr, 1972). Examination of the fine structure of bean seedlings grown under continuous far-red irradiation shows that they do not contain normal chloroplasts, and the plastids have parallel thylakoid sheets with no grana and only a small amount of thylakoid appression where adjacent thylakoid sheets overlap (De Greef, Butler and Roth, 1971). In mustard cotyledons, prolonged far-red irradiation results in plastids with crystalline prolamellar bodies and parallel thylakoid sheets with a substantial amount of thylakoid appression, to give what may be described as paired primary thylakoids rather than grana (Häcker, 1967; Kasemir, Bergfeld and Mohr, 1975). The latter workers also showed that in etioplast development in the cotyledons of dark-grown mustard the prolamellar bodies at first possessed a crystalline form which subsequently was lost. However, crystallisation of the prolamellar body occurred in response to either prolonged far-red or short-red irradiation, and to some extent in response to short far-red irradiation, and was inferred to result from the action of phytochrome. Berry and Smith (1971) found that short red-light treatments induced prolamellar body crystallisation in dark-grown barley leaves but as there was no far-red induced reversal they inferred that phytochrome was not involved. In the primary leaves of dark-grown *Phaseolus,* the prolamellar bodies retain their crystalline form, at least up to 30 days of dark growth, when senescence and death intervene (Bradbeer *et al.,* 1974a). Thus we have not been able to observe irradiation-induced prolamellar body crystallisation in bean, but it is possible that irradiation at an early stage of etioplast development might induce such a crystallisation.

In angiosperms, normal chloroplasts have so far been found only in plants which have been exposed to substantial periods of irradiation with white light. Since chloroplast development is clearly not wholly controlled by phytochrome, it is evident that the process requires both phytochrome and other photoreceptors.

The complexity of a situation involving more than one photoreceptor, together with the time-consuming nature of the quantitative investigation of fine structure which will be required to resolve the situation, has resulted in very slow progress in this area. As an initial investigation, we studied the effects of short exposures of dark-grown beans to irradiation of different wavelengths. The following broad bandwidth illumination treatments were used.

Blue: 20 min exposure to 50 μW cm^{-2}, equivalent to 230 nE cm^{-2} per irradiation; peak transmission, 450 nm; bandwidth at half peak transmission, 47 nm.

Red: 5 min exposure to 178 μW cm^{-2}, equivalent to 280 nE cm^{-2} per irradiation; peak transmission, 634 nm; bandwidth at half peak transmission, 57 nm.

Far-red: 20 min exposure to the following distribution of radiation: <700 nm of 6 μW cm^{-2} (40 nE cm^{-2}), 700–750 nm of 32 μW cm^{-2} (230 nE cm^{-2}), 750–800 nm of 112 μW cm^{-2} (870 nE cm^{-2}).
White: 5 min exposure to 900 μW cm^{-2}, equivalent to 1300 nE cm^{-2}, from Ekco 'Double Light' fluorescent tubes.

Dark-grown beans were exposed to these irradiations, supplied either singly or in certain combinations, once on each of days 12, 13 and 14. They were maintained in darkness between the irradiation treatments and fixed for analysis on day 15, exactly 24 h after the preceding irradiation (Bradbeer, 1971; Bradbeer *et al.,* 1974c). The results of the short illumination treatments may be compared with those of investigations of the effects of continuous white light on chloroplast development (Bradbeer, 1969; Bradbeer *et al.,* 1969, 1974b).

Table 15.1 shows the effects of the various short irradiation treatments on several parameters of leaf and plastid development. For comparison, the effects of 3 days' continuous irradiation with white light (1.6 mW cm^{-2}) on 12-day-old dark-grown beans are also included.

None of the measurements recorded in *Table 15.1* show any difference between the short blue irradiation and the dark control. Both the increases in leaf area and in fresh and dry weights satisfied the operational requirements for control by Pfr, being promoted by red, reversed by brief far-red and scarcely affected by brief far-red light alone. A similar observation was reported by Downs (1955). Short white irradiation was more effective than red in increasing leaf area, apparently by promoting the development of intercellular spaces. It is possible that this latter aspect of leaf development is not a phytochrome response. In continuous white irradiation, the leaf area was close to that found after short white irradiations but the fresh and dry weights were much higher, possibly resulting from the photosynthetic activity of the leaf in continuous light increasing the dry matter content of the leaf. There was some light-induced cell division which was clearly under the operational control of phytochrome without any evidence of the participation of another photoreceptor. *Table 15.1* also shows that plastid division, plastid expansion and the formation of plastid membrane were all under the operational control of phytochrome but that when red irradiation was followed by blue there was a statistically significant reduction of the red-induced promotion. This result differs somewhat from that of Possingham (1973), who found that chloroplast replication in cultured leaf discs of spinach required either a high intensity of continuous white light (6.5 mW cm^{-2}) or high irradiations from red or blue laser light. On the other hand, Pfr has been reported to control chloroplast replication in the germinating spores of *Polytrichum* (Kass and Paolillo, 1974). Kasemir, Bergfeld and Mohr (1975) have shown that plastid expansion is controlled by Pfr. As the plastid membrane content increased much more under continuous illumination than under short illumination, it is evidently under the control of another photoreceptor in addition to phytochrome. The total leaf protein content was also under the operational control of phytochrome but the red-induced increase was stimulated further by

Table 15.1 The 'slow' effects of the irradiation of 12-day-old dark-grown *Phaseolus vulgaris* L. cv. Alabaster seedlings. Short irradiation treatments given on days 12, 13 and 14. Leaves analysed after 15 days' dark growth, 24 h after the preceding short light treatment.

Treatment	Plastid diameter (μm)	Per primary leaf						Plastid membrane (cm²)
		Leaf area (mm²)	Fresh weight (mg)	Dry weight (mg)	Total protein (mg)	$10^{-6} \times$ cell number	$10^{-7} \times$ plastid number	
12 days dark	2.3	300	34	6.5	2.1	15	23	189
15 days dark	2.8	309	37	6.8	2.3	15	28	327
5 min red daily	3.7	708	50	8.4	2.8	21	51	757
20 min far-red daily	2.9	354	41	7.4	2.4	17	35	391
5 min red/20 min far-red daily	2.7	321	41	7.2	2.2	16	27	213
20 min blue daily	2.8	344	38	7.2	2.3	16	28	289
5 min red/20 min blue daily	3.1	659	51	8.7	3.1	23	41	503
5 min white daily	3.8	1072	52	8.7	3.3	24	58	729
Continuous white light	4.3	1461	155	16.7	6.5	23	58	2000[1]

[1] 14 days' dark growth, 3 days' illumination

following red with blue irradiation. In this case, continuous illumination was more effective than short illumination. *Table 15.2* summarises our conclusions about the slow effects of phytochrome.

Table 15.2 Summary of photoreceptors considered to be involved in various 'slow' reactions in leaf and chloroplast development in *Phaseolus vulgaris* L. cv. Alabaster

Reaction	Pfr	Continuous white irradiation more effective than short red treatment	Red/blue interaction
Leaf expansion	+	+[1]	0
Increase in leaf dry weight	+	+	0
Cell division	+	0	0
Plastid division	+	0	−
Plastid expansion	+	+	−
Plastid membrane formation	+	+	−
Grana formation	0	+	0
Accumulation of leaf protein	+	+	+

[1] Also short white irradiation.

In an attempt to elucidate the mechanism of some fine structural changes not discussed in this chapter, an investigation of the 'rapid' effects of various irradiation treatments was carried out by taking samples at intervals during the 3–h period immediately after the irradiation of 12-day-old dark-grown beans. In addition, a sample was fixed after 24 h. Although the results of these latter experiments have been discussed in detail elsewhere (Bradbeer *et al.*, 1974c), they are summarised in *Table 15.3.*

All of the irradiation treatments except blue gave complete prolamellar body transformation from the crystalline state to the reacted state within 30–40 min of the commencement of illumination. Continuous white irradiation gave a similar effect. The average volume of prolamellar body per plastid also fell in response to all of the irradiations, although in this case blue again was least effective. Within 3 h of the commencement of irradiation, prolamellar body recondensation had commenced in the case of the far-red, the red/far-red and the blue treatments. This result suggests that prolamellar body recondensation may be inhibited by Pfr while, in contrast, the crystallisation of the mustard prolamellar body has been reported to require Pfr (Kasemir, Bergfeld and Mohr, 1975). The area of the thylakoids increased in response to all irradiation treatments which included a red component and fell in response to all irradiation treatments which did not include red (far-red and blue). This effect indicates that thylakoid formation is dependent on a red-absorbing photoreceptor other than phytochrome. Kasemir, Bergfeld

Table 15.3 The 'rapid' effects on etioplast fine structure of the exposure of 12-day-old dark-grown beans to short irradiation treatments

Treatment	Percentage of prolamellar bodies transformed within 1 h of illumination	Maximum percentage loss of prolamellar body volume within 3 h of illumination	Occurrence of prolamellar body recondensation within 3 h of illumination	Maximum percentage change of area of thylakoids within 3 h of illumination
5 min red	100	90	0	+23
20 min far-red	100	90	+	−19
5 min red/20 min far-red	100	75	+	+35
20 min blue	37	55		−31
5 min red/20 min blue	100	100	0	+23
5 min white	100	80	0	+44

and Mohr (1975) suggest that the formation of primary thylakoids may result from the conversion of protochlorophyllide to chlorophyllide.

The rapid responses in the etioplast which seem to be phytochrome controlled are prolamellar body recondensation (*Table 15.3*), the crystallisation of the prolamellar body (Kasemir, Bergfeld and Mohr, 1975), the stimulation of the Shibata shift (Jabben and Mohr, 1975) and the release of gibberellin-like substances from etioplast membranes (Evans, Chapter 10). Of these, at least the first three are presumably located in the prolamellar body, but only the first two are fine-structural changes and only the first one has been seen in etiolated beans, which already possess crystalline prolamellar bodies.

Wellburn and Wellburn (1973a) reported a 'rapid' effect of red light (which fulfilled the operational criteria for phytochrome) on the fine structure of etioplasts both *in vivo* and *in vitro*. However, the result was obtained by the application of their index of chloroplast development, which represents the transformation and loss of the prolamellar body in one dimension and the elaboration of the thylakoids in the other dimension (Wellburn and Wellburn, 1973b). It appears to us that the measurement of two distinct processes on the same scale is both liable to give misleading results and certain to fail to define the exact nature of the photo-responses. Although prolamellar body loss may well induce further thylakoid development (Bradbeer *et al.*, 1974b), it is clear from our data (*Table 15.3* and Bradbeer *et al.*, 1974c) that these two processes are separate and are probably controlled differently. Henningsen (1967) found that the process of vesicle dispersal was controlled by a photoreceptor with an action spectrum consisting of a sharp peak at 450 nm. Vesicle dispersal is an artefact resulting from the use of permanganate fixation in electron microscopy but the stage is probably equivalent to prolamellar body dispersal which is seen in glutaraldehyde/OsO_4-fixed material. Unfortunately, Henningsen's experiments have not been repeated with glutaraldehyde/OsO_4 fixation and the nature of the response that he studied has not been resolved. At least the process does not seem to involve phytochrome. The investigation of the 'rapid' effects of light on etioplasts clearly requires further investigation.

Effect of Light Intensity on Chloroplast Development

The published electron micrographs of chloroplasts from plants grown under different light intensities show that at higher light intensities there appears to be less thylakoid membrane per chloroplast and a smaller proportion of the thylakoid membrane occurs in the grana. This is evident for soybeans grown under 22 and 9 mW cm^{-2} irradiation (Ballantine and Forde, 1970) and for *Atriplex patula* grown under 20, 6.3 and 2 mW cm^{-2} irradiation (Björkman *et al.*, 1972). In soybean, the number of chloroplasts in a leaf cross-section was approximately the same at each light intensity but for *Atriplex patula* the light micrographs show more chloroplasts with increasing light intensity. The extreme development of chloroplasts in plants grown in low light intensities is found in the deep shade plant *Alocasia macrorrhiza*, which receives an average irradiation of 0.1 mW cm^{-2} and which contains massive chloroplasts, each with a large amount of thylakoid membrane and very large grana (Anderson, Goodchild and Boardman, 1973).

As no quantitative determinations of the amounts of thylakoid membrane in leaves grown under different light intensities had been reported, we investigated the development of 14-day-old dark-grown maize leaves under the following light intensities: 5 (high light), 1.6 (intermediate light) and 0.16 mW cm^{-2} (low light). The highest of these intensities was the highest available in our growth cabinets (Controlled Environments Ltd. Model E7, Winnipeg, Canada). The *Zea mays* (cv. First of All) was sown and grown for 14 days in the dark at 23 °C before being illuminated at 25 °C. The chloroplast development was studied in a section of the first leaf (15–35 mm below the leaf apex) which had completed its elongation in 14-day-old plants. The section was not removed from the plant until the time of fixing the material for electron microscopy. The number of plastids per leaf section and plastid diameter were determined by light microscopy and the plastid membrane was measured on electron micrographs as described previously (Bradbeer *et al.*, 1974a; Montes and Bradbeer, 1974).

Maize, like most other C$_4$ plants, shows dimorphic plastids with those of the bundle sheath that are distinct from those of the mesophyll sheath. The bundle sheath is a layer of stout thick-walled cells surrounding each vascular bundle and this in turn is surrounded with a single layer of mesophyll cells. Most of the chloroplasts of the leaf occur in these two sheaths of cells. The etioplasts, which occur in the dark-grown leaves, are also distinct. The mesophyll etioplasts contain complex prolamellar bodies, which are built up of membrane tubules, while the prolamellar bodies of the bundle sheath etioplasts are smaller and simpler (*Figures 15.1a,b*). *Figure 15.1a* also shows that the mesophyll etioplasts often show closely parallel arrangements of thylakoids while those of the bundle sheath tend to be more distant from each other (*Figure 15.1b*). In both etioplasts and chloroplasts, *Figure 15.1* also shows a peripheral reticulum, an anastomosing system of tubules immediately within the plastid envelope, which is characteristic of C$_4$ plants.

Figure 15.1c shows a bundle sheath chloroplast and a mesophyll chloroplast in adjacent cells in a maize leaf grown under diurnal illumination. The bundle sheath chloroplast contains parallel thylakoids which show only a very small amount of appression with adjacent thylakoids. As the leaves had been destarched, by placing them in the dark for several hours prior to fixation for electron microscopy, there are no starch grains to be seen in the bundle sheath chloroplasts. The mesophyll chloroplasts contain prominent grana and they do not normally possess starch grains.

Irradiation of dark-grown leaves with either the high or intermediate light intensities would produce chloroplasts similar to those shown in *Figure 15.1c*. However, irradiation at low light intensities for 48 h gives plastids like those in *Figures 15.2a,b*. In the mesophyll plastid (*Figure 15.2a*), some of the thylakoids occur in a closely parallel arrangement with the remainder less close and there is very little thylakoid appression. In contrast the bundle sheath plastid (*Figure 15.2b*) is very similar to normal except that there is a small amount of the closely parallel arrangement of thylakoids. Transfer of these plants from low light to high light, after 48 h in the former conditions, resulted in a return towards the normal condition. *Figure 15.2c* shows a mesophyll plastid 12 h after transfer. Granal formation had occurred although it was not as extensive as in the controls which had received the higher light intensities after 14 days of dark growth. The transferred bundle sheath plastids (*Figure 15.2d*) were not qualitatively different from the normal bundle sheath chloroplasts.

The amount of thylakoid membrane per maize leaf section is shown in *Table 15.4*. Over the range 0.16–5 mW cm^{-2}, the amount of thylakoid membrane per leaf section in both bundle sheath and mesophyll increased with increasing irradiation. It can be calculated that the percentage of mesophyll thylakoid membrane in grana increased from zero at the low intensity to 56 per cent at the intermediate intensity and 65 per cent at the high intensity. When the net CO_2 exchange of the leaves was determined at 11 mW cm^{-2} after the plants had been irradiated for 48 h at their appropriate light intensities, the low-intensity plants showed a very small net uptake with the intermediate- and high-intensity plants yielding successively greater net uptakes. When the low-intensity plants were transferred to high intensity conditions for 24 h, there was a considerable increase of the rate of CO_2 uptake together with the development of grana in the mesophyll plastids. The transferred plants had 82 per cent of their mesophyll thylakoid membrane in grana although there had been a decrease of their total thylakoid membrane during the 24 h at the high light intensity. This loss of thylakoid membrane presumably reflects the onset of senescence in the first leaf of maize. When maize plants are grown for a prolonged period in the dark and then illuminated, the completion of greening between 48 and 72 h after the commencement of illumination is immediately followed by the onset of senescence.

Another peculiar feature of the low-intensity plants is that the bundle sheath chloroplasts developed deep pockets containing mitochondria, as shown in *Figure 15.3a* (Montes and Bradbeer, 1975). The pockets occurred near the periphery of the chloroplasts and serial sectioning established that they were always in communication with the surrounding cytoplasm. Since about 25 per cent of the sections of the bundle sheath plastids contained at least one mitochondrial pocket, it can be estimated that each plastid contains one or two such pockets after this treatment. The phenomenon is not confined to maize, as shown in the micrograph of a *Hyptis suaveolens* chloroplast (*Figure 15.3b*) which comes from a plant grown under a light intensity of less than 0.3 mW cm^{-2}. There was no difference between the appearance of the mitochondria in the pockets and that of the free mitochondria. On transfer

Table 15.4 Thylakoid membrane content of leaf sections (1.5-3.5 cm from the tip of the first leaf) and the rate of net CO_2 exchange (at 25 °C with 11 mW cm^{-2} irradiation) of 14-day-old dark-grown *Zea mays* subjected to various illumination treatments

Illumination treatment	Thylakoid membrane (cm^{-2} leaf section)		Thylakoid membrane in grana (cm^{-2} leaf section)[1]		Net CO_2 exchange in light (μmol h^{-1} dm^{-2})
	Bundle sheath	*Mesophyll*	*Bundle sheath*	*Mesophyll*	
None	7.7	7.9	0	0	−341
48 h at 5 mW cm^{-2}	85	132	~0.1	86	+850
48 h at 1.6 mW cm^{-2}	41	73	~0.1	41	+660
48 h at 0.16 mW cm^{-2}	19	44	~0.1	0	+35
48 h at 0.16 mW cm^{-2} followed by 24 h at 5 mW cm^{-2}	17	33	~0.2	27	+432

[1] Amount of appressed thylakoid membrane. Typical grana are not found in bundle sheath plastids

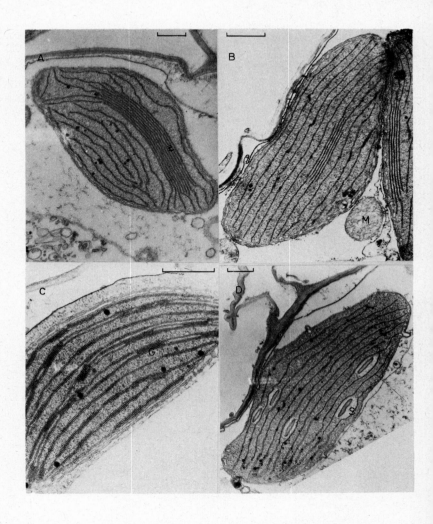

Figure 15.2   Plastids of 14-day-old dark-grown plants of Zea mays *after 48 h of illumination at 0.16 mW cm⁻² (A and B) and after a further 12 h at 5 mW cm⁻² (C and D).   The scale lines represent 0.5 μm.   S, starch grain; other letters as in* Figure 15.1. * A, mesophyll plastid greened in low light, ✕ 15 000; B, bundle sheath plastid greened in low light, ✕ 20 000; C, mesophyll plastid 12 h after transfer to high light, ✕ 27 000; D, bundle sheath plastid 12 h after transfer to high light, ✕ 13 500*

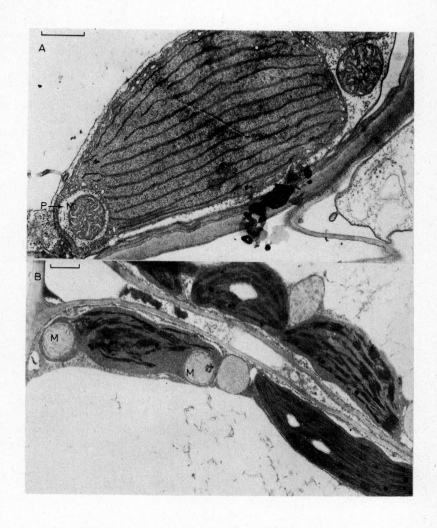

Figure 15.3 Plastids in leaves of Zea mays *and* Hyptis suaveolens *after growth in low light intensities. The scale lines represent 0.5 μm. P, pocket; other letters as in* Figure 15.1. *A, bundle sheath plastid of 14-day-old dark-grown* Zea mays *plant after 48 h illumination at 0.16 mW cm^{-2},* × *22 500; B, plastid of* Hyptis suaveolens *grown at* <*0.3 mW cm^{-2} in a glasshouse,* × *15 000*

of the low-light grown maize to the high light intensity, the mitochondria were lost from the pockets, which retained only their cytoplasmic contents.

Our investigation of the effects of light intensity on plastid development differs from those of other workers in that we have investigated the continuous illumination of dark-grown seedlings while they investigated the diurnal illumination of seedlings germinated and grown under diurnal alternation of light and dark (Ballantine and Forde, 1970; Björkman *et al.*, 1972). Consequently, the investigations are not strictly comparable. However, it might be speculated that, up to a certain light intensity, the amount of thylakoid membrane in a leaf may increase with increasing intensity, while at higher light intensities the amount of thylakoid membrane in a leaf may decrease with increasing intensity.

Conclusion

Chloroplast development is affected by several photoreceptors of which phytochrome seems to be concerned in the largest number of responses, although most of these are slow responses which take more than 3 h after irradiation to be detected. Of the more rapid responses, only the prevention of prolamellar body recondensation, the promotion of prolamellar body recrystallisation, the control of the Shibata shift and gibberellin release seem to be controlled by Pfr. On the other hand, prolamellar body transformation and prolamellar body dispersal and thylakoid formation do not appear to be controlled by Pfr. Photoreceptors other than phytochrome, concerned with chloroplast development, may be a red-absorbing one (possibly protochlorophyllide or chlorophyllide), a blue absorbing one and possibly the whole photosynthetic system. As light intensity also has considerable effects on chloroplast development, further investigations in which both light quality and intensity are studied together should be carried out so that there can be further elucidation of the mechanism of chloroplast development.

Acknowledgements

Acknowledgements are due to the Consejo Nacional de Investigaciones Cientificas y Tecnológicas (CONICIT) Caracas, Venezuela (grant and a studentship to G.M.), to the Science Research Council (electron microscope), to the Wellcome Foundation (ulta-microtome) and to Professor Renata Wulff (*Hyptis suaveolens* seeds).

References

ANDERSON, J.M., GOODCHILD, D.J. and BOARDMAN, N.K. (1973). *Biochim. Biophys. Acta,* **325,** 573

BALLANTINE, J.E.M. and FORDE, B.J. (1970). *Am. J. Bot.,* **57,** 1150

BERRY, D.R. and SMITH, H. (1971). *J. Cell. Sci.,* **8,** 185

BJÖRKMAN, O., BOARDMAN, N.K., ANDERSON, J.M., THORNE, S.W., GOODCHILD, D.J. and PYLIOTIS, N.A. (1972). *Carnegie Inst. Washington Yearb.,* **71,** 115

BRADBEER, J.W. (1969). *New Phytol.,* **68,** 233

BRADBEER, J.W., GYLDENHOLM, A.O., WALLIS, M.E. and WHATLEY, F.R. (1969). In *Progress in Photosynthesis Research,* Vol.1, pp.272-279. Ed. Metzner, H. Tübingen

BRADBEER, J.W. (1971). *J. Exp. Bot.,* **22,** 382

BRADBEER, J.W., IRELAND, H.M.M., SMITH, J.W., REST, J. and EDGE, H.J.W. (1974a). *New Phytol.,* **73,** 263

BRADBEER, J.W., GYLDENHOLM, A.O., IRELAND, H.M.M., SMITH, J.W., REST, J. and EDGE, H.J.W. (1974b). *New Phytol.,* **73,** 271

BRADBEER, J.W., GYLDENHOLM, A.O., SMITH, J.W., REST, J. and EDGE, H.J.W. (1974c). *New Phytol.,* **73,** 281

DE GREEF, J., BUTLER, W.L. and ROTH, T.F. (1971). *Plant Physiol.,* **47,** 457

DOWNS, J. (1955). *Plant Physiol.,* **30,** 468

HÄCKER, M. (1967). *Planta,* **76,** 309

HENNINGSEN, K.W. (1967). In *Biochemistry of Chloroplasts,* Vol.2, pp.453-457. Ed. Goodwin, T.W. Academic Press, London

JABBEN, M. and MOHR, H. (1975). *Photochem. Photobiol.,* **22,** 55

KASEMIR, H., BERGFELD, R. and MOHR, H. (1975). *Photochem. Photobiol.,* **21,** 111

KASS, L.B. and PAOLILLO, D.J. Jr. (1974). *Plant Sci. Lett.,* **3,** 81

MOHR, H. (1972). *Lectures on Photomorphogenesis.* Springer, Berlin

MONTES, G. and BRADBEER, J.W. (1974). In *Proceedings of the Third International Congress on Photosynthesis,* pp.1867-1876. Ed. Avron, M. Elsevier, Amsterdam

MONTES, G. and BRADBEER, J.W. (1976). *Plant Sci. Lett.,* **6,** 35

POSSINGHAM, J.V. (1973). *J. Exp. Bot.,* **24,** 1247

WELLBURN, A.R. and WELLBURN, F.A.M. (1973a). *Ann. Bot., N.S.,* **37,** 11

WELLBURN, F.A.M. and WELLBURN, A.R. (1973b). *New Phytol.,* **72,** 55

16

THE EFFECT OF LIGHT ON RNA METABOLISM IN DEVELOPING LEAVES

D. GRIERSON
S.N. COVEY
A.B. GILES

Department of Physiology and Environmental Studies, University of Nottingham, School of Agriculture, Sutton Bonington, Loughborough, Leics. LE12 5RD, U.K.

Introduction

Many of the important physiological and biochemical events that take place during leaf development are regulated by light. These events include cell division and expansion (Dale and Murray, 1969) and a variety of ultrastructural and biochemical changes associated with the development of the photosynthetic apparatus (Bradbeer and Montes, Chapter 15; Kirk, 1970). In many cases, changes in the metabolism of either macromolecules or plant growth substances are implicated in these responses. Despite the fact that a great deal is known about perception of the light stimulus and the nature of the light-induced modifications in leaf development, very little is known about the biochemical control mechanisms that govern these processes. Such regulatory mechanisms might be expected to operate by the modification of

(1) RNA synthesis and metabolism;
(2) protein synthesis;
(3) enzyme activation or
(4) by allosteric interaction of small metabolites with enzyme systems already formed in the dark.

It is highly probable that more than one type of control system operates but it is of considerable interest to establish the relative importance of these mechanisms in regulating development. In order to obtain further information concerning the operation of processes (1) and (2), a detailed knowledge of RNA metabolism is necessary. Observations following the application of inhibitors of RNA synthesis are of little value except in confirming the belief that gene expression is in some way related to development. The most interesting type of RNA from the point of view of gene expression, messenger-RNA (mRNA), has always been the most difficult to study because it is present only in

229

small amounts and is difficult to purify and characterise. More recent developments in the field of RNA synthesis have made it possible to study mRNA metabolism with greater precision. This chapter describes the results of preliminary experiments to characterise the various types of RNA synthesised in developing leaves and to assess the effect of light on the metabolism of different RNA fractions.

In most experiments, plants were grown under sterile conditions at 25 °C either in darkness or illuminated by warm white fluorescent lights. Manipulation and labelling of dark-grown plants were carried out under a green safelight. All experiments were carried out with intact plants unless otherwise stated in the figure legends. Labelling of RNA was by the direct application of solutions of radioisotopes to the leaves or occasionally the roots. Relevant experimental details are included in the legends to the figures and tables. Further details can be obtained from the papers by Grierson (1972, 1974) and Grierson and Loening (1972, 1974).

Ribosomal RNA Synthesis in Developing Leaves

During germination of seeds of *Phaseolus aureus* in the dark, the number of cells in each primary leaf increases from about 0.25×10^6 to 2.5×10^6 during the first 5 days and then remains constant. Cell division is stimulated when seedlings are grown in cycles of 12 h white light and 12 h darkness and the number of cells in each leaf increases to approximately 8×10^6 over the same period (*Figure 16.1a*). The light-stimulated increase in cell number is first detected between the second and third day after germination although plants grown in the dark for longer periods before illumination are also capable of responding in a similar way. Corresponding changes in leaf fresh weight are associated with these increases in cell number (*Figure 16.1b*). During the cell division phase of leaf development in the light, the amount of nucleic acid increases from approximately 6 to 80 μg per leaf. There is a similar increase in the nucleic acid content of leaves of dark-grown plants although this is generally initiated approximately 1 day later than in the light (Grierson *et al.*, 1970). Polyacrylamide gel electrophoresis of the nucleic acid enables the chloroplast and cytoplasmic ribosomal RNA (rRNA) molecules to be distinguished from the low molecular weight RNA and DNA (*Figure 16.2*).

A comparison of the RNA samples from leaves of different ages shows that there are changes in the relative amounts of each type of RNA during leaf development. These changes can be quantified by measuring the areas under each RNA peak after gel electrophoresis and relating them to the total amount of RNA present in each sample. Changes in the amount of chloroplast and cytoplasmic rRNA during leaf development are shown in *Figure 16.3*. Both types of rRNA are synthesised in the dark and in the light. The main effect of light is to hasten the onset of RNA accumulation and the synthesis of

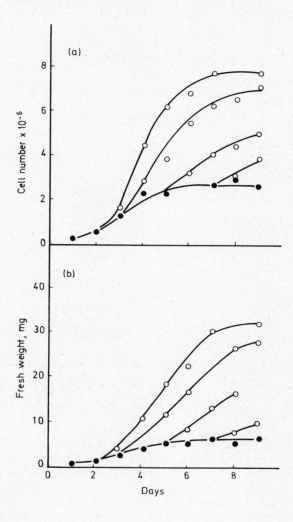

Figure 16.1 Changes in leaf cell number and fresh weight during development.
Phaseolus aureus *plants were grown in total darkness (●) or in a 12 h dark/12 h*
light photoperiod (○). Further batches of plants were grown in darkness and
introduced into the light on successive days. (a) Leaf cell number; (b) fresh
weight. (From Grierson and Covey, 1975)

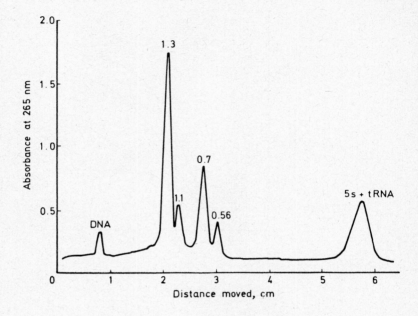

Figure 16.2   Fractionation of total nucleic acid from leaves of P. aureus *by polyacrylamide gel electrophoresis.   Nucleic acids were extracted and purified from
leaves of 6-day-old plants in the presence of* Mg^{2+}*, which was also present during
all subsequent purification procedures.   Electrophoresis was carried out for 3 h in
a 2.6 per cent acrylamide gel using a Tris–HCl buffer containing* Mg^{2+}*.   (After
Grierson, 1974)*

cytoplasmic rRNA is preferentially stimulated.   After 5 days in the
light, under the growth conditions used, each primary leaf of approximately 8×10^6 cells contains 40 μg of cytoplasmic rRNA.   The
accumulation of chloroplast rRNA occurs less rapidly and reaches a
maximum of about 18 μg per leaf after 6 days.   These values corres-
pond to an average of 1.5×10^6 cytoplasmic ribosomes and $0.75 \times
10^6$ chloroplast ribosomes in each cell.   At the completion of the cell
division phase, a gradual decline in rRNA content takes place, in contrast to the rapid decline in mRNA content described later.

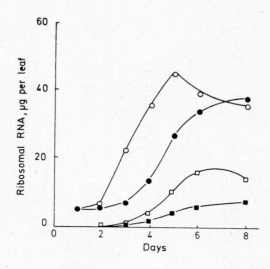

Figure 16.3 Changes in chloroplast and cytoplasmic rRNA during leaf develop-ment in Phaseolus aureus. ○, Cytoplasmic rRNA; □, chloroplast rRNA in the light; ●, cytoplasmic rRNA; ■, chloroplast rRNA in the dark. (From Grierson and Covey, 1975)

Mechanism of Synthesis of rRNA

The synthesis of rRNA can be studied by pulse-labelling experiments. Gel electrophoresis of rapidly labelled RNA from plants shows a number of peaks of RNA superimposed upon a background of polydisperse RNA (Rogers, Loening and Fraser, 1970; Leaver and Key, 1970; Grierson *et al.*, 1970). These peaks of RNA are macromolecular precursors to rRNA and various processing intermediates. The pattern of RNA mole-cules synthesised in leaves is slightly different to that found in roots (Grierson and Loening, 1972) and is shown in *Figure 16.4*. Cell frac-tionation experiments have shown that rapidly labelled RNAs with mole-cular weights of approximately $2.5-2.6 \times 10^6$, 1.45×10^6 and 1.0×10^6 daltons are synthesised in the nucleus (Grierson and Loening, 1974). A slightly larger RNA with a molecular weight of 2.9×10^6 daltons is synthesised in the chloroplasts together with smaller RNAs, which may represent precursors of chloroplast rRNA, and an unknown com-ponent with a molecular weight of approximately 0.48×10^6 daltons (Hartley and Ellis, 1973; Grierson and Loening, 1974). These rapidly labelled RNA peaks are metabolically unstable and it follows that they

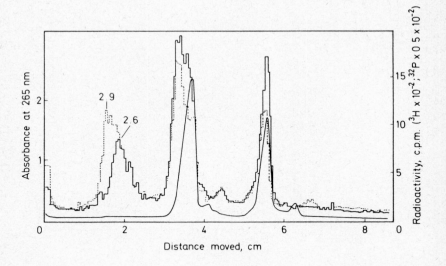

*Figure 16.4   Gel electrophoresis of a mixture of mung bean leaf and root RNA.
Seedlings were grown under sterile conditions in sand in glass tubes at 22 °C.
For labelling of the leaves, seedlings were germinated in the dark for 3 days and
then illuminated (warm white fluorescent tubes) for 24 h, and 120 µCi of
[^{32}P]phosphate in 10 µl of sterile distilled water was applied directly to each of
the primary leaves of the intact plants (about 3 × 10^6 cells per leaf).   After
incubation for 1.5 h the leaves were cut off, washed with water and mixed with
labelled roots for RNA extraction.   For labelling of the roots, seedlings were
germinated for 30 h, when the roots were about 1.5 cm long, and 1 mCi of
[^{3}H]uridine in 2 ml of sterile water was injected into the sand around the roots.
After 1.5 h, the seedlings were washed with water and the apical 1 cm of each
root was cut off and mixed with the leaves.   Nine roots and four leaves were
ground in detergent and extracted with phenol mixture.   The nucleic acids were
precipitated and washed with ethanol and the DNA was digested with 10 µg ml^{-1}
of electrophoretically purified DNAase (Sigma) in MES buffer, pH 7.0, containing
2 mM MgCl$_2$, for 10 min at 0 °G.   Sodium acetate (0.15 M) and SDS (0.5 per
cent) were added and the RNA was precipitated with ethanol.   The precipitate
was dissolved in 100 µl of half-concentrated E buffer (E buffer is 0.036 M Tris,
0.030 M NaH$_2$PO$_4$, 1 mM EDTA and 0.2 per cent SDS) containing 20 per cent
of sucrose and 50 µl were layered on a gel.   Electrophoresis was carried out in
a 9 mm diameter, 2.4 per cent polyacrylamide gel for 4.5 h at 50 V and 11 mA.
The gel was scanned in a Joyce Loebl Polyfrac ultraviolet scanner, frozen in dry-
ice and cut into 0.5–mm slices on a Mickle gel slicer.   The slices were heated at
60 °C for 4 h in 10 per cent piperidine–1 mM EDTA and dried.   The slices were
swollen, the hydrolysed RNA was dissolved in 0.4 ml of water for 1 h and 10
ml of scintillation fluid were added (60 per cent toluene, 40 per cent methoxy-
ethanol, 0.5 per cent butyl-PBD).   Radioactivity was determined in a Packard Tri-
Carb spectrometer and the ^{3}H counts were corrected for 5 per cent spill-over
from the ^{32}P.   ——, Absorbance at 265 nm; ⌐⌐, ^{32}P; ⌐⌐, ^{3}H.   Molecular
weights are indicated, assuming values for the rRNA of 1.3 × 10^6 and 0.7 × 10^6,
which were obtained by comparison with E. coli rRNA of assumed molecular
weight 1.1 and 0.56 × 10^6.   (From Grierson and Loening, 1972)*

must either be degraded after synthesis or converted into other types of RNA. The nucleotide composition (calculated by the ^{32}P labelling method) of the nuclear RNAs is very similar to that of cytoplasmic rRNA labelled for a similar time (*Table 16.1*). This evidence suggests that these molecules are involved in the synthesis of cytoplasmic rRNA. The nuclear RNA with a molecular weight of 2.5–2.6 × 10^6 daltons is sufficiently large to contain one sequence of each of the cytoplasmic

Table 16.1   Base composition of pulse-labelled [^{32}P] RNA fractions

RNA from leaves labelled for 1.5 or 3 h was fractionated on gels which were cut into 0.5–mm slices. Two or three slices were analysed for each radioactive RNA peak. The results are the average of several determinations, the number of determinations being given in parentheses. (From Grierson and Loening, 1974).

RNA fraction (mol. wt. × 10^{-6})	^{32}P-labelled nucleotides (mol/100 mol)				$\dfrac{GMP}{AMP}$
	CMP	AMP	GMP	UMP	
2.9 (4)	20.8	28.4	28.0	22.9	1.0
2.5 (2)	20.7	28.4	28.1	22.7	1.0
1.45 (3)	20.4	28.1	30.9	20.2	1.1
1.3 (2)	21.25	28.55	30.4	19.8	1.1
0.7 (2)	20.65	26.9	28.5	24.0	1.1
3.0–0.5 (5)[1]	18.4	32.7	24.4	24.3	0.74

[1] The base composition of the heterogeneous RNA varies with the molecular weight. The figures shown represent average values for five heterogeneous RNA fractions from different regions of the gel

rRNAs (molecular weights 1.3 and 0.7 × 10^6 daltons) together with some excess of non-ribosomal RNA. This would represent a polycistronic transcription unit, similar to those found in animal cells, containing RNA sequences transcribed from alternating cistrons of 18S and 26S ribosomal DNA. The results of competitive hybridisation experiments between *P. aureus* DNA and radioactive rRNA precursor, using unlabelled cytoplasmic rRNAs as competitors, support this interpretation (*Figure 16.5*). The scheme shown in *Figure 16.6* summarises current views about the formation of cytoplasmic rRNA. The 2.5–2.6 × 10^6 daltons RNA synthesised in the nucleus appears to be the first detectable transcription product of the rRNA genes. It contains one sequence of each of the two high molecular weight rRNAs and is processed to smaller intermediates before mature rRNA is finally produced. Although this scheme is only tentative and requires confirmation by RNA sequencing techniques, it is consistent with most of the available evidence. We have not investigated the function of the rapidly labelled chloroplast RNAs, but Head and Hartley (personal communication) have shown that similar molecules in spinach chloroplasts are involved in the synthesis of chloroplast rRNA. It therefore seems probable that the 2.9 × 10^6 daltons RNA detected in *P. aureus* chloroplasts is a polycistronic precursor to chloroplast rRNA. In young leaves of plants grown in the

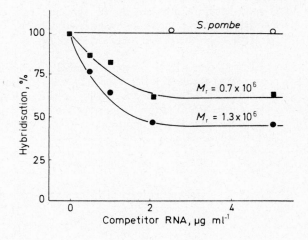

Figure 16.5   Competition hybridisation of pre-rRNA with cytoplasmic rRNA.
Total leaf DNA was immobilised on nitrocellulose filters (5 μg per filter) and
incubated at 70 °C in 2 × standard saline citrate containing either unlabelled
Schizosaccharomyces pombe *RNA (o) or* $M_r = 1.3 \times 10^6$ *(•) or 0.7 × 10^6 (■)*
rRNA from Phaseolus aureus *at the concentration shown.   After 2 h the filters*
were washed, treated with ribonuclease and incubated together in ^{32}P-labelled
pre-rRNA.   The radioactive RNA was prepared by labelling the leaves of 60-h-old
plants grown in the dark with [^{32}P]orthophosphate for 2 h.   The actual labelling
was performed in the light because the leaves take up more [^{32}P]orthophosphate
under these conditions.   RNA was extracted and fractionated on 2.4 per cent
gels and the radioactivity located by Cerenkov counting of the wet gel slices.
RNA from the appropriate slices was recovered as described by Grierson (1974).
The specific radioactivity of the pre-rRNA was not determined.   The total
Cerenkov radiation from the $M_r = 2.6$‑2.4×10^6 *RNA fraction from 20 leaves*
was 23 000 counts min^{-1}.   Hybridisation was carried out at 66 °C in 6 × stan-
dard saline citrate containing 0.06 per cent phenol for 1.5 h.   The filters were
then washed, treated with ribonuclease and counted.   100 per cent hybridisation
was 400 counts min^{-1}.   (From Grierson and Loening, 1974)

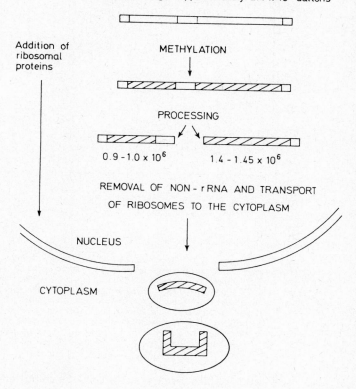

RIBOSOMAL RNA PRECURSOR
molecular weight approximately 2.4 x 10⁶ daltons

Addition of
ribosomal
proteins

METHYLATION

PROCESSING

0.9 - 1.0 x 10⁶   1.4 - 1.45 x 10⁶

REMOVAL OF NON - rRNA AND TRANSPORT
OF RIBOSOMES TO THE CYTOPLASM

NUCLEUS

CYTOPLASM

Figure 16.6   Synthesis of rRNA.   This scheme outlines current views about the post-transcriptional modification of the ribosomal RNA precursors in plants.   The molecular weights of the RNA molecules are shown in millions.   In P. aureus the precursor to cytoplasmic rRNA has a molecular weight of about 2.5 × 10⁶ daltons.   In other plants, the molecular weight of the precursor may be slightly different and in some species there are apparently two precursors

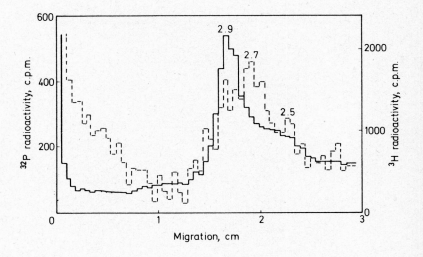

Figure 16.7   Comparison of high-molecular-weight pulse-labelled RNA from leaves at different stages of development.   Plants were germinated in the dark.   After 60 h the leaves of one group were labelled for 1.5 h with [^{32}P]orthophosphate and total nucleic acid extracted.   Another group of similar plants was grown for a further 24 h under continuous illumination and then labelled for 1.5 h with [5-^{3}H]uridine prior to nucleic acid extraction.   The samples were mixed, treated with deoxyribonuclease and fractionated on 2.4 per cent gels for 4.5 h.   Only the first 3 cm of the gel are shown.   About 6 per cent of the ^{32}P was registered in the ^{3}H channel.   Solid line, ^{3}H; dashed line, ^{32}P.   (From Grierson and Loening, 1974)

dark, very little of this RNA is synthesised in comparison with the amount produced in the light (*Figure 16.7*).   This result is consistent with the observation that young leaves contain very little chloroplast rRNA.

Effect of Light on rRNA Synthesis

The results in *Figure 16.3* show that cytoplasmic rRNA accumulation proceeds more rapidly in the light than in the dark.   The question arises as to how rapidly this accumulation occurs and by what mechanism.   Labelling experiments show that there is an increased incorporation of radioactivity into cytoplasmic rRNA within a few hours of introducing plants into the light.   However, in these experiments there is also a parallel increase in the rate of uptake of radioisotope by the

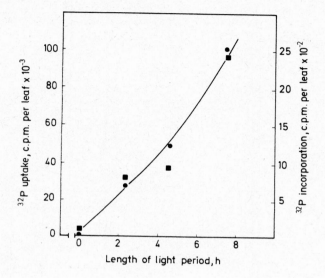

Figure 16.8   Effect of light on the uptake and incorporation of [³²P]orthophos-phate.   Dark-grown plants of Phaseolus aureus *were introduced into the light for various times and then returned to darkness and incubated with [³²P]orthophos-phate for 2 h together with dark controls.   ³²P uptake into the leaves was deter-mined by washing plants with phosphate buffer, grinding the leaves in the medium used for RNA extraction and counting a sample of the supernatant from the centrifuged homogenate prior to phenol deproteinisation of the nucleic acids. Samples of purified RNA equivalent to that from two leaves were fractionated by gel electrophoresis and the radioactivity incorporated into rRNA was deter-mined by summing the amount of radioactivity associated with each peak.   (From Grierson, 1972)*

leaves (*Figure 16.8*).   It remains possible, therefore, that an increase in the rate of labelling of rRNA arises solely by virtue of increased uptake of the radiochemical supplied and does not necessarily indicate a net increase in RNA synthesis.   In consequence, it is not possible to con-clude from this type of experiment how rapidly the increase in RNA synthesis occurs.

However, it is possible to demonstrate a rapid effect of light on the processing of the rRNA precursor to mature rRNA by using a different experimental approach.   When the RNA of plants grown under con-trolled conditions is labelled for a short time and then purified and fractionated by gel electrophoresis, a characteristic pattern is observed. For example, with young leaves of dark-grown plants, after a labelling period of 1 h, the rRNA precursors are prominent and there is a certain amount of radioactivity associated with the mature rRNAs which

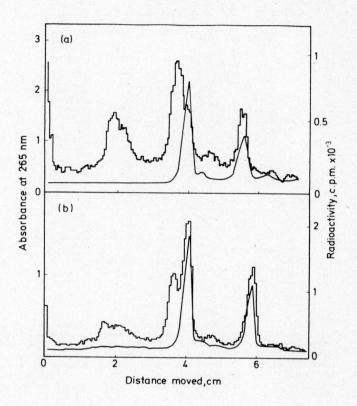

Figure 16.9 Effect of light on processing of pre-rRNA in dark-grown leaves of Phaseolus aureus. (a) Plants grown for 60 h and labelled for 1 h in total darkness by application of $[^{32}P]$orthophosphate to the leaves. (b) Leaves of plants grown for 60 h were introduced into the light for 5 h and labelled for the fifth hour with $[^{3}H]$uridine. Electrophoresis was carried out in 2.4 per cent gels for 4.5 h. (From Grierson, 1972)

represents that proportion of precursor molecules which have been processed since the start of the labelling period (*Figure 16.9a*). The ratio
of precursors to products is independent of the amount of radioactivity
taken up by the plants. When plants are illuminated and then labelled
for 1 h, the amount of radioactivity incorporated into RNA increases,
as expected, but in addition there is an increased proportion of radioactivity present in the mature rRNA relative to that in the precursor
molecules (*Figure 16.9b*). This suggests that an increase in the rate of
processing of the rRNA precursor occurs when plants are illuminated.
By comparing the ratio of radioactivity in the rRNA precursors to
that in the processing products it is possible to detect altered rates of
precursor processing. The results in *Table 16.2* show that with three
batches of dark-grown plants, each labelled for 1 h, this ratio is very

Table 16.2 Effect of light on the processing of precursors to rRNA
Leaves of plants grown in darkness or introduced into the light for
a short time were incubated with $[^{32}P]$orthophosphate for 1 h. RNA
was extracted and DNA removed with deoxyribonuclease. Samples
were separated on 2.4 per cent gels and the amounts of radioactivity
present in the rRNA precursors (mol. wt. 2.9-2.5×10^6) and processing products (mol. wt. 1.45, 1.3, 1.0 and 0.7×10^6) were determined.
The results are expressed as the ratio of radioactivity in the rRNA
precursors compared to the products, after incubation for 1 h in
$[^{32}P]$. (From Grierson, 1972).

	Dark(a)	Dark(b)	Dark(c)	Light 1 h	Light 1.8 h	Light 3.3 h	Light 4 h
Ratio precursors: products	0.56	0.52	0.53	0.50	0.45	0.36	0.38

similar. However, when plants of the same age, grown under identical
conditions, are introduced into the light for a short time and then
labelled for 1 h, the ratio changes. The results indicate that an
increased rate of processing of the rRNA precursor occurs within 2 or
3 h of the onset of light treatment (*Table 16.2*).
There are two complicating factors to be considered when interpreting this type of experiment. Firstly, it has been suggested that prominent peaks of radioactivity associated with the rRNA precursors are due
to an impaired processing of these molecules caused by the high levels
of radioisotope employed (Jackson and Ingle, 1973). That this is not
always the case is clearly demonstrated by the experiments of Cox and
Turnock (1973), who detected prominent rRNA precursor peaks with
very low levels of radioisotope. Nevertheless, in some circumstances,
precursor processing may be affected by high levels of radioisotopes.
The results shown in *Table 16.2* were obtained with plants from the
same batch, germinated, grown and labelled under identical conditions.
The only factor that varied was the length of the light period and

therefore the observed effect of light on precursor processing is assumed to be independent of any isotope effect.

A second problem is that it is difficult, without resorting to cell fractionation procedures, to assess the contribution of chloroplast RNA to the labelling pattern. It is certainly true that in older leaves large amounts of the 2.9×10^6 daltons chloroplast rRNA are synthesised. However, it is clear from the results in *Figures 16.3* and *16.7* that with leaves from plants germinated in the dark for 60 h very little chloroplast RNA is synthesised. As the plants used to obtain the results shown in *Figure 16.9* and *Table 16.2* were used 60 h after germination it seems reasonable to conclude that the rapid effect of light is predominantly on the processing of the precursors to cytoplasmic rRNA.

It is possible that, when the rate of rRNA synthesis is low, part of the pool of rRNA precursor molecules is degraded without being processed and that in the light a larger number of these molecules are converted into mature rRNA. An alternative possibility is that light stimulates the rate of processing, which in turn would tend to reduce the pool of rRNA precursors. In order to maintain an increased rate of production of rRNA by this mechanism, a corresponding increase in RNA transcription would be expected within a short time. Further studies will be necessary in order to determine whether a transcriptional or post-transcriptional mechanism operates.

Fractionation of mRNA on the Basis of Poly(adenylic) Acid Content

A number of animal mRNAs have been shown to contain poly(adenylic) acid sequences at the $3'$-end of the molecule (Matthews, 1973; Brawerman, 1974) and this facilitates their purification by procedures that select for poly(A) sequences, such as affinity chromatography on columns of oligo(dT)-cellulose or poly(U)-Sepharose. Poly(A) sequences have also been shown to be present in plant RNA (Sagher, Edelman and Jakob, 1974) and the mRNA for the plant protein leghaemoglobin has been partially purified on the basis of poly(A) content (Verma, Nash and Schulman, 1974).

Figure 16.10 illustrates the fractionation of pulse-labelled plant RNA into two components by oligo(dT)-cellulose chromatography. Gel electrophoresis of these labelled RNA fractions from leaves of *P. aureus* shows that the column specifically binds polydisperse RNA but does not retain rRNA or rRNA precursors (compare *Figure 16.11b* with *16.11c*). Part of the polydisperse RNA lacks poly(A) and therefore does not bind to the column. The bound RNA is composed of a large number of molecules with different molecular weights, which accounts for the broad nature of the peak after gel electrophoresis (*Figure 16.11c*). Each molecule contains a short sequence of 100–250 residues of adenylic acid (Sagher, Edelman and Jakob, 1974) attached to an RNA sequence from several hundred to a few thousand nucleotides in length.

Although historically the idea has developed that mRNA is polydisperse in size and is rapidly labelled, these criteria alone are insufficient to identify a particular RNA fraction as mRNA. However, additional evidence shows that poly(A)-containing RNA is associated with polyribosomes and directs the synthesis of proteins. *Figure 16.12* illustrates the fractionation by sucrose gradient centrifugation of polyribosomes prepared from leaves of *P. vulgaris*. If polyribosomes are extracted from leaves in which the RNA has previously been pulse labelled, the nature of the rapidly labelled RNA associated with the polyribosomes may be determined. Part of the polysome-associated RNA does indeed bind to oligo(dT)-cellulose. This fraction has, not surprisingly, a very high content of AMP (*Table 16.3*) and is different in nucleotide composition from other types of leaf RNA (compare *Table 16.1* with *Table 16.3*). Gel electrophoresis of the poly(A)-containing RNA from polyribosomes shows that it is composed of molecules which vary considerably in size (*Figure 16.13*). This is to be

Table 16.3 Base composition of pulse-labelled [^{32}P] RNA fractions from polyribosomes separated by oligo(dT)-cellulose chromatography

Leaves of 7-day-old dark-grown plants were introduced into white light for 16 h and labelled with [^{32}P] orthophosphate for 4 h. Polyribosomes were prepared and the purified RNA fractionated into poly(A)-containing and poly(A)-minus RNA by oligo(dT)-cellulose chromatography. The number of determinations is shown in parentheses.

RNA fraction	^{32}P-labelled nucleotides (mol/100 mol)				$\frac{GMP}{AMP}$
	CMP	AMP	GMP	UMP	
Peak 1 (3) [Poly(A)-minus]	18.8	27.6	29.9	23.6	1.08
Peak 2 (6) [Pol(A)-plus]	16.6	39.0	21.6	22.7	0.55

expected of an mRNA population derived from cells known to be synthesising a large number of different proteins. There is a tendency for this RNA to aggregate, which makes accurate measurement of the molecular weight more difficult. After treatment with 8 M urea, which disrupts aggregates, the average size of the RNA (7×10^5 daltons) corresponds to about 2 100 nucleotides. Assuming the poly(A) sequence to be 150 nucleotides long, the average coding capacity of each mRNA (1 950 nucleotides) is equivalent to about 650 amino acids. This corresponds to a protein with a molecular weight of approximately 6.5×10^4 daltons. Further evidence that this fraction is composed of mRNA comes from the observation that purified poly(A) containing RNA from leaf polyribosomes directs the synthesis of proteins when added to an *in vitro* protein synthesising system prepared from wheat embryos (*Figure 16.14*).

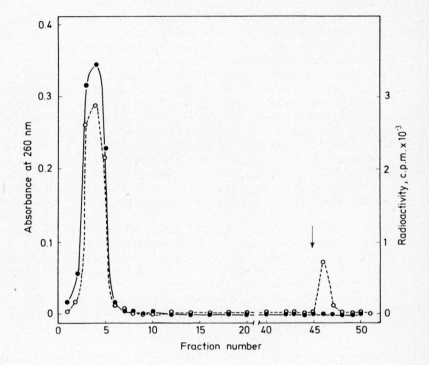

Figure 16.10   Fractionation of RNA from leaves of Phaseolus aureus *by oligo(dT)-cellulose chromatography.   Solid curve, A$_{260nm}$; dotted curve, radioactivity.   The RNA was extracted from leaves labelled with [^{3}H]uridine for 3 h.   (From Grierson and Covey, 1975)*

Oligo(dT)-cellulose does not bind all the polydisperse RNA from polyribosomes and it should be emphasised that not all mRNA molecules contain poly(A).   However, experiments with the polydisperse RNA purified from polyribosomes of sycamore cells grown in suspension culture have shown that at least 40 per cent of the mRNA molecules contain poly(A) sequences (Covey and Grierson, in preparation) and it therefore seems reasonable to conclude that we are studying a representative fraction of leaf mRNA.   However, it is possible that measurements of poly(A)-containing mRNA relate only to the cell cytoplasm and not to chloroplasts and mitochondria.   Poly(A) is absent from yeast mitochondria (Groot *et al.,*   1974) and chloroplasts from spinach (Hartley, personal communication).

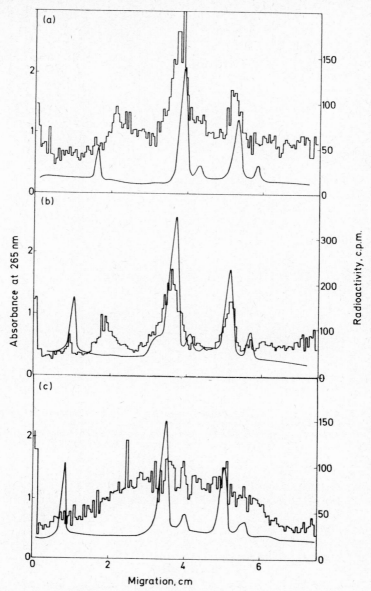

Figure 16.11   Gel electrophoresis of RNA fractions from an oligo(dT)-cellulose column.   Leaves of Phaseolus aureus were labelled for 2 h with [^{32}P]orthophosphate and RNA extracted and fractionated by oligo(dT)-cellulose chromatography.   Samples of the unbound (b) and bound (c) RNA were collected and compared with total RNA (a) by electrophoresis in 2.4 per cent acrylamide gels for 3 h.   Smooth curve, A_{265}nm; histogram, radioactivity.   (From Grierson and Covey, 1975)

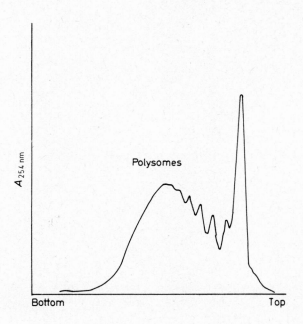

Figure 16.12   Fractionation of polyribosomes from leaves of Phaseolus vulgaris.
*Polyribosomes were prepared by the method of Davies, Larkins and Knight (1972)
and fractionated on a 30-60 per cent sucrose gradient at 36 000 r.p.m. in the
SW 50.1 rotor of a Beckman ultracentrifuge for 75 min at 4°C*

Quantitative Estimation of the Amount of Poly(A) and Poly(A)-associated RNA in Leaves in the Dark and in the Light

The fact that all mRNA molecules that contain poly(A) will bind to
oligo(dT)-cellulose provides a means of quantitatively estimating the
amount present in different RNA samples.   It is difficult to accom-
plish this accurately by estimating the amount of bound and unbound
RNA by optical methods because poly(A)-containing RNA represents
only a very small percentage of the total RNA and trace amounts of
UV-absorbing contaminants introduce errors in the measurements.   This
difficulty can be overcome by using radioactive RNA.   In principle,
it is necessary to know the specific radioactivity of each RNA fraction
from one sample, but in practice it is much simpler if the specific

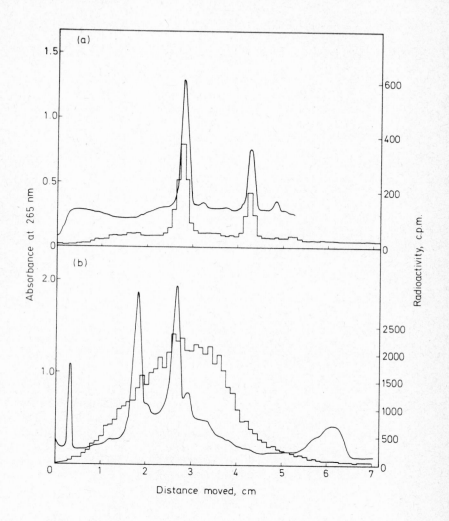

Figure 16.13 Gel electrophoresis of RNA from polyribosomes of leaves of
Phaseolus vulgaris.   Leaves were labelled for 4 h with [^{32}P]orthophosphate and
polyribosomes extracted and purified.   RNA was prepared from the polyribosomes
and fractionated in 2.4 per cent acrylamide gels (a).   A separate aliquot was
further fractionated by oligo(dT)-cellulose chromatography and the bound [poly(A)-
containing] RNA was collected, unlabelled carrier RNA added and precipitated
with ethanol.   This sample was then dissolved in 8 M urea and heated for
3 min at 60 °C, before fractionation in a 2.4 per cent gel (b)

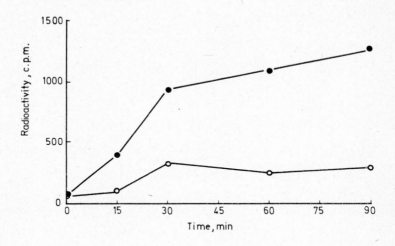

Figure 16.14   Activity of poly(A)-containing RNA from Phaseolus vulgaris *in directing the incorporation of radioactive leucine into protein in vitro. The in* vitro *protein-synthesising system from wheat germ described by Roberts and Paterson (1973) was used.* ●, *4 µg of poly(A)-containing RNA per reaction mixture;* ○, *no RNA added*

radioactivity of each fraction is equal. For this to be achieved, the plants from which the labelled RNA is obtained must have been previously incubated in radioisotope for several days. After appropriate purification, the percentage of radioactive RNA from a particular sample that binds to oligo(dT)-cellulose represents the percentage of mRNA in the sample. If the amount of RNA in each leaf is known, then the total amount of poly(A)-containing mRNA in each leaf can be calculated.

When this is done, the percentage of poly(A)-containing mRNA present in light-grown *P. aureus* leaves at different stages of development is found to vary from 0.5 to 2 per cent. Changes in the actual amount of this RNA during leaf development are shown in *Figure 16.15*. The rapid increase in mRNA content coincides with the onset of cell division and rRNA synthesis (*Figures 16.1* and *16.3*) and occurs at the same time as a dramatic increase in the protein content of the leaves (*Table 16.4*). The maximum mRNA content (approximately 0.7 µg per leaf) occurs between day four and day five (*Figure 16.15*). Assuming an average molecular weight for each mRNA of approximately 0.7×10^6 daltons, this corresponds to approximately 7.5×10^4

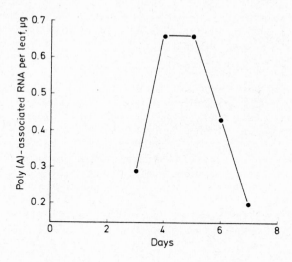

Figure 16.15   Changes in the amount of poly(A)-containing RNA during leaf development.   Phaseolus aureus plants were germinated and grown in the light with [³H]uridine present in the sand in which they were growing (approximately 40 μCi ml⁻¹), RNA was extracted and purified, the DNA removed with deoxyribonuclease and the RNA fractionated by oligo(dT)-cellulose chromatography.   The bound and unbound RNA was precipitated with 5 per cent (final concentration) trichloroacetic acid and trapped and counted on membrane filters.   The total amount of RNA in each leaf was calculated from the absorbance of the original RNA sample at 260 nm.   The percentage of the total RNA that bound to oligo(dT)-cellulose was used to calculate the amount of poly(A)-containing RNA in each leaf.   (From Grierson and Covey, 1975)

molecules of poly(A)-containing mRNA per cell, or one molecule for every twenty cytoplasmic ribosomes. When the plants are grown for a longer period, the amount of leaf mRNA declines rapidly, in marked contrast to the much slower decline in rRNA content (compare *Figures 16.15* and *16.3*).

Table 16.4 Changes in total protein content during development of primary leaves of *Phaseolus aureus*

Age (days)	Protein per leaf (μg)
3	310
4	800
5	900
6	1600

There are a number of limitations to the use of radioactive RNA for the estimation of amounts of poly(A)-containing RNA. Apart from difficulties associated with the variation in specific activity of different types of cellular RNA after short labelling times, and the need to take account of differences in isotope uptake *and* incorporation when comparing different treatments, such an approach excludes the possibility of measuring amounts of previously synthesised or stored mRNA. For these reasons, an alternative approach was investigated. The method used involves challenging unlabelled RNA samples with high specific radioactivity synthetic [^{3}H] poly(uridylic) acid. The poly(U) rapidly forms a nucleic acid hybrid structure with any poly(A) in the sample. Under appropriate conditions, hybridised poly(U) is insensitive to ribonuclease and can therefore be estimated and related to the amount of poly(A) present. Model experiments have shown that the method is capable of detecting as little as 5×10^{-9} g of poly(A), which is equivalent to the amount present in a few million leaf cells.

Changes in the amount of poly(A) during leaf development are shown in *Figure 16.16*. The results show a pattern of change in poly(A) content similar to that found for the poly(A)-associated RNA (*Figure 16.15*). The actual amounts are different because the poly(A) region represents only a small proportion of each molecule. There is evidence for some poly(A) in the leaves of ungerminated seeds but a period of synthesis occurs during the rapid phase of leaf development, which is followed by a rapid decline in poly(A) content. The results also show that there is a period of poly(A) synthesis in leaves of plants grown in total darkness. It occurs slightly later and results in the accumulation of approximately half the amount of poly(A) produced in the light. However, taking into account the fact that in the dark fewer cells are present, the results show that there is actually more poly(A) present per cell in the dark than in the light. It is possible that, as with rRNA synthesis, light may stimulate the synthesis of poly(A) but there is clearly no absolute requirement for light. In view of the fact that after growth for several days in the dark leaves retain the capacity to respond to light by cell division, we investigated the possibility that light induced the synthesis of additional poly(A). The results in *Table 16.5* indicate that there is no change in poly(A) content detectable within 6 h of bringing plants into the light.

Table 16.5   Poly(A) content of leaves of plants grown in darkness for 6 days and illuminated with white light for short periods of time. (From Grierson and Covey, 1975)

	Period of illumination (h)			
	0	2	5	6
Poly(A) (ng per 100 μg total RNA)	27	31	29	33

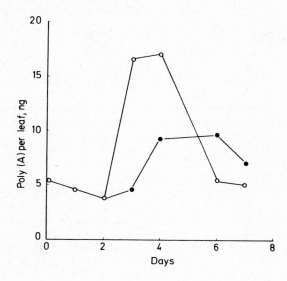

Figure 16.16   Changes in poly(A) content during leaf development.   Poly(A) content of RNA from Phaseolus aureus *leaves of different ages expressed as the amount per leaf.   ○, Light-grown plants; ●, dark-grown plants.   Each point is the average of three measurements.   (After Grierson and Covey, 1975)*

Conclusions

During primary leaf development, there is a period of rapid synthesis of cytoplasmic and chloroplast rRNA and poly(A)-containing mRNA a few days after germination.   These changes are associated with a phase of cell division and expansion and an increase in leaf protein content. Both types of rRNA and poly(A)-containing mRNA accumulate in the dark and the light.   There is no evidence for a specific effect of light on the type of RNA synthesised.   When expressed on a per leaf basis, light appears to hasten the onset of accumulation of cytoplasmic rRNA and poly(A)-containing RNA and also causes the formation of more chloroplast rRNA than appears in the dark.   However, when the difference in leaf cell number is taken into account, the results show that in leaves of plants grown in the dark each cell actually synthesises more of each type of RNA than cells of leaves grown in the light.   It is possible, however, that light may either modify the rate of RNA synthesis or processing or affect the metabolic stability of RNA.

The synthesis of poly(A)-containing mRNA takes place at about the time that cell division occurs. The comparatively large amounts involved, and the fact that histone mRNAs are reported to lack poly(A) (Adesnick *et al.*, 1972), suggest that this mRNA is not necessarily directly related to cell division. This conclusion is strengthened by the observation that the comparatively high poly(A) content of leaves in the dark results only in modest cell division, whereas after light treatment, which stimulates cell division, there is no rapid change in poly(A) content.

The apparently rapid decline in the amount of poly(A)-containing RNA in leaves in the light suggests that it is rapidly degraded at the conclusion of the cell division phase, in contrast to rRNA, which is much more stable. However, this result needs confirmation by using other techniques because a similar result would have been obtained if the mRNA simply lost its poly(A) region.

These experiments take no account of that portion, perhaps 50 per cent, of mRNA that lacks poly(A). With this limitation in mind and considering the fact that the mRNA changes actually measured are probably related to the synthesis of hundreds of proteins, it is clear that changes in transcription of a comparatively small number of genes would remain undetected. On the other hand, it is equally clear that light does not stimulate leaf development by regulating the transcription of a large number of genes. A more precise assessment of this problem could be made following the appropriate competition hybridisation experiments. In the meantime, further studies on post-transcriptional mechanisms for regulating leaf development should prove fruitful.

References

ADESNICK, M., SALDITT, M., THOMAS, W. and DARNELL, J. (1972). *J. Mol. Biol.*, **71**, 21

BRAWERMAN, G. (1974). *Annu. Rev. Biochem.*, **43**, 621

COX, B.J. and TURNOCK, G. (1973). *Eur. J. Biochem.*, **37**, 367

DALE, J.E. and MURRAY, D. (1969). *Proc. R. Soc. B*, **173**, 541

DAVIES, E., LARKINS, B.A. and KNIGHT, R.H. (1972). *Plant Physiol.*, **50**, 581

GRIERSON, D. (1972). *Ph.D. Thesis*, University of Edinburgh

GRIERSON, D. (1974). *Eur. J. Biochem.*, **44**, 509

GRIERSON, D. and COVEY, S. (1975). *Planta*, **127**, 77

GRIERSON, D. and LOENING, U.E. (1972). *Nature New Biol.*, **235**, 80

GRIERSON, D. and LOENING, U.E. (1974). *Eur. J. Biochem.*, **44**, 501

GRIERSON, D., ROGERS, M.E., SARTIRANA, M.-L. and LOENING, U.E. (1970). *Cold Spring Harbor Symp. Quant. Biol.*, **35**, 589

GROOT, G.S.P., FLAVELL, R.A., VAN OMMEN, G.J.B. and GRIVELL, L.A. (1974). *Nature, Lond.*, **252**, 167

HARTLEY, M.R. and ELLIS, R.J. (1973). *Biochem. J.*, **134**, 249

JACKSON, M. and INGLE, J. (1973). *Biochem. J.*, **131**, 523

KIRK, J.T.O. (1970). *Annu. Rev. Plant Physiol.*, **21**, 11

LEAVER, C.J. and KEY, J.L. (1970). *J. Mol. Biol.*, **49**, 671

MATTHEWS, M.B. (1973). *Essays Biochem.*, **9**, 59

ROBERTS, B.E. and PATERSON, B.M. (1973). *Proc. Nat. Acad. Sci. U.S.A.*, **70**, 2330

ROGERS, M.E., LOENING, U.E. and FRASER, R.S.S. (1970). *J. Mol. Biol.*, **49**, 681

SAGHER, D., EDELMAN, M. and JAKOB, K.M. (1974). *Biochim. Biophys. Acta*, **349**, 32

VERMA, D.P.S., NASH, D.T. and SCHULMAN, H.M. (1974). *Nature, Lond.*, **251**, 74

IV

PHYSIOLOGICAL ASPECTS OF PHYTOCHROME ACTION

17

PHYTOCHROME ACTION AS A THRESHOLD PHENOMENON

H. MOHR
H. OELZE-KAROW
Institut für Biologie II, Universität Freiburg, D-78 Freiburg i. Br., Schänzlestrasse 9-11, West Germany

Introduction

Threshold (all-or-none) phenomena are known to every biologist, and the elucidation of the molecular processes that govern the all-or-none phenomena is a great challenge in present-day biology. All-or-none processes seem to be the basis for the proper functioning of many biological structures in sensory, nerve and muscle physiology (Adam, 1970, 1973). As classical examples of threshold phenomena we shall mention only the contraction of a skeletal muscle upon direct electrical excitation and the appearance of an action potential in an isolated nerve fibre (axon) caused by an externally applied electrical stimulus. The principle (*Figure 17.1*) is that the electrical stimulus must have reached a certain threshold level before the response takes place. As long as the extent of the stimulus remains below the threshold level, no response occurs; if the stimulus is increased beyond the threshold level, the response (e.g. action potential) does not show any further increase. Most investigations of all-or-none behaviour have been carried out on the non-myelinated giant axon of the squid. A significant step forward in the understanding of this phenomenon was achieved when Cole and co-workers (*see* Cole, 1968) developed the voltage clamp of the squid axon and thereby reduced, by rigorous electrical characterisation of the squid axon, the difficult phenomenon of an action potential to the conceptually simpler aspect of axon excitation. Subsequently, Hodgkin and Huxley (for a review, *see* Hodgkin, 1964) gave a physicochemical description of the ion fluxes at excitation. In addition, they advanced a phenomenological mathematical description of the observations in question. This work has had a strong influence on nerve physiology. The physico-chemical mechanism of axon excitation has received much attention, and sophisticated theoretical treatments have appeared (e.g. Adam, 1970).

257

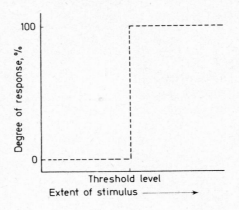

Figure 17.1   A principal presentation of a threshold (all-or-none) response

It is generally agreed that threshold phenomena are characteristics of highly organised biological systems. In addition, it is agreed that the basis of threshold responses is a sharp change of membrane properties, e.g. co-operative transitions of subunits after some external stimulus has reached the threshold level (e.g. Changeux *et al.*, 1967). In the nerve membrane, for instance, the external stimulus might be a change of the membrane potential, the elementary subunit a proteolipid complex of the membrane, and the mechanism of interaction between the external stimulus and the elementary subunits an ion exchange within a specific site of the subunit (Adam, 1973). A more general model is that the subunits of a matrix ('membrane') which carries the receptor sites for some ligand perform a transition with a high degree of co-operativity after the level of the external ligand has reached the threshold level. This model does not necessarily imply positive co-operativity in the binding of the stimulating ligand. Experimental evidence indicates that Michaelian binding curves of ligands to membrane receptors is common (Levitzki, 1974).

In the phytochrome-mediated threshold response described in this paper, the main features seem to be the following: the binding of the stimulating ligand (Pfr) to the receptor sites (X) of the matrix ('membrane') is non-co-operative (this implies that binding is linear, at least at low receptor occupancy); the response of the matrix occurs within a very narrow concentration range of the stimulating ligand (Pfr); and the full biological signal is achieved at low receptor occupancy (Oelze-Karow and Mohr, 1976). These requirements can be met by and large by recently developed theoretical models (Levitzki, 1974).

The Phytochrome System

The following considerations are based on the phytochrome model presented in *Scheme 17.1*. This model accounts for the major features of the phytochrome system as it occurs in the cotyledons and in the hypocotylar hook of the mustard seedling (*Sinapis alba* L.). The model is the result of a systems analysis approach and concomitant spectrophotometry in the cotyledons and hook of the mustard seedling (Schäfer and Mohr, 1974; Schäfer, Schmidt and Mohr, 1973). The so-called 'dark reversion' of Pfr (Pfr $\xrightarrow{\text{dark}}$ Pr) has been ignored in the construction of the minimum model. The justification for this decision can only be given later in this chapter.

$$Pr' \xrightarrow{^{o}k_{s}} Pr \underset{k_{2}}{\overset{k_{1}}{\rightleftarrows}} Pfr \xrightarrow{k_{d}} Pfr'$$

Scheme 17.1   Simplified model of the phytochrome system in mustard cotyledons and hypocotylar hook. Pr' $\xrightarrow{^{o}k_{s}}$ Pr represents the zero-order de novo synthesis; Pr $\overset{k_{1}}{\underset{k_{2}}{\rightleftarrows}}$ Pfr represents the first-order light reactions; Pfr $\xrightarrow{k_{d}}$ Pfr' represents the first-order destruction process. Only k_{1} and k_{2} depend on light. While the elements of the system and the orders of reaction are the same in cotyledons and hook, the numerical values of the rate constants may differ considerably in the two organs (cf. Figure 17.3). (After Schäfer and Mohr, 1974; Schäfer, Schmidt and Mohr, 1973)

Biophysically, an important feature of the phytochrome system is that the absorption spectra of Pr and Pfr overlap throughout the visible range of the spectrum. This overlap is the reason that photostationary states are characteristic of the status of the phytochrome system in solution as well as in the cell under conditions of saturating irradiations. *Figure 17.2* gives some information on how much of the total phytochrome is present as Pfr if the photostationary state is established in the hypocotylar hook of the mustard seedling by monochromatic light of the indicated wavelengths.[1]

A second important biophysical feature of the phytochrome system in the mustard seedling (relevant to the present topic) is that the phytochrome system shows the characteristics of a steady state. According to the phytochrome model in *Scheme 17.1*, any change of total phytochrome can be described by

$$\frac{d}{dt}[\text{Ptot}] = {}^{o}k_{s} - {}^{1}k_{d}[\text{Pfr}]$$

[1]For convenience, the following phytochrome symbols are used in this chapter: [Ptot], total phytochrome ([Pr] plus [Pfr]); square brackets denote relative concentrations; $\phi_{\lambda} = ([\text{Pfr}]/[\text{Ptot}])_{\lambda}$ = photostationary state of phytochrome at wavelength λ.

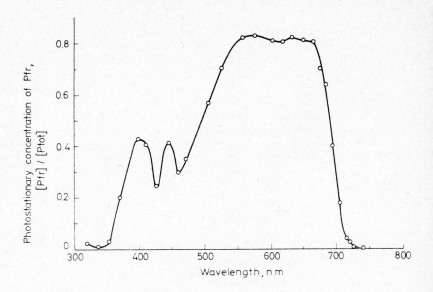

Figure 17.2 The fraction of [Pfr] at photoequilibrium (photostationary state) as a function of wavelength (in vivo measurements at 25°C with hypocotylar hooks of mustard seedlings, Sinapis alba *L.) The data were obtained by K.M. Hartmann and C.J.P. Spruit. (From Hanke, Hartmann and Mohr, 1969)*

In the steady state, i.e. when $\dfrac{d}{dt}[\text{Ptot}] = 0$,

$$^{0}k_{s} = {}^{1}k_{d}[\text{Pfr}] = {}^{1}k_{d} \cdot \phi_{\lambda}[\text{Ptot}]$$

Hence, in the steady state, [Pfr] is a function only of the synthesis of Pr and the rate of destruction of Pfr. [Pfr] steady state does not depend on λ or on ϕ_{λ} (the level of total phytochrome, [Ptot] steady state, does depend on ϕ_{λ} in addition to $^{0}k_{s}$ and $^{1}k_{d}$).

This brief treatment shows that any type of light (except when ϕ_{λ} is too small, cf. Schäfer, 1975) is capable of establishing a steady state of the phytochrome system. The question is only how long the transient phase will be before the steady state is actually established after the onset of light. We have found empirically that 'standard far-red light', which is equivalent (as far as the photostationary state of phytochrome is concerned) to the wavelength 718 nm, will rapidly establish a photo steady state of the phytochrome system, [Ptot] = constant, in the cotyledons and hypocotylar hook of the mustard seedling, if the onset of light is 36 h after sowing (*Figure 17.3*) (Oelze-Karow and Mohr, 1973).

Since the $^{0}k_{s}$ values differ by a factor of 3.2 in the two organs (cotyledons and hypocotylar hook), the ϕ_{λ} values must differ by the

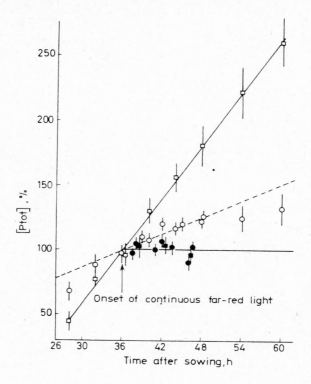

Figure 17.3 Time courses of change of total phytochrome ([Ptot] = [Pr] + [Pfr]) in the cotyledons and in the hypocotylar hook of the mustard seedling in the dark and under continuous far-red light. The deviations of the measured points from the regression line (hook) below 30 h and above 48 h can be accounted for by changes in the system of reference ('hook'). □, Cotyledons, dark; ■, cotyledons, far red; o, hook, dark; •, hook, far red. (From Schäfer, Schmidt and Mohr, 1973)

same degree ($\phi_{far-red, hook} = 0.023$; $\phi_{far-red, cotyledons} = 0.074$) (Schäfer, Schmidt and Mohr, 1973). The important question of whether or not the rate of *de novo* synthesis of Pr is constant and of zero order irrespective of the level of Ptot has been repeatedly investigated. *Figure 17.4a,b* documents the fact that there is indeed no detectable 'feedback' from [Ptot] to the rate of Pr synthesis. The rate is the same irrespective of the level of Ptot. The phytochrome system *per se (Scheme 17.1)* does not include threshold reactions or any other co-operative reaction. The photochemical transformations are strictly first order, and the destruction of Pfr also follows first-order kinetics:

$$- \frac{d}{dt}[Pfr] = {}^1k_d[Pfr]$$

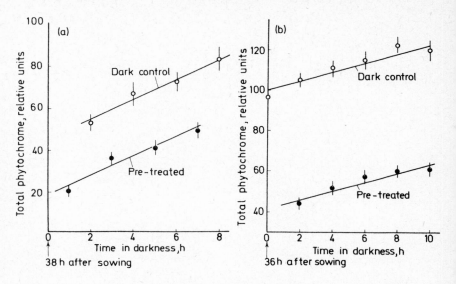

Figure 17.4 (a) Increase in total phytochrome, [Ptot], in mustard seedling cotyledons in the dark. The increase is due to de novo synthesis of Pr (cf. Scheme 17.1). o, Control (no treatment); •, treatment of seedlings with 36 h dark, 1 h red, 5 min far-red. Obviously the rate of synthesis of Pr (^0k_s) does not depend on the amount of Ptot. (From Drumm, Wildermann and Mohr, 1975). (b) Increase in total phytochrome, [P_{tot}], in mustard seedling hypocotylar hook in the dark. The increase is due to de novo synthesis of Pr (cf. Scheme 17.1). o, Control (no treatment); •, treatment with 36 h dark, 5 min red, 55 min dark, 5 min red, 55 min dark, 5 min far-red. (From Oelze-Karow, Schäfer and Mohr, 1976)

at least in the well investigated dicotyledonous seedlings. This finding implies that once a particular Pfr molecule has been formed, its chance of being destroyed compared with the other molecules of the total Pfr population is random. We shall see later, however, that although degradation of Pfr is always a first-order process the type of binding to a particular receptor site (X, X′) decides in some instances whether or not the Pfr molecules are available at all to the degradative 'agent'.

A Phytochrome–receptor Model

All considerations in this chapter about phytochrome–receptor inter-actions will be based on an 'open phytochrome–receptor model' advanced by Schäfer (1975) to explain the 'high irradiance reaction' of photomorphogenesis on the basis of phytochrome (*Scheme 17.2*). The phytochrome receptors (X, X′) are considered to be integral

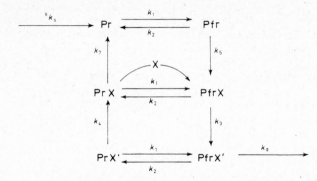

Scheme 17.2 Diagrammatic representation of the open phytochrome-receptor model advanced by Schäfer (1975). The model considers de novo synthesis of Pr as well as destruction of Pfr and includes an interaction of Pfr and Pr with a receptor X or $X^{\prime}$. The model was originally developed to account for the binding properties of phytochrome to an operationally defined 'particulate fraction' isolated from maize coleoptiles or squash hypocotylar hooks (cf. Quail, Marmé and Schäfer, 1973). It was suggested that the receptor X to which phytochrome rapidly binds exists in two forms (X and $X^{\prime}$) and that the transition $X \rightarrow X^{\prime}$ is mediated by Pfr

constituents of matrices (some of them possibly 'membranes'). In the case of phytochrome-mediated threshold responses, the pre-existing 'membrane' is thought to be capable of performing fully reversible conformational transitions with a high degree of co-operativity. Pfr has the properties of a ligand capable of binding to the receptor sites (X, $X^{\prime}$). The model implies that the binding of Pfr to the receptor sites is non-co-operative [linear binding at low receptor occupancy is strongly indicated by physiological data (Drumm and Mohr, 1974)]. However, it further implies that the availability of receptor sites is not limiting for binding even at a photostationary state as high as $\phi_{red} = 0.8$. This assumption is also supported by experimental data (Drumm and Mohr, 1974).

The Experimental System

We have studied the control by phytochrome (Pfr) of the increase of the level (= extractable activity)[1] of the enzyme lipoxygenase in the cotyledons of the mustard seedling (*Sinapis alba* L.). Lipoxygenase

---

[1]The term 'synthesis' is used for convenience to denote any increase in enzyme level although (for experimental reasons) *de novo* synthesis of lipoxygenase has not so far been rigorously demonstrated during the period of our experimentation. The usual inhibitor experiments have led to the conclusion that intact RNA and protein synthesis is required for an increase in the lipoxygenase level (Oelze-Karow *et al.*, 1970). Lipoxygenase 'synthesis' is only used as a tool in the present investigation. The interpretation of the threshold reaction of phytochrome is not influenced by the kind of interpretation of the operationally measured 'increase of lipoxygenase level'

(LOG)[1] is a plant enzyme which catalyses the oxidation of unsaturated fatty acids containing a methylene-interrupted multiple unsaturated system in which the double bonds are all *cis,* such as linoleic, linolenic and arachidonic acids, to the conjugated *cis, trans*-hydroperoxides.[2] Although enzymes grouped under the general definition of lipoxygenases occur widely in the plant kingdom, and some enzymes of this type (e.g. in legume and cereal seeds) have been known and studied for many years, the physiological role of lipoxygenases is not fully understood at present. However, this shortcoming does not hamper the use of LOG in the present studies. LOG is localised in the mustard seedling exclusively in the laminae (95 per cent) and petioles (5 per cent) of the cotyledons. No LOG can be extracted either from the hook region of the seedling or from the lower parts of the hypocotyl and the tap root (cf. *Figure 17.5*). By definition (*see* Oelze-Karow and Mohr, 1973), the total phytochrome level, [Ptot], at 36 h after sowing in complete darkness at 25 °C is equal to unity (or 100 per cent) in the cotyledons as well as in the hypocotylar hook of the mustard seedling.

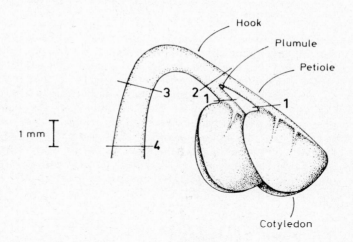

Figure 17.5 Upper parts of the mustard seedling 36 h after sowing (25 °C). Points of excision are indicated by the numbers (1 to 4). The experiment which involves sectioning of the seedling is described in the section 'Inter-organ correlation in the LOG Response'). (From Oelze-Karow and Mohr, 1974)

[1]LOG = lipoxygenase (E.C. 1.13.11.12)
[2]The enzyme assay and the enzymological results obtained in the development of the enzyme assay were described in detail previously. As in the original paper, LOG activity will be expressed in relative units in this chapter. For each single experiment a control experiment was performed (at time zero = 36 h after sowing, or under continuous far-red light), the result of which was taken as the point of reference (100 per cent) (for details and justification of the procedure, *see* Oelze-Karow and Mohr, 1973)

In this chapter we shall consider only the threshold control by Pfr of LOG 'synthesis', although other developmental processes are also known to be controlled by phytochrome through an all-or-none mechanism, e.g. hypocotyl lengthening (Schopfer and Oelze-Karow, 1971). It is hoped that the insight obtained regarding the LOG response will facilitate further research on the threshold control by Pfr of growth phenomena.

The Basic Observations

Figure 17.6 shows that the increase in LOG activity ('synthesis') in the mustard cotyledons is arrested immediately after the onset of standard far-red light, i.e. after the formation of a relatively low ($\phi_{\text{far-red}} = 0.023$ in the hypocotylar hook[1]) but approximately stationary concentration of the effector molecule Pfr in the seedling (cf. *Figure 17.3*). The inhibition can be maintained over at least 11 h. *Figure 17.6* shows in addition the kinetics of LOG activity after a saturating red or far-red light pulse given at time zero ($\phi_{\text{red}} = 0.8$; $\phi_{\text{far-red}} = 0.023$). If 80 per cent of [Ptot] is transferred to Pfr by a 5-min red light pulse given to a dark-grown mustard seedling at time zero (= 36 h after sowing), a delay period of 4.5 h is found in darkness before enzyme synthesis is resumed. A corresponding far-red light pulse which gives 2.3 per cent Pfr at time zero leads to a 40-min delay (by delay we define the duration of time between the end of irradiation and the resumption of activity increase). The delay period is also about 40 min following exposure to far-red light for 8 h before returning to darkness. This is not surprising since 2.3 per cent Pfr ($\phi_{\text{far-red}} = 0.023$) remains in the mustard seedling when it is transferred to darkness after 8 h of continuous far-red light. We recall that [Ptot] remains approximately constant in the mustard cotyledons and hypocotylar hook for at least 11 h after the onset of continuous far-red light (cf. *Figure 17.3*).

Figure 17.6 also shows that the operational criteria for the involvement of phytochrome in this response are fulfilled. If a red light pulse at time zero is followed immediately with a far-red light pulse, the result is that the red effect is completely reversed. The two curves, 5 min far-red → dark and 5 min red + 5 min far-red → dark, are virtually identical.

The results suggest a threshold mechanism for the action of Pfr in suppressing the increase of LOG activity. Further, the precision of the response indicates that the cells of the mustard cotyledons which produce the enzyme form a fully synchronised cell population within the seedling.

[1]The reason why the ϕ_{λ} values characteristic of the hypocotylar hook (rather than for the cotyledons) are used throughout this chapter can be given only later (*see* section 'Inter-organ Correlation in the LOG Response').

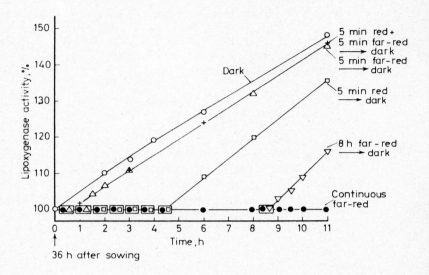

Figure 17.6   Time courses of lipoxygenase synthesis after a light pulse (5 min at time zero) with standard red, far-red or red followed by far-red light.   The time courses for continuous far-red light (onset at 36 h after sowing) and for the programme 8 h far-red light → dark are also given.   The identical delay period before resumption of enzyme synthesis in the cases 5 min far-red → dark and 8 h far-red → dark support the notion that a photo steady state of the phytochrome system is maintained under standard far-red light (cf. Figure 17.3).   (After Oelze-Karow and Mohr, 1973; Oelze-Karow, Schopfer and Mohr, 1970)

On the basis of *Figure 17.6*, one can calculate the level of Pfr that remains at the end of the delay period.   Using the constants justified previously, namely first-order destruction of Pfr, half-life of Pfr = 45 min, ϕ_{red} = 0.8 and $\phi_{far\text{-}red}$ = 0.023, it was found (Oelze-Karow and Mohr, 1973) that the level of Pfr at the point of resumption of LOG synthesis is the same in both cases (i.e. after a red as well as after a far-red light pulse), namely 1.25 per cent Pfr based on [Ptot] at time zero = 100 per cent.

A Threshold (All-or-none) Mechanism

Scheme 17.3 summarises a major result of more than 6 years of experimental work on the control of LOG synthesis by phytochrome (Oelze-Karow and Mohr, 1976; Oelze-Karow, Schopfer and Mohr, 1970). A threshold (all-or-none) mechanism for the action of Pfr was postulated.   The data we have obtained over the years are in accordance

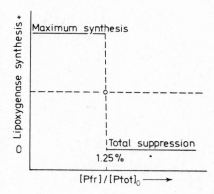

Scheme 17.3 Scheme to describe the concept of a threshold regulation of lipoxygenase synthesis in the mustard cotyledons by Pfr. [Ptot]₀ is the total phytochrome at time zero (36 h after sowing); this value is a constant. The amount of Pfr, [Pfr], is always expressed as a fraction or percentage of [Ptot]₀. Expressed in this way, the threshold level, [Pfr]ₜₕ, is approximately 1.25 per cent (0.0125). (From Mohr, 1972; Mohr and Oelze-Karow, 1973)

with the photometrically determined half-life of 45 min for Pfr between 36 and 48 h after sowing (Schäfer, Schmidt and Mohr, 1973), provided that a level of 1.25 per cent Pfr, based on [Ptot] at time zero (36 h after sowing) = 100 per cent, is a threshold level for LOG synthesis. If the actual level of Pfr exceeds the threshold level, the increase in LOG activity is immediately and completely arrested. If the actual level of Pfr decreases below the threshold level, the increase in LOG activity is immediately resumed at full speed.

The conspicuous symmetry of the threshold model has been justified by several experimental approaches, one of which will be described briefly (*Figure 17.7*). As soon as the threshold level of Pfr (1.25 per cent, based on [Ptot] at time zero = 100 per cent) is exceeded (onset of red light at time zero), the LOG synthesis is suppressed. The corresponding release from suppression occurs with the same rapidity. After 90 min of continuous red light, the relative level of Ptot is so low (approximately 32 per cent based on [Ptot] at time zero = 100 per cent; Schäfer, Schmidt and Mohr, 1973) that a shift from red to far-red light will lead to a Pfr level of 0.74 per cent, which is far below the threshold level (1.25 per cent). As a result, LOG synthesis is resumed immediately after the shift from red to far-red light. If the seedlings are transferred back to red light after 3.5 h of continuous far-red light, LOG synthesis is again suppressed since the Pfr level is far

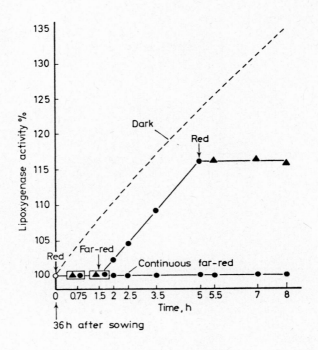

Figure 17.7   Time courses of lipoxygenase synthesis in the intact mustard seed-
ling in darkness (---; cf. Figure 17.6*), under continuous far-red light (●) and*
under the irradiation sequence 1.5 h red light (▲), 3.5 h far-red light (●), 3 h
red light (▲).   (After Oelze-Karow and Mohr, 1974)

above the threshold level under red light.   *Figure 17.7* shows that the
threshold control occurs repeatedly and without a detectable time lag
in both directions (suppression and release from suppression).

The sharp threshold (*Scheme 17.3*) can be understood only if one
assumes that somewhere in the causal chain between Pfr and the on-off
response a reaction step is strongly co-operative.   We have interpreted
the strong co-operativity implied by the steepness of the threshold in
terms of co-operativity of the 'primary reaction matrix' (Mohr and
Oelze-Karow, 1973).   According to this concept, Pfr would be analogous
to a ligand and the primary reactant (or receptor) of Pfr (designated as
X) would be an integral constituent of a pre-existing matrix (possibly a
'membrane'), capable of performing fully reversible conformational transitions with a high degree of co-operativity.

However, a different kind of experimental approach was required to
confirm (or even prove) the concept that the 'primary reaction of Pfr'

is indeed the site where the threshold reaction occurs. A successful experimental approach became feasible when we tried to understand some sets of data that had troubled us since 1969 (Oelze-Karow, Schopfer and Mohr, 1970).

Troublesome Data

In experiments in which the threshold level was slowly approached from below, i.e. with a small k_1 (cf. *Scheme 17.1*), it was found that the law of reciprocity is valid over a wide range of irradiances with red or far-red light (*Figures 17.8* and *17.9; Table 17.1*). The question arose of why reciprocity holds even under conditions of irradiation where the photochemical rate constant k_1 is very low compared with the light-independent rate constant of Pfr destruction (remember that half-life of Pfr is 45 min in cotyledons and hypocotylar hook under our experimental conditions).

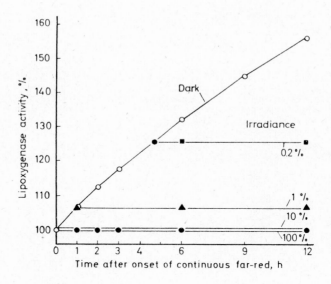

Figure 17.8 Test for reciprocity with continuous far-red light in synthesis of lipoxygenase. Standard irradiance (3.5 W m^{-2}) = 100 per cent. Onset of light at 36 h after sowing. The results indicate that irrespective of irradiance a given number of quanta (einsteins) per square metre is required to reach the threshold level of Pfr. (After Oelze-Karow, Schopfer and Mohr, 1970)

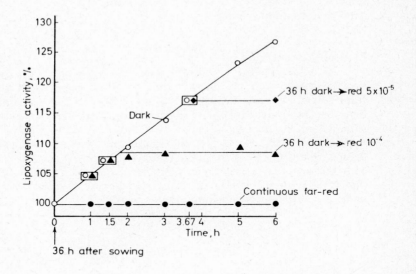

Figure 17.9   Test for reciprocity with continuous red light ('red') in synthesis of lipoxygenase.   Standard irradiance (0.675 W m^{-2}) = unity.   Onset of light at 36 h after sowing.   The results indicate that irrespective of irradiance a given number of quanta (einsteins) per square metre is required to reach the threshold level of Pfr.   (After Oelze-Karow and Mohr, 1976)

Table 17.1   Results from *Figure 17.8* rearranged so as to show directly the validity of the law of reciprocity (after Oelze-Karow, Schopfer and Mohr, 1970)

Irradiation time, t(s)	Irradiance,[1] I (pE cm^{-2} s^{-1})	I.t (pE cm^{-2})
36	2 180	78 500
360	218	78 500
3 360	22	73 900
16 800	4.5	75 600

[1] Light source: standard far-red source (Mohr, 1966)

There are two hypotheses to explain *Figures 17.8* and *17.9:*

(1) The Pfr molecules originating from the photochemical transformation are somehow 'counted' before they undergo destruction, and this information is stored until the threshold level (1.25 per cent, based on [Ptot] at time zero = 100 per cent) is reached.

(2) The Pfr molecules are not subject to destruction as long as the amount of Pfr remains below the threshold level.

Since the small level of 1.25 per cent Pfr cannot be measured spectrophotometrically with sufficient accuracy, alternatives (1) and (2) were checked by physiological means. In addition, the idea had to be considered that only a fraction of total phytochrome is involved in the control of LOG synthesis. This implies that only a fraction of the total Pfr reacts with the Pfr receptors at the particular 'matrix' which is connected to LOG synthesis. It was found previously (e.g. Drumm and Mohr, 1974) that in the case of phytochrome-mediated anthocyanin synthesis in the mustard seedling no threshold reaction is involved, and the conclusion was drawn that there must exist within a plant several or even many response-specific 'matrices' to which Pfr can bind.

Experiments to Decide Between Hypotheses (1) and (2)

One of the original experiments is illustrated in *Figure 17.10*. Red light is applied for 1.5 h at 10^{-4} standard irradiance, which yields close to 1 per cent Pfr. At this point, a saturating pulse (5 min) with 746–nm light (ϕ_{746} = 0.25 per cent) is applied and immediately afterwards irradiation with 10^{-4} standard red light is continued.

If the 'counting mechanism' [hypothesis (1)] is correct, the threshold level must be reached after a further 20 min and suppression of LOG synthesis would begin at this point. If, however, hypothesis (2) is correct (no destruction of Pfr below the threshold value), approximately 90 min in 10^{-4} standard red light will be required after the 746–nm pulse to increase the Pfr level from 0.25 to 1.25 per cent. The experimental result (*Figure 17.10*) is consistent only with hypothesis (2). We conclude that the Pfr molecules do not undergo destruction as long as the level of Pfr is below the threshold level.

Does destruction of Pfr cease if the threshold level is reached from *above*? An answer to this question is given by *Figure 17.11*. At 36 h after sowing (= time zero), a saturating far-red light pulse (5 min) is applied. The mustard seedling is then left in darkness for 2 h 55 min and, at 3 h, irradiation with continuous 10^{-4} standard red light is commenced. The rationale behind this treatment is as follows. The far-red pulse establishes a photostationary state with 2.3 per cent Pfr ($\phi_{far-red}$ = 0.023). It requires a period of 40 min to lower the Pfr level by first-order destruction with a half-life of 45 min from 2.3 per cent to the threshold level (1.25 per cent). At this point (45 min after time zero), suppression of LOG synthesis is abolished, and LOG synthesis is resumed at the full rate (cf. *Figure 17.6*). If Pfr destruction (at the 'matrix' relevant for the LOG response) continues after the threshold level is reached from above, one can predict that at 3 h the Pfr level will be close to 0.016 per cent. To reach the threshold level (1.25 per cent Pfr) again from this low level, approximately 1 h

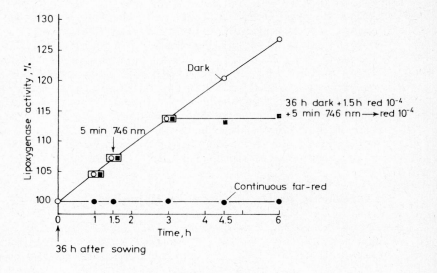

Figure 17.10   Test experiment to decide between the alternatives 'counting mechanism' or 'no destruction of Pfr below the threshold level'.   The rationale behind the experiment is described in the text.   Irradiance of red light: 0.675 W m⁻².10⁻⁴.   (After Oelze-Karow and Mohr, 1976)

35 min in continuous red light at 10^{-4} standard irradiance would be required.   However, if there is no significant destruction of Pfr below the threshold level, the red light (onset at 3 h) will immediately suppress LOG synthesis.   The data in *Figure 17.11* are consistent only with the concept that there is no Pfr destruction below the threshold level.

If a saturating pulse with 746-nm light is applied immediately prior to the resumption of the red light, approximately 90 min of red light at 10^{-4} standard irradiance is required to reach the threshold level (*Figure 17.11*).   This is the irradiation time required to lift the Pfr level from 0.25 per cent (ϕ_{746} = 0.25 per cent) to the threshold level (1.25 per cent).

In conclusion, all data available (Oelze-Karow and Mohr, 1976) support hypothesis (2), which states that the Pfr molecules involved in the LOG response do not undergo destruction if the level of Pfr is below the threshold level.   It does not matter whether or not destruction of Pfr has already occurred in the system, the only decisive factor is whether the actual level of Pfr is above or below 1.25 per cent (based on [Ptot] at time zero = 100 per cent).

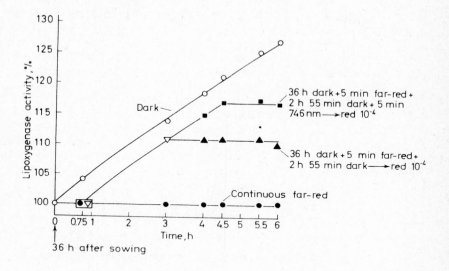

Figure 17.11   Test experiment to answer the question of whether or not the destruction of Pfr ceases if the threshold level is reached from above.   The rationale behind the experiment is described in the text.   Irradiance of red light: 0.675 W m⁻². 10⁻⁴.   (After Oelze-Karow and Mohr, 1976)

The LOG Response and the 'Open Phytochrome-receptor Model' (*SCHEME 17.2*)

The data referred to in the previous section show that the threshold reaction in the case of the LOG response is indeed located at the site where binding and destruction of Pfr take place.   The 'open phytochrome–receptor model' advanced by Schäfer (1975) to explain the 'high irradiance reaction' of photomorphogenesis on the basis of phytochrome (*Scheme 17.2*) offers a convenient framework for the data obtained with phytochrome and LOG.   We emphasise only three features of the model: any Pfr molecule formed from Pr is rapidly bound to the receptor X (the availability of X is not limiting for binding); the transition PfrX → PfrX′ is a rapid process; and Pfr destruction starts from PfrX′.

The LOG threshold response fits into this model as follows.   As long as the level of Pfr is below the threshold level, the Pfr remains bound to X but the transition PfrX → PfrX′ does not occur.   Only

after the threshold level has been reached does the transition
PfrX → PfrX$'$ become possible, and destruction comes into play.
If the level of Pfr falls below the threshold level, destruction stops
since the remaining PfrX$'$ is rapidly returned to PfrX. As long as
the actual level of Pfr is below the threshold level, only the form X
of the phytochrome receptor is stable; if the actual level of Pfr is
above the threshold level, only the form X$'$ of the phytochrome
receptor is stable. Since Pfr destruction starts from PfrX$'$, there is
no destruction of Pfr as long as the PfrX → PfrX$'$ transition is
blocked, i.e. as long as [Pfr] is below the threshold level. In brief,
the destruction process itself is a threshold response.

The model implies that the threshold reactions in connection with
Pfr destruction and with respect to the LOG response reflect the same
event (PfrX ⇌ PfrX$'$).   Suppression of LOG synthesis depends on
the presence of PfrX$'$.   However, PfrX$'$ is stable only above the
threshold level of [Pfr]. Hence the simplest formulation of the actual
threshold reaction in the LOG response is

$$PfrX \qquad \rightleftharpoons \qquad PfrX'$$

State at [Pfr] < 1.25%	State at [Pfr] > 1.25%
No Pfr destruction	Pfr destruction (^1k_d)
LOG synthesis unimpaired	LOG synthesis suppressed

The 'primary reaction of Pfr' in the case of the LOG response must
be written as

$$Pfr + X \rightleftharpoons PfrX \rightleftharpoons PfrX'$$

The reversible threshold reaction is thus an integral part of the 'primary'
reaction' occurring at the 'matrix' specific for the LOG response.

Schäfer's model (*Scheme 17.2*) does not consider the aspect that the
receptor sites X may be integral constituents of different matrices
which differ, e.g. in the capability of performing fully reversible conformational transitions with a high degree of co-operativity. As pointed
out previously (Drumm and Mohr, 1974), there is no escape from the
conclusion that Pfr-mediated control of anthocyanin synthesis in the
mustard seedling (no threshold, no co-operativity) and Pfr-mediated
control of LOG synthesis differ in principle even at the level of the
'primary reaction'. In the case of anthocyanin induction by phytochrome, the 'primary reaction' does not involve any threshold reaction.
The transfer of the signal from Pfr to the response is linear. It seems
that there is no common matrix of the Pfr + X → PfrX → PfrX$'$
reaction which could account for both phenomena.

As far as *Scheme 17.2* is concerned, the data obtained with the LOG
response are fully consistent with Schäfer's model since this model
deals only with phytochrome *per se* and with the receptor site(s).
Phytochrome action, in particular the problem of the 'multiple action',
has not been considered in connection with *Scheme 17.2,* because
Schäfer's model was derived from binding data (involving phytochrome

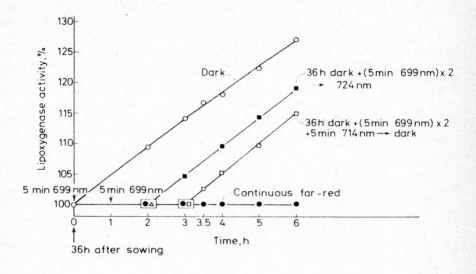

Figure 17.12   Test experiments to check the hypothesis that there is no signifi-cant (rapid) dark reversion of Pfr even though the photostationary state ϕ_λ established by a saturating light pulse is below 0.28. Spectrophotometric data (Schmidt and Schäfer, 1974) had led to the concept that at least most of the Pfr in the mustard seedling undergoes dark reversion rather than destruction if an initial photostationary state below approximately 0.28 is established. The following photostationary states were used: $\phi_{699} = 0.27$; $\phi_{714} = 0.039$; $\phi_{724} = 0.013$ (cf. Figure 17.13). The rationale behind the experiment with 724 nm is as follows. If the two light pulses at 699 nm (applied at time zero and 1 h) lead to 'normal' Pfr destruction and thus to a significant decrease in [Ptot], one would expect that continuous 724-nm light could no longer suppress LOG synthesis. Without the 699-nm pre-treatment, continuous 724-nm light suppresses LOG synthesis during the whole experimental period. (After Oelze-Karow, Schäfer and Mohr, 1976)

and some operationally defined 'particulate fraction' from maize coleop-tiles and pumpkin hypocotylar hooks) (Quail, Marmé and Schäfer, 1973) rather than from physiological data.

Schäfer's model (Scheme 17.2) does not contain any site from which the previously postulated rapid Pfr dark reversion (Pfr $\xrightarrow{\text{dark}}$ Pr) (Marmé, Marchal and Schäfer, 1971; Schmidt and Schäfer, 1974) could occur. This conspicuous feature is also consistent with our findings. PfrX as a potential site of rapid dark reversion is excluded by the data on reciprocity (Figures 17.8 and 17.9); PfrX' is excluded by many previous data (spectrophotometric and physiological) which clearly require that dark reversion and destruction do *not* compete for the same Pfr pool (e.g. Oelze-Karow and Mohr, 1973; Schäfer, Schmidt and Mohr, 1973). Moreover, recent experimental data are consistent only with the assumption that dark reversion (Pfr $\xrightarrow{\text{dark}}$ Pr) does not come into play in connection with the LOG response (Oelze-Karow, Schäfer and Mohr, 1976). One argument will be presented briefly (*Figure 17.12*). If we irradiate with a 714‒nm light pulse at time zero, we know that the delay (Δt) before LOG synthesis is resumed is 1 h 18 min (cf. *Figure 17.13*). Assuming that Pfr will disappear by irreversible destruction only, one can predict that two consecutive pulses with 699‒nm light (applied at time zero and at 1 h) will decrease the level of Ptot within 2 h by approximately 26 per cent. Instead of 3.9 per cent Pfr, a 714‒nm light pulse applied at 2 h would then form only 2.9 per cent Pfr, based on [Ptot] at time zero = 100 per cent. This would have the consequence that the delay (Δt) before LOG synthesis is resumed after the 714‒nm light pulse applied at 2 h is shortened by approximately 20 min. According to *Figure 17.12,* this is precisely the case.

We suggest (Oelze-Karow, Schäfer and Mohr, 1976) that the notion of a 'rapid dark reversion of Pfr' in dicotyledonous seedlings should be re-examined from the point of view that possibly the spectrophotometric signals on which this notion was based deserve a new interpretation.

Inter-organ Correlation in the LOG Response

EXPERIMENTS LEADING TO THE HYPOTHESIS OF AN INTER-ORGAN
SIGNAL TRANSMISSION

The LOG response was used to determine physiologically the photostationary states, ϕ_λ, which are established by different wavelengths in the red and far-red range of the spectrum. The physiological approach is, in principle, as follows (*Figure 17.13*). At time zero, a brief saturating light pulse is applied to the mustard seedling with different wavelengths, and the delay period (Δt) before enzyme synthesis is resumed is measured by extrapolating the kinetic curve for the increase in LOG activity to the base-line. It was found that wavelengths above 727 nm have no detectable suppressive effect on LOG synthesis, so that the threshold level of Pfr cannot be reached with such light. Using the constants justified previously (first-order destruction of Pfr, half-life of Pfr = 45 min and $\phi_{\text{standard red light}}$ = 0.8), a threshold value was calculated which is close to 1.25 per cent Pfr based on [Ptot] at time zero = 100 per cent. On the basis of the threshold value of 1.25 per cent (= 0.0125), the experimental Δt values and the half-life of Pfr (45 min), the initial photostationary state (ϕ_λ) can be extrapolated for every wavelength, including the standard far-red light source, which turns out to be equivalent to 718 nm as far as the phytochrome system in the mustard seedling is concerned (*Figure 17.14*).

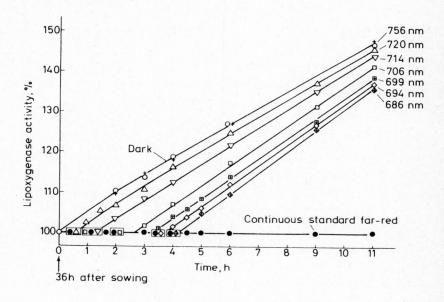

Figure 17.13 Time courses of lipoxygenase synthesis after a single, brief but saturating irradiation (5 min) at time zero with light of different wavelengths (DEPIL interference filters. AL interference filter for 756 nm). The Δt values can be extrapolated with high accuracy. (After Oelze-Karow and Mohr, 1973)

If the ϕ_λ values obtained this way are plotted as a function of wavelength, it becomes obvious (*Figure 17.15*) that they agree with the ϕ_λ values as determined spectrophotometrically by Hartmann and Spruit in mustard seedling hypocotylar hooks (*Figure 17.2*) but not with the data obtained by Schäfer, Schmidt and Mohr (1973) with mustard seedling cotyledons (cf. *Figure 17.3*). On the other hand, we recall that LOG is localised exclusively in the cotyledons and that LOG synthesis takes place in these organs, as shown in experiments with isolated cotyledons. The conclusion was drawn (Oelze-Karow and Mohr, 1973) that the suppression of LOG synthesis in the cotyledons is controlled by phytochrome located in the hypocotylar hook, in other words that an inter-organ signal transfer is involved. This bold but inevitable conclusion has been substantiated by both spectrophotometric (Schäfer, Schmidt and Mohr, 1973) and physiological (Oelze-Karow and Mohr, 1974) evidence.

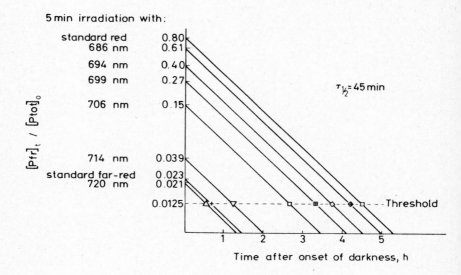

Figure 17.14   Using a threshold of 0.0125 ([Pfr]t/ [Ptot]$_{zero\ time}$), the photo-stationary states established by a 5–min irradiation at time zero can be extrapolated for every wavelength investigated, including standard far-red light.   The [Pfr]/[Ptot] value for standard far-red light (= ϕ_{fr}) is estimated to be 0.023. This confirms the value reported previously (Oelze-Karow, Schopfer and Mohr, 1970).   (After Oelze-Karow and Mohr, 1973)

EXPERIMENTS TO JUSTIFY THE CONCEPT OF A RAPID INTER-ORGAN SIGNAL TRANSMISSION

The experiments included mechanical segmentation of cotyledons and hypocotylar hook (cf. *Figure 17.5*) and the following results were obtained (Oelze-Karow and Mohr, 1974).   The lower parts of the hypocotyl and the radicle (the remaining seedling below excision point 4) have no influence on LOG synthesis during the experimental period; if the excision is performed at point 2, no control of LOG synthesis by continuous far-red light can be detected.   Rather, the increase of LOG activity is precisely the same in light and dark; if more than half of the hook is attached to the cotyledons (excision point 3), the normal control by far-red light of LOG synthesis in the cotyledons is observed.   A major experiment will be discussed more in detail since it documents in an impressive manner the precision and rapidity of the inter-organ signal transfer in the mustard seedling.   We recall from *Figure 17.6* that following a 5–min pulse with standard red light at

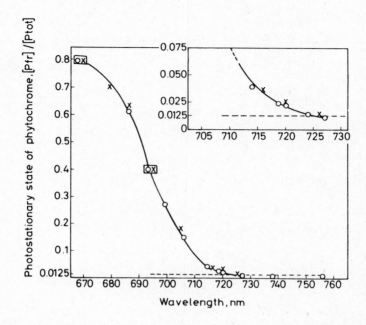

Figure 17.15   Values for the [Pfr]/[Ptot] ratios (φλ values) as obtained by the procedure described in Figure 17.14 *plotted as a function of wavelength (○).   In addition, the [Pfr]/[Ptot] ratios measured spectrophotometrically by Hartmann and Spruit (cf.* Figure 17.2) *above 670 nm are indicated (x).   For the wavelengths 738 and 756 nm, the [Pfr]/[Ptot] ratios are below the threshold level. The actual threshold value (dashed line) is obviously located between the [Pfr]/[Ptot] ratio established by wavelength 727 nm and the [Pfr]/[Ptot] ratio of 0.0130 (established by wavelength 724 nm) (inset).   (From Oelze-Karow and Mohr, 1973)*

time zero, a delay period of 4.5 h must pass before LOG synthesis is resumed.   This effect was used in the following excision experiments (*Figure 17.16*):   5 min of red light was applied at time zero and the excision of the cotyledons at excision point 2 was performed 1.5 h after the red light pulse.   A nearly instantaneous resumption of LOG synthesis was observed in the excised cotyledons.

These results indicate that the signal transmitted from the hypocotylar hook to the cotyledons is not conserved to any significant extent in the cotyledons.   As soon as the cotyledons are separated from the hook, the suppression of LOG synthesis by Pfr is completely lost.   As long as the hook is attached to the cotyledons (excision point 4), the excision procedure has no influence on the increase of LOG activity in the cotyledons (Oelze-Karow and Mohr, 1974).

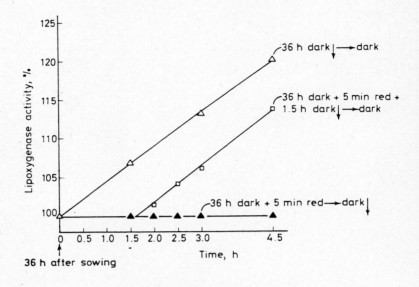

Figure 17.16   Time courses of lipoxygenase synthesis in cotyledons (excision point 1 in Figure *17.5) after a red light pulse (5 min) applied at time zero (= 36 h after sowing).* ▲, *Excision of the cotyledons just before enzyme extraction;* □, *excision of the cotyledons 1.5 h after time zero;* △, *excision of the cotyledons at time zero (no light treatment).   The arrows indicate times of excision.   (From Oelze-Karow and Mohr, 1974)*

The data suggest that transmission of the signal from the hypocotylar hook to the cotyledons is rapid and that the signal cannot be stored in the cotyledons, so that a rapid and precise interorgan signal transmission must occur.   It appears that this type of communication between neighbouring organs cannot be accounted for by hormones. Rather, the signal seems to be transmitted from one organ to the next by means of a biophysical system.

To summarise, the experimental data obtained with the LOG response require a threshold (all-or-none) reaction on the level of the 'primary reaction of Pfr'.   They further require a high degree of synchronism on the cellular and organismal level and a very precise communication system between the hypocotylar hook and the cotyledons.

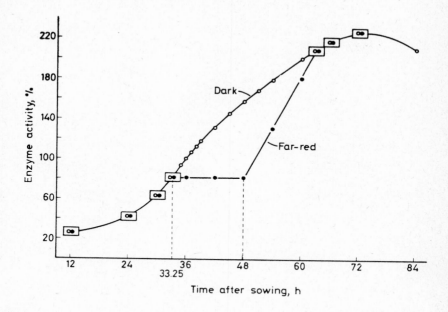

Figure 17.17 Time courses of lipoxygenase synthesis in mustard seedling cotyledons in continuous darkness and under continuous standard far-red light (equivalent to 718–nm light). The onset of far-red light was at the time of sowing of the seeds (zero = time of sowing). o, Dark; •, continuous far-red (onset 0 h). (From Oelze-Karow and Mohr, 1970)

The Time Course of Competence

For the developmental biologist, the problem has been to ascertain whether LOG synthesis responds to Pfr throughout the whole period of development of the mustard seedling, i.e. up to 84 h after sowing at 25 °C. If one irradiates with standard far-red (or red) light from the time of sowing, there is no control of LOG synthesis up to approximately 33 h (*Figure 17.17*). At this point, the full suppression of LOG synthesis by far-red (or red, cf. Mohr, 1972) light (i.e. by a Pfr level above the threshold level) suddenly comes into play, while at 48 h after sowing, the LOG synthesis of the seedling suddenly and completely escapes from control by Pfr. LOG synthesis is resumed even under continuous far-red (or red) light. The kinetics of LOG level after resumption of increase return to the dark kinetics.

Exactly the same temporal pattern was observed in a number of experiments in which different levels of Pfr were used (Oelze-Karow

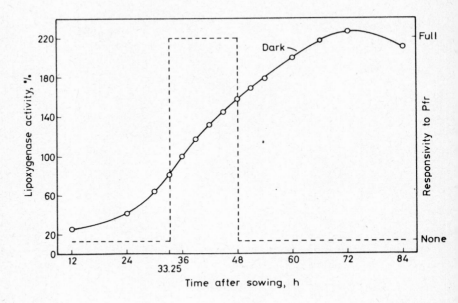

Figure 17.18   Summary of the experimental information regarding the control of lipoxygenase synthesis by light (i.e. by an amount of Pfr above the threshold level). Note that the sensitivity of the system towards an above-threshold level of Pfr changes abruptly in the form of an all-or-none response (dashed line). This is true for the onset (33.25 h) as well as for the termination (48 h) of the period of sensitivity. (After Oelze-Karow and Mohr, 1970)

and Mohr, 1970).   Under all circumstances the LOG-producing system escapes from Pfr control 48 h after sowing (with 25 °C and standard conditions).   It was concluded that the temporal pattern of response must be determined by changes in the system on which Pfr acts rather than by Pfr.   The basic experiment reported in *Figure 17.17* was repeated over several years with seed samples which differed in origin (different seed growers) and in age.   While the on-off pattern of sensitivity (*Figure 17.17*) was always found, the duration of time during which the LOG-producing system is sensitive ('competent') for Pfr differed (e.g. *see Figure 152* in Mohr, 1972).   According to our experience, the duration of time of sensitivity becomes shorter in the course of ageing of the seed material (kept at 4 °C).

It must be emphasised that the onset and duration of the period of sensitivity ('competence') for Pfr is response-specific.   As an example, phytochrome-induced anthocyanin synthesis in the mustard seedling starts 27 h after sowing (B. Steinitz, personal communication) and phytochrome-mediated synthesis of ribulose bisphosphate carboxylase starts only at 42 h after sowing (at 25 °C and with standard conditions) (Brüning, Drumm and Mohr, 1975).

Returning to the original experiments (Oelze-Karow and Mohr, 1970), *Figure 17.18* summarises the experimental information regarding the control of LOG synthesis by continuous far-red (or red) light (i.e. by Pfr at an amount above the threshold level). In the course of development of the seedling, the sensitivity of the LOG-producing system towards Pfr changes abruptly in the manner of an all-or-none response. This is true for the onset (at 33.25 h) as well as for the termination (at 48 h) of the period of sensitivity.

Hence, not only the LOG response *per se* shows the characteristics of an all-or-none response, as the processes of acquiring and losing 'competence' (for Pfr with respect to suppression of LOG synthesis) strictly follow the pattern of all-or-none responses.

The pattern of competence in space and time (the pattern of 'primary differentiation', *see* Mohr, 1972) is *not* influenced by phytochrome. This important statement is supported by many facts, for example it was found that neither the beginning nor the end of the period of sensitivity for Pfr of the LOG-producing system can be influenced by phytochrome (Oelze-Karow and Mohr, 1970).

Conclusion

The LOG response indicates that all-or-none responses and rapid signal transfer also occur in biological systems that do not possess a nervous system. This indicates that threshold responses are *general* biological phenomena, not necessarily connected with the organisation of nerve or muscle fibre and not necessarily correlated with sophisticated electrophysiological phenomena. [Phytochrome-mediated electrical signals could not so far be detected in hypocotylar hooks of bean seedlings, while blue light-mediated electrical signals ('action potentials') can easily be measured with this organ (Hartmann, 1974)] In our opinion, a general approach to explain the mechanism of all-or-none reactions should not be related to the marginal conditions of an axon. In any case, the LOG response shows that threshold responses are not only an indispensable attribute of nerve and sensory physiology, but rather all-or-none responses must also be considered in developmental physiology.

Acknowledgements

This work was supported by grants from the Deutsche Forschungsgemeinschaft (SFB 46). We thank Miss Inge Schneider for 6 years of skilful and competent technical assistance.

References

ADAM, G. (1970).   In *Physical Principles of Biological Membranes,* p.35. Ed. Snell, E., Wolken, J., Iverson, G.J. and Lam, J.   Gordon and Breach, New York

ADAM, G. (1973).   In *Synergetics: Cooperative Processes in Multicomponent Systems,* p.220.   Ed. Haken, H.   Teubner, Stuttgart

BRÜNING, K., DRUMM, H. and MOHR, H. (1975).   *Biochem. Physiol. Pflanz.* **168,** 141

CHANGEUX, J.P., THIÉRY, J., TUNG, Y. and KITTEL, C. (1967).   *Proc. Nat. Acad. Sci. U.S.A.,* **57,** 335

COLE, K.S. (1968).   *Membranes, Ions and Impulses.   A Chapter of Classical Biophysics.*   University of California Press, Berkeley

DRUMM, H. and MOHR, H. (1974).   *Photochem. Photobiol.,* **20,** 151

DRUMM, H., WILDERMANN, A. and MOHR, H. (1975).   *Photochem. Photobiol.,* **21,** 269

HANKE, J., HARTMANN, K.M. and MOHR, H. (1969).   *Planta,* **86,** 235

HARTMANN, E. (1974).   In *Proceedings of the Annual European Symposium on Plant Photomorphogenesis,* Vol.8, p.8.   Ed. De Greef, J.A.   State University Centre, Antwerp

HODGKIN, A.L. (1964).   *The Conduction of the Nervous Impulse,* University Press, Liverpool

LEVITZKI, A. (1974).   *J. Theor. Biol.,* **44,** 367

MARMÉ, D., MARCHAL, B. and SCHÄFER, E. (1971).   *Planta,* **100,** 331

MOHR, H. (1966).   *Z. Pflanzenphysiol.,* **54,** 63

MOHR, H. (1972).   *Lectures on Photomorphogenesis.*   Springer, Berlin, Heidelberg, New York

MOHR, H. and OELZE-KAROW, H. (1973).   *Biol. Unserer Zeit,* **5,** 137

OELZE-KAROW, H. and MOHR, H. (1970).   *Z. Naturforsch.,* **25b,** 1282

OELZE-KAROW, H. and MOHR, H. (1973).   *Photochem. Photobiol.,* **18,** 319

OELZE-KAROW, H. and MOHR, H. (1974).   *Photochem. Photobiol.,* **20,** 127

OELZE-KAROW, H. and MOHR, H. (1976).   *Photochem. Photobiol.,* **23,** 61

OELZE-KAROW, H., SCHÄFER, E. and MOHR, H. (1976).   *Photochem. Photobiol.,* **23,** 55

OELZE-KAROW, H., SCHOPFER, P. and MOHR, H. (1970).   *Proc. Nat. Acad. Sci. U.S.A.,* **65,** 51

QUAIL, P.H., MARMÉ, D. and SCHÄFER, E. (1973).   *Nature New Biol.,* **245,** 189

SCHÄFER, E. (1975).   *J. Math. Biol.,* **2,** 41

SCHÄFER, E. and MOHR, H. (1974).   *J. Math. Biol.,* **1,** 9

SCHÄFER, E., SCHMIDT, W. and MOHR, H. (1973).   *Photochem. Photobiol.,* **18,** 331

SCHMIDT, W. and SCHÄFER, E. (1974).   *Planta,* **116,** 267

SCHOPFER, P. and OELZE-KAROW, H. (1971).   *Planta,* **100,** 167

18

RAPID EFFECTS OF RED LIGHT ON HORMONE LEVELS

P.F. WAREING
A.G. THOMPSON
Department of Botany, University College of Wales, Aberystwyth, U.K.

Gibberellin Effects

A number of rapid effects of irradiation with red light on endogenous hormone levels have been reported in recent years. Following his earlier finding (Köhler, 1965) that exposure of etiolated pea seedlings to continuous low-intensity red light caused a 10-fold increase in endogenous gibberellin levels, Köhler (1966a) showed that intermittent short periods of red light (2.5 min h^{-1}) also increased the gibberellin levels, and that the effects of red could be reversed by far-red light and *vice versa*, indicating phytochrome control. He concluded, however, that effects of phytochrome on gibberellin production have a lag-phase of about 1 day after light exposure. Köhler (1966b) also reported that increased gibberellin levels could be detected in lettuce seed 60 min after irradiation with red light had commenced, but he is reported to have been unable to repeat these results (Black, 1968).

The first unequivocal evidence of rapid increases in gibberellin levels was provided by the work of Reid, Clements and Carr (1968), who showed that when etiolated barley leaves were exposed to 30 min of red light, extractable gibberellins reached a maximum after a further period of 15 min in the dark, but declined rapidly thereafter. In view of the recent demonstrations that the etioplasts are the site of these rapid changes (*see below*), it is of interest to note that the extraction procedure of these workers was to homogenise the tissue after red light or dark treatments, then to centrifuge, discard the pellet and partition the supernatant only against ethyl acetate at two pH values. Their procedure would have meant, therefore, that gibberellins contained in the plastids were probably not completely extracted and that the observed gibberellin changes related only to the cytoplasmic fraction.

The effects of inhibitors of gibberellin biosynthesis were tested by infiltrating the leaves with CCC or Amo 1618 before exposure to red light or continuous dark. It was found that both substances strongly inhibited the red light-promoted increases in gibberellins. In further experiments (Reid and Clements, 1968), the studies were extended to the effects of inhibitors of nucleic acid and protein synthesis, including actinomycin D (an inhibitor of DNA transcription), chloramphenicol (an inhibitor of protein synthesis in chloroplasts) and cycloheximide (an inhibitor of DNA and RNA synthesis). All of these substances strongly inhibited the increase in gibberellin following exposure to red light and the authors concluded that RNA and protein synthesis are prerequisites of the red light-induced gibberellin production. These results suggested that the increase in gibberellin levels is dependent not only on gibberellin biosynthesis, but also upon enzyme synthesis.

Rapid increases in extractable gibberellins following red light treatment were also demonstrated by Beevers *et al.* (1970) for etiolated wheat leaves, where it was found that large increases in extractable gibberellins could be detected after only 5 min of red light and 10 min of darkness. These studies were extended by Loveys and Wareing (1971a,b) whose extraction procedure involved the freezing and homogenisation of the tissue, followed by extraction in 80 per cent methanol a procedure which might be expected to extract both cytoplasmic and chloroplast gibberellins, although Stoddart (1968) found that after extraction of chloroplasts with methanol, the residue still contained considerable amounts of gibberellin which could be released by ultrasonication. Loveys and Wareing (1971a) partitioned the extracts into acidic and neutral ethyl acetate-soluble fractions, followed by separation by thin-layer chromatography (TLC). It was found that exposure to red light resulted in an increase in the level of a relatively non-polar gibberellin in the acidic fraction and a concomitant decrease in the level of activity in more polar material in the neutral fraction. Thus, the red light appeared to be promoting the release of a non-polar gibberellin from a conjugated or 'bound' form. This conclusion was further supported by the finding that treatment of the gibberellin peak in the neutral fraction with pronase led to an increase in the more non-polar zone, but whether this latter component was identical with that promoted by red light was not shown. In further experiments, etiolated wheat leaves were infiltrated with Amo 1618 before red light treatment and extraction. Amo 1618 did not prevent the usual decrease in the polar fraction and increase in the non-polar fraction following 5 min red light and 10 min dark, although the changes were not as great as in the water controls. Similar effects were observed when the light treatments used by Reid and co-workers (30 min red light and 10 min dark) were applied, but pre-treatment with Amo 1618 appeared to cause some reduction in gibberellin activity. The results of experiments with actinomycin D were difficult to interpret, since this inhibitor had the unexpected effect of increasing the levels of the non-polar gibberellin in the dark controls.

Pre-treatment with abscisic acid (ABA), on the other hand, completely inhibited any increase in gibberellins in response to red light (Beevers *et al.*, 1970; Loveys, 1970). The results of these experiments of Loveys and Wareing, in which the treatment periods (5 min red light + 10 min dark) were shorter than those used by Reid, Clements and Carr (1968), seemed to suggest that the effects of red light involved conversion of a polar into a non-polar gibberellin, rather than gibberellin biosynthesis, but it is possible that biosynthesis is involved when longer periods of treatment are used.

More recently, Reid *et al.* (1972) made the interesting discovery that the red light-promoted gibberellin changes could be obtained in crude homogenates of etiolated barley leaves. Moreover, when $[^3H]GA_9$ was added to the homogenate, its metabolism into at least two other gibberellins was promoted by red light. Further studies on leaf homogenates were carried out simultaneously by Cook and Saunders at Aberystwyth and Evans and Smith at Sutton Bonington. Cooke and Saunders (1975) prepared an etioplast fraction from etiolated wheat leaves and showed that exposure of this fraction to red light resulted in a marked increase in two gibberellins separated by TLC and assayed by the lettuce hypocotyl test. Evans and Smith (1975) likewise isolated etioplast preparations from etiolated barley leaves and showed that the gibberellin activity extractable from the preparation by 75 per cent methanol is increased 3-fold by exposure to 5 min of red light. The effect of red is reversible by far-red light. Ultrasonication of etioplasts maintained continuously in the dark resulted in a gibberellin activity three times that which could be extracted from preparations maintained throughout in the dark. Evans and Smith (1975) interpret these results to indicate that exposure to red light causes a change in permeability of the etioplast membranes, permitting the movement of gibberellins from inside the etioplasts to the ambient medium. The same effect is apparently achieved by ultrasonication of etioplasts maintained continuously in the dark.

These are clearly very interesting and significant results, but before the hypothesis can be regarded as established, the following further questions need to be answered:

(1) Is the gibberellin fraction released by ultrasonication identical with that which increases in response to red light?
(2) Why is the observed increase in gibberellins so transient? Presumably the rapid decline in gibberellins after exposure to red light is due to inactivation or breakdown, but the high rate of metabolism is, at first sight, surprising.
(3) Why does pre-treatment of etiolated wheat leaves with ABA completely inhibit the *in vivo* increase in gibberellins in response to red light? It is possible that ABA treatment inhibits the increased permeability of the etioplast membranes in response to red light since it is known to affect the phytochrome-controlled adhesion of root tips to glass (Tanada, 1973a,b), but other interpretations are possible.

Rapid effects of red light on gibberellin levels have also been reported for other plant material. Kopcewicz and Porazinski (1973) showed that increased levels of gibberellins could be detected after 10–15 min of red light treatment in extracts of etiolated seedlings of *Pinus sylvestris.* Smolenska and Lewak (1971) have shown that repeated exposure (six 30-min irradiations at 15-min intervals) of apple seeds to red light leads to a 3-fold increase in the levels of GA_4. The effect of red is reversed by far-red light. Pre-treatment of the seeds with Amo 1618, which inhibits their germination, does not affect the red light-promoted increase in GA_4.

Seeds of certain provenances of *Picea sitchensis* are light-requiring and, if such seeds are exposed to red light for 30 min and then immediately extracted, it is found that there is a 2-fold increase in the levels of extractable gibberellins (Taylor and Wareing, unpublished work).

So far, we have considered only reports of rapid increases in endogenous gibberellin levels in etiolated tissues and in seeds. Similar effects appear to occur also in green leaves. Thus, Railton and Wareing (1973) studied changes in endogenous gibberellin levels in leaves of the potato species *Solanum andigena,* in which tuberisation is promoted by short days. Levels of gibberellins in the leaves are higher under long days than under short days. If a long night is interrupted by 30 min of red light, a significant increase in gibberellin levels can be detected 20 min after the end of the red irradiation.

Cytokinin Effects

Rapid effects upon endogenous cytokinin levels, paralleling these effects on gibberellin levels, have been reported for several systems in recent years. Van Staden and Wareing (1972a) found that if the light-requiring seeds of *Rumex obtusifolius* are exposed to 10 min of red light and then immediately frozen and extracted, large increases could be demonstrated both in the butanol-soluble fraction, which includes the free bases and ribosides, and the aqueous fraction, which contains the cytokinin ribotides. The effects of 10 min of red light were completely reversed by 20 min of far-red light, clearly indicating phytochrome control. As with the gibberellin changes in barley and wheat leaves, the levels of extractable cytokinins rapidly decline during a further period of darkness after exposure to red light.

Barzilai and Meyer (1964) had earlier reported that extractable cytokinin levels increase in lettuce seed 24 h after exposure to light. The changes in lettuce seed have been further studied by Van Staden (1973), who found that although the total cytokinin activity was not markedly affected by exposure to red light, the levels of the butanol-soluble cytokinins increase while those in the aqueous fraction decline in response to 20 min of red light. The effect of red light is not completely reversed by far-red light. It was suggested that exposure to red light promotes the conversion of nucleotide cytokinins into the nucleosides.

Van Staden, Olatoye and Hall (1973) studied the cytokinin changes in the seed of *Spergula arvensis,* in relation to light and exogenous ethylene. This seed shows little germination in either light or dark in the absence of ethylene but in its presence germination is greatly increased in the light. Seeds kept in light or dark in the presence or absence of ethylene were extracted 48 h after sowing, at which time there was little germination. There was some increase in the butanol-soluble cytokinins in the light and in the presence of ethylene, but the effects were much greater in the aqueous fraction. These effects were observed after 48 h of continuous illumination and it is not possible to say how rapidly they occur in response to light, but they are of interest in showing, once again, the effects of light on cytokinin levels.

Rapid effects of red light on cytokinin levels have also been observed in green leaves of poplar (*Populus robusta*) (Hewett and Wareing, 1973). Detached leaves were collected from the field during the night (at 03.30 h) and exposed to 1, 5, 10 or 30 min of red light. They were then immediately frozen, extracted and assayed for cytokinin activity. A maximum level of butanol-soluble cytokinins was obtained after 5 min of red light treatment. If leaves were exposed to 5 min of red light at the end of the main dark period and then maintained in darkness for a further period, cytokinin levels reached a maximum after 30 min of darkness. If given 5 min red light at the beginning of the dark period, cytokinin levels were continuing to rise 60 min after exposure to red light. As in the other systems we have considered, the increases in hormone levels are transient and there is a rapid decline during a further period of darkness following exposure to red light.

Poplar leaves collected during the main photoperiod contain a complex of several cytokinin-active fractions but these decline to low levels during darkness. Leaves exposed to 5 min of red light and harvested after 10 min of darkness were found to contain a single main peak of cytokinin activity, which has been identified as 6-(*o*-hydroxybenzyl)-adenosine, i.e. as the riboside of the base (*o*-hydroxybenzyl)adenine (Horgan *et al.,* 1973), which is an active cytokinin in several tests. This was the first report of a natural cytokinin that has an aromatic side-chain, instead of the isoprene side-chain characteristic of zeatin and its derivatives.

It has been possible to apply the single ion current monitoring (SICM) technique with combined gas chromatography–mass spectrometry (GCMS) to determine the levels of (*o*-hydroxybenzyl)adenosine (HBA), focusing on the ion of m/e 661, which was shown to be unique for extracts of poplar leaves. Moreover, by adding the unnatural isomer (*p*-hydroxybenzyl)adenosine to the extract as an internal standard, it is possible to estimate the percentage recovery in the final purified fractions and to make the necessary corrections for losses during purification.

Using this technique, it has been possible to show that the endogenous levels of HBA in the leaves of plants growing in the field are low during the night, but at dawn they rise rapidly and then decline, in a manner similar to the changes seen following a period of 5 min of red light (Thompson, Horgan and Heald, 1975). This decline is followed later by a slow rise in HBA during the day.

It has been possible to demonstrate clearly red/far-red reversibility using the SICM technique. Moreover, the 'escape time' for far-red reversibility was found to be between 5 and 10 min. A cell-free 'chloroplast' preparation was obtained from poplar leaves by chopping the leaves, filtering and centrifuging at 1 000 g. The pellet so obtained, containing a high proportion of chloroplasts, was re-suspended in buffer and used for studies on the effects of red light and sonication. This 'chloroplast' preparation was shown to contain significant amounts of HBA but neither red light nor sonication appeared to have any significant effect upon the distribution of HBA between the chloroplasts and the supernatant. Thus the findings reported for gibberellins in cell-free preparations of barley and wheat etioplasts do not appear to hold for HBA present in green chloroplasts obtained from poplar leaves.

Effects on Auxin and Ethylene

The effects of red light on levels of other types of endogenous hormones are less striking and less rapid. Thus, Briggs (1963) observed that red light inhibits auxin production in maize coleoptiles. Similarly, Fletcher and Zalik (1964) found that red light reduced the auxin content of etiolated plants of bean (*Phaseolus vulgaris*). Kang and Ray (1969) investigated changes in ethylene and CO_2 production in relation to the red light-promoted opening of the bean hypocotyl hook. In this system, ethylene inhibits and CO_2 promotes opening of the hook. It was found that exposure to red light leads to a reduction in the output of endogenous ethylene after 6 h and to an increase in CO_2 output after 2 h. Thus, these phytochrome-controlled effects are much less rapid than is the case for the effects on gibberellin and cytokinin levels.

There has, as yet, been no unequivocal evidence that ABA levels are under phytochrome control.

Discussion

The question naturally arises as to what is the functional significance of the rapid changes in gibberellin and cytokinin levels following treatment with red light and whether the growth responses to red light are brought about by the hormone changes. Certainly it would seem to be significant that many of the phytochrome-controlled growth responses can be effected in dark-grown tissue by application of exogenous hormones. Thus, the effect of red light can be simulated by application of exogenous gibberellins and cytokinins in seed and spore germination, in leaf expansion, internode extension, two-dimensional growth in moss protonemata, flowering, etc.

At first sight, therefore, it seems logical to conclude that the hormone changes effected by red light play an important role in the growth responses. However, numerous studies on the interaction between the effects of red light and exogenous gibberellins and cytokinins in various tests have, in general, indicated that the effects of the two factors are independent and additive, as has been shown for the growth of coleoptiles, internodes and leaves (*see,* for example, various papers in Withrow, 1959). However, before conclusions can be drawn from experiments with gibberellins such as GA_3, or unnatural cytokinins such as kinetin and benzyladenine, it is necessary to identify the endogenous hormones which vary in response to red light and to test the effect of the specific hormones on the growth responses of the parent plant material. Preliminary studies of this kind have been carried out with the unrolling responses of wheat leaves (Loveys and Wareing, 1971b). Red light causes unrolling of etiolated leaf sections within 24 h, and the effect of red light can be partially reproduced by treating dark-grown sections with GA_3 and, more effectively, with kinetin. When chromatograms of extracts of etiolated wheat leaves were assayed simultaneously by both the lettuce hypocotyl test and by the effects upon the unrolling of etiolated wheat leaf sections, only one narrow zone of the chromatogram was effective in stimulating leaf unrolling, whereas a much wider zone showed gibberellin activity in the lettuce hypocotyl assay.

On the other hand, in a detailed discussion of the role of endogenous hormones in the light responses of lettuce seeds, Black (1968) concluded that the effect of red light on germination is not mediated through increases in either gibberellins or cytokinins. Studies on the interaction between red light and exogenous GA_3 in lettuce seed germination indicate that, if sub-optimal amounts of red light and of GA_3 are given, a strong synergism in their effects can be demonstrated. Black concluded that the effects of Pfr and GA_3 must be independent.

Following the report of Van Staden of quantitative changes in lettuce seed after red light, Black, Bewley and Fountain (1974) examined the possibility that these changes might promote germination. The approach was based upon the fact that cytokinins can overcome the inhibitory effects of ABA on lettuce seed germination. It was found that exposure to red light did not enhance the ability of the seeds to overcome the effects of ABA, which would have been expected if the primary effect of red light is mediated through increases in endogenous cytokinin levels. Thus the question as to whether the germination responses of lettuce seed to a single short exposure to red light are mediated through endogenous hormone changes must remain open for the present.

It is also important to consider the implications of these phytochrome-controlled transient hormone effects in green tissue under natural light.

As we have seen, a short period of red light applied during the dark period leads to rapid increases in both cytokinin levels in poplar leaves and gibberellin levels in potato leaves (pp. 288 and 289). It is well known that a short period of red light ('night break') nullifies the effect of a long dark period in photoperiodic flowering responses. Tuberisation in *Solanum andigena* and bud dormancy in poplar are promoted by short days, so that these two species are photoperiodically sensitive, although the 'night break' effect has apparently not been directly demonstrated for their respective short-day responses. Moreover, cytokinin levels in the leaves, stems and roots of *Xanthium strumarium* fall dramatically after exposure to a single long dark period, and this effect is nullified by a night break (Van Staden and Wareing, 1972b; Henson and Wareing, 1974). Thus, the finding that a short period of red light applied during the dark period leads both to a profound modification of the photoperiodic responses and to a marked fluctuation in the hormone levels in the leaves clearly calls for further investigation of the relationship between these two phenomena.

These rapid and transient effects of phytochrome conversion are of considerable interest from two points of view. Firstly, as Evans and Smith (1975) point out, they provide a very promising model system for studying the mode of phytochrome action. Secondly, they seem likely to provide a clue to the way in which various phytochrome-controlled growth responses, such as internode elongation and bud dormancy, are mediated. Although it is probable that phytochrome acts primarily by affecting membrane properties, and thereby affects a wide variety of responses of which growth responses are only one aspect, nevertheless these rapid fluctuations in hormone levels are among the earliest detectable metabolic changes following phytochrome conversion. Thus, although it would appear that responses such as light-promoted germination of seeds cannot be explained solely in terms of rapid increases in endogenous hormone levels, nevertheless these latter changes may form an essential part of several processes initiated by phytochrome conversion, which lead to germination. In addition to these transient effects of short periods of irradiation, there is also evidence that the quality of the light to which plants are exposed during the main light period of photosynthesis (i.e. at high intensities for prolonged periods) has effects on the hormonal levels in the plant.

Thus, in some preliminary experiments by Hewett and Wareing (1973), it was found that whereas extracts of poplar leaves taken from the field contained considerable cytokinin activity in several distinct fractions, similar extracts from poplar plants maintained under fluorescent lighting in a growth room showed very low levels of activity. This observation suggested that the difference in cytokinin levels was probably due to differences in the quality of the light to which the same leaves had been exposed. Since light from white fluorescent lamps is deficient in red and far-red light compared with daylight, poplar plants were maintained for 3 weeks under either fluorescent lamps alone or fluorescent supplemented with tungsten filament lamps, and then extracted and assayed for cytokinins. The cytokinin levels in the leaves, apices and sap were appreciably higher in the plants which had received supplementary lighting from tungsten filament lamps. Both the increment in stem height during the experiment and the leaf area were increased by supplementation with incandescent light. In a further experiment, all plants were maintained under fluorescent lamps for 16 h per day, followed by illumination from fluorescent, incandescent, red or far-red light for a further 4 h. All of the additional light treatments resulted in higher cytokinin levels, but the greatest effect was with red light, which produced a 20-fold increase in cytokinin levels compared with those in the dark controls.

Although these results from very simple, preliminary experiments are clearly of a tentative nature, they seem to indicate that the quality of the light (presumably the Pfr:Ptot ratio) to which plants are exposed has a marked effect upon the cytokinin levels. Moreover, the irradiances and duration of light used in these experiments suggest that the 'high irradiance reaction' may be involved. There is circumstantial evidence suggesting that endogenous gibberellin levels may also be affected by light quality. Thus, the growth responses to light of varying quality may be mediated through variations in endogenous hormone levels, including both cytokinins and gibberellins.

References

BARZILAI, E. and MAYER, A.M. (1964). *Aust. J. Biol. Sci.,* **17**, 798

BEEVERS, L., LOVEYS, B.R., PEARSON, J.A. and WAREING, P.F. (1970). *Planta,* **90**, 286

BLACK, M. (1968). *Symp. Soc. Exp. Biol.,* **23**, 193

BLACK, M., BEWLEY, J.D. and FOUNTAIN, D. (1974). *Planta,* **117**, 145

BRIGGS, W.R. (1963). *Am. J. Bot.,* **50**, 196

COOKE, R.J. and SAUNDERS, P.F. (1975). *Planta,* **123**, 299

EVANS, A. and SMITH, H. (1975). *Proc. Nat. Acad. Sci. U.S.A.,* in the press

FLETCHER, R.A. and ZALIK, S. (1964). *Plant Physiol.,* **39**, 328

HENSON, I.E. and WAREING, P.F. (1974). *Physiol. Plant.,* **32**, 185

HEWETT, E.W. and WAREING, P.F. (1973). *Planta,* **114**, 119

HORGAN, R., HEWETT, E.W., PURSE, J. and WAREING, P.F. (1973). *Tetrahedron Lett.,* **30**, 2827

KANG, B.G. and RAY, P.M. (1969). *Planta,* **87**, 193

KÖHLER, D. (1965). *Planta,* **67**, 44

KÖHLER, D. (1966a). *Planta,* **69**, 27

KÖHLER, D. (1966b). *Planta,* **70**, 42

KOPCEWICZ, J. and PORAZINSKI, Z. (1973). *Bull. Acad. Pol. Sci.,* **21**, 383

LOVEYS, B.R. (1970). *Ph.D. Thesis,* University of Wales

LOVEYS, B.R. and WAREING, P.F. (1971a). *Planta,* **98**, 109

LOVEYS, B.R. and WAREING, P.F. (1971b). *Planta,* **98**, 117

RAILTON, I.D. and WAREING, P.F. (1973). *Physiol. Plant.,* **29,** 430

REID, D.M. and CLEMENTS, J.B. (1968). *Nature, Lond.,* **219,** 607

REID, D.M., CLEMENTS, J.B. and CARR, D.J. (1968). *Nature, Lond.,* **217,** 580

REID, D.M., TUING, M.S., DURLEY, R.C. and RAILTON, I.D. (1972). *Planta,* **108,** 67

SMOLENSKA, G. and LEWAK, St. (1971). *Planta,* **99,** 144

VAN STADEN, J. and WAREING, P.F. (1972a). *Planta,* **104,** 126

VAN STADEN, J. and WAREING, P.F. (1972b). *Physiol. Plant.,* **27,** 331

VAN STADEN, J. (1973). *Physiol. Plant.,* **28,** 222

VAN STADEN, J., OLATOYE, S.T. and HALL, M.A. (1973). *J. Exp. Bot.,* **24,** 662

STODDART, J.L. (1968). *Planta,* **81,** 106

TANADA, T. (1973a). *Plant Physiol.,* **51,** 150

TANADA, T. (1973b). *Plant Physiol.,* **51,** 154

THOMPSON, A.G., HORGAN, R. and HEALD, J.K. (1975). *Planta,* in the press

WITHROW, R.B. (1959). *Am. Assoc. Adv. Sci. Publ.,* No.55

19

PHYTOCHROME-MEDIATED INTER-ORGAN DEPENDENCE AND RAPID TRANSMISSION OF THE LIGHT STIMULUS

J.A. DE GREEF
R. CAUBERGS
J.P. VERBELEN
E. MOEREELS
Department of Biology, University of Antwerp, B-2610 Wilrijk, Belgium

Introduction

Botanists have long known that apical dominance and flower induction are phenomena of plant development that depend on the interaction between different organs. Dominance of apical over lateral growth can easily be demonstrated by removing the apical bud from a plant. Decapitation usually releases the lateral bud(s) from correlative inhibition by the apex and in its turn the lateral bud(s) nearest the apical bud will quickly establish dominance over the remaining buds. Most of our knowledge of apical dominance comes from studies in which the apices of plants were removed and replaced by hormones (Thimann and Skoog, 1933; Thimann, 1937). It was therefore suggested that auxins may play a major role in the problems of apical dominance. Since then, however, it has been found that all of the known plant growth regulators have an extremely wide spectrum of action and interact quantitatively to control many growth responses.

Since Garner and Allard (1920) formulated the principle of photoperiodism in flowering, much work has been aimed at establishing which part of the plant receives the photoperiodic stimulus. The organs that received most attention were the leaves and buds. Knott (1934) demonstrated that in spinach, a long-day plant, the leaves are the perceptive site of the photoperiodic stimulus. He postulated that 'something' is produced in the leaves in response to a photoinductive cycle and translocated to the apical tips, causing the initiation of floral primordia. Overwhelming evidence for leaves being the organs of perception in the flowering response to photoinductive cycles has since been obtained. Giving inductive photoperiods to a single leaf, while the remainder of the plant is on a non-inductive cycle, is sufficient to cause flowering (Hamner and Bonner, 1938; Long, 1939). Grafting leaves from a photoinduced plant to a plant on non-inductive photoperiods promotes flowering of the defoliated receptor plant (Heinze, Parker and Borthwick, 1942; Naylor, 1953).

Although there are now substantial experimental data for both apical dominance and flower induction, the reactions involved remain uncertain on present-day biochemical maps, in spite of the many research efforts made in recent decades in these fields.

In most aspects of plant morphogenesis, photoreactions in organs have been evoked by *in situ* illumination. In most cases of photomorphogenic induction, we are dealing with long-term effects. When there is a long period before we can measure the expression of a growth effect after its induction, then it is more probable that secondary effects will emerge, making the interpretation of primary reaction(s) more difficult. In this way, we are looking at consequences that are too close to the target but too far from the primary events of growth responses.

In relation to the above line of thought, we decided to search for inter-organ dependent phenomena of plant development during the transition from heterotrophic to autotrophic growth in etiolated seedlings. For this purpose, it seemed logical to experiment with those plant parts which are sensitive to and responsible for primary growth under light control. At the same time, we looked for fast reactions evoked by photoinduction and related to macrophysiological effects. The first hint we had in this respect was the absence of red/far-red light effects in the vegetative growth of *Marchantia thalli* when the apical meristems were excised or shielded from the light (Fredericq and De Greef, 1968).

The *Phaseolus* System and Methodology

Seedling development of *Phaseolus vulgaris* L. in complete darkness is characterised by a progressive evolution in its general morphology. Uniform development of the seedlings and their physiological age are mainly dependent on the temperature and relative humidity of the ambient atmosphere and on the sowing depth. Therefore, the growth conditions were kept rigorously constant during plant cultivation. Under the conditions used (21 °C, 80 per cent relative humidity and 4 cm sowing depth), the etiolated seedlings were taken into the experimental runs between days 8 and 11. At these physiological ages the plants had reached a morphological development as shown in *Figure 19.1*. In this developmental stage, the plumule, the primary leaves, the hook region and the cotyledons were all grouped together but these organs were at sufficient distances from each other to be manipulated separately. The organs of these seedlings were very sensitive to light stimuli.

To demonstrate light-controlled inter-organ dependence, we illuminated one type of organ by shielding the other organs of the same plant and measured a photoreaction in one of the shielded organs. In the following sections 'selective light treatment' means that only the plant part indicated was exposed to light. Light shielding was

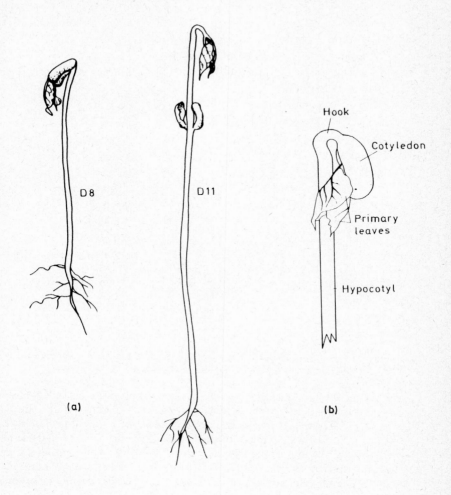

Figure 19.1   Morphological stages in the development of etiolated bean seedling. (a) In the experiments described in this chapter plants were used between eight (D8) and eleven days (D11) of etiolation.   (b) Detailed picture of eight-day-old seedling top

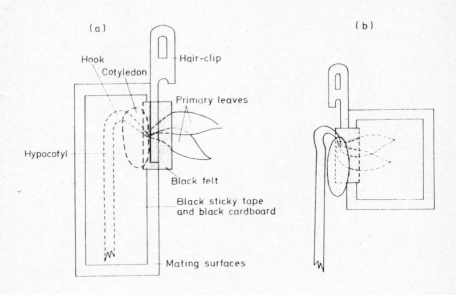

Figure 19.2   Light-shielding of the embryonic axis (a) and of the primary leaves (b) for the studies of greening and hook opening

performed by a black-body effect. The organs were wrapped in black adhesive tape lined with soft black cardboard (*Figure 19.2*). In the study of leaf expansion, a black box was used (*Figure 19.3*). The primary leaves were pulled through light-tight slits to the outside of the box while the remainder of the seedling was kept inside the box. At the top of the box red and far-red light sources were built in so as to provide the possibility of giving light flashes to the embryonic axis.

Extent of the Embryonic Axis

In our work, the term 'embryonic axis' is used to denote the highly meristematic hook tissue, wherever the hook may be located during the morphological development of the seedlings.

Epicotyl elongation of 11-day-old etiolated seedlings was followed in order to determine the extent of the embryonic axis. The length of the (remaining) epicotyl region was measured 1 week after cutting different segments of the epicotyl, as indicated in *Figure 19.4*. It can

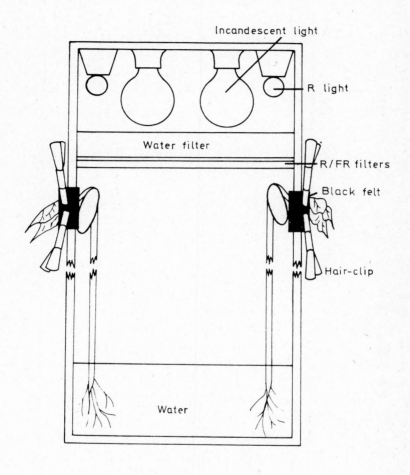

Figure 19.3   Black box for shielding the embryonic axis in studies of leaf expansion

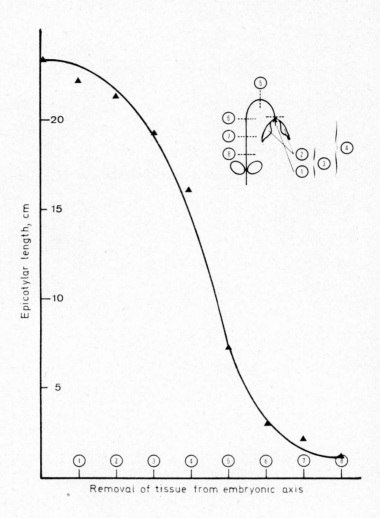

Figure 19.4 Extent of the embryonic axis. Effect of removal of tissue of the embryonic axis on epicotylar growth in darkness. The figures indicate the segments removed in subsequent steps: (1) plumule; (2) primary leaves; (3) plumule + primary leaves; (4) cut underneath the node of the primary leaves; (5), (6), (7) and (8) increasing parts of the elbow region

be seen that two thirds of the increment of epicotyl growth is due to
the hook region itself. When the apex is cut just underneath the node
of the primary leaves, epicotyl elongation is inhibited by one third.
Segments a few centimetres underneath the top of the hook elbow did
not contribute to stem growth. In order to obtain efficient shielding
of the hook region and to facilitate manipulations of the material, we
cut the tops of etiolated seedlings 4 cm under the elbow in some
experiments. In agreement with our results, Powell and Morgan (1970)
found that the lower stem and root tissues have little or no effect on
the opening of bean hooks.

Phytochrome-mediated Inter-organ Dependence in Classical Long-term Phenomena

Leaf greening, hook opening and leaf expansion are classical parameters
in plant development under phytochrome control. Using these parameters, an attempt was made to demonstrate the existence of a light-
triggering system between different organs of the embryonic axis
(De Greef and Caubergs, 1972, 1973).

GREENING OF PRIMARY LEAVES

With increasing age, etiolated seedlings develop a so-called lag-phase of
chlorophyll synthesis when they are brought under continuous white
light (Liro, 1909). Under the cultivation conditions used, 8-day-old
etiolated bean seedlings showed a lag-phase of 2 h before rapid chlorophyll formation started. This lag-phase can be prevented by pre-irradiation with a small dose of 660-nm light and subsequent dark incubation
for 5–15 h before transfer to continuous white light (Withrow, Wolff
and Price, 1956). The stimulatory effect of red light can be nullified
by subsequent treatment with far-red light (Price and Klein, 1961).
Studying the action spectrum for elimination of the lag-phase of chlorophyll synthesis, Virgin (1961) demonstrated a maximum of photomorphogenic induction at 660 nm. The antagonistic effect of red and
far-red light and the position of the action spectrum peak indicated
that phytochrome is the responsible photoreceptor. It is generally
assumed that the reaction centre for the photomorphogenic induction
of leaf greening is situated in the leaf itself.

We studied leaf greening by a selective pre-illumination of the
embryonic axis in comparison with the normal greening of intact
plants and excised leaves, either pre-treated with red light or red
followed by far-red light, or not pre-treated. *Figure 19.5* summarises
the results of these different experiments. Maximal chlorophyll synthesis occurred when intact etiolated seedlings were entirely pre-exposed
to 5 min of red light, then kept in darkness for 12 h and subsequently
transferred to continuous white light illumination. The kinetics are

shown in *Figure 19.5*, path (1), side a. The amount of chlorophyll a synthesised within 8 h of white light was taken as 100 per cent. For all other treatments, pigment values are expressed in relation to this maximum effect. When tops of seedlings were treated in the same way as the intact plants mentioned above, there was only a slight quantitative difference [path (1), side b]. Intact etiolated seedlings or tops of the same physiological age but not pre-illuminated with red light (dark controls) greened under white light as in path (3), side b. Irradiation with far-red light following the red-light treatment decreased the red effect of intact plants and tops to the level of the dark controls, but they showed higher chlorophyll values [path (3), side a] owing to the photochemical conversion of protochlorophyll(ide) initially present.

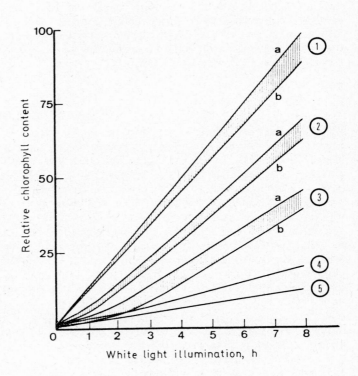

Figure 19.5 Chlorophyll synthesis under white light in primary leaves of etiolated bean plants pre-irradiated with red light, and red followed by far-red light, respectively. Pre-illuminations were given either to whole plants or plant tops, or selectively to leaves or embryonic axis, or to excised leaves (all or not incubated in Hoagland sucrose medium). The results were compared with those of etiolated plants of the same age but not pre-treated with light. For further explanation, see text

A selective pre-treatment with red light given either to the leaves of intact plants or to excised leaves incubated in a Hoagland sucrose solution stimulates chlorophyll synthesis under white light after a dark period of optimal length [path (2), side a]. The effect is obviously smaller than was found in intact plants or tops. When the embryonic axis was selectively pre-treated with red light, greening proceeded in a similar way as was found for a selective pre-exposure of the leaves to red light [path (2), side b]. Again the lower pigment values of the latter experiment are explained by the lack of photoconversion of protochlorophyll(ide). In both experiments on photoinduction by selective light pre-treatments, far-red reversibility was equally evident [path (3), sides a and b]. Excised leaves incubated in water do not demonstrate a red/far-red antagonism [path (4)]. Compared with their dark controls [path (5)], the higher chlorophyll values in red and red/far-red pre-treated leaves are the result of protochlorophyll(ide) conversion.

From these experiments, we conclude that intact bean plants demonstrate an obvious inter-organ dependence for greening of the primary leaves. This co-operation is phytochrome mediated. The increased capacity of chlorophyll metabolism can find its expression only by an adequate supply of substrate(s). In intact plants, this metabolic interplay between organs must be controlled by phytochrome.

HOOK OPENING

Straightening of the hypocotyl hook of etiolated bean seedlings was found to be under phytochrome control (Klein, Withrow and Elstad, 1956; Withrow, Klein and Elstad, 1957). In these earlier investigations, excised hooks were used so that only the effect of phytochrome present in the hook region could be examined. In accordance with the greening experiments, the effects of short red and red/far-red pre-irradiations selectively administered to the primary leaves or to the embryonic axis were compared with dark and white light controls. All of these data were submitted to analysis of variance and to a Student–Newman–Keuls multiple range test (Woolf, 1968). In *Figure 19.6* the results are presented schematically.

After 24 h of white light, unbending of the hook in intact control plants was complete. At that time the degree of hook opening was measured in the other experimental series. A 5-min red light pre-illumination of the whole plant, or selectively of the hook, followed by a 24-h dark period, caused the same degree of unbending, but it was smaller than that of the white light controls. Selective red light pre-treatment of the leaves also induced hook opening but to a lesser extent than when the hook was selectively illuminated. Full far-red reversibility could be obtained only either when red and far-red light were applied to the whole plant or when the whole plant was pre-exposed to red light, and far-red light was subsequently applied to the

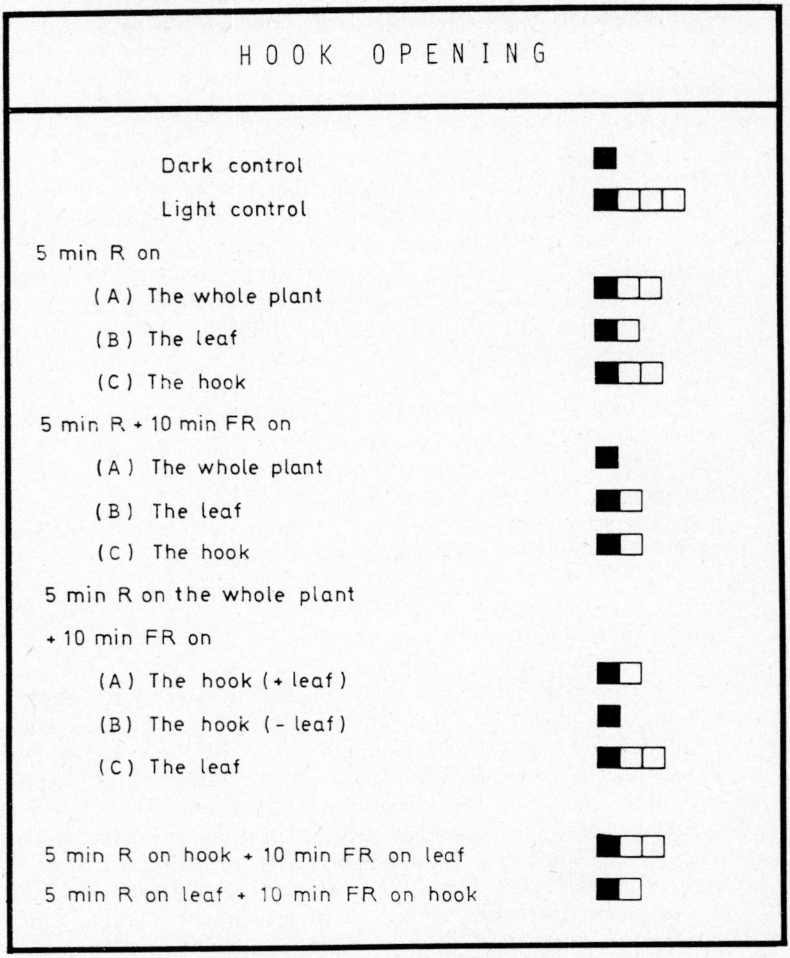

HOOK OPENING

Dark control

Light control

5 min R on

    (A) The whole plant

    (B) The leaf

    (C) The hook

5 min R + 10 min FR on

    (A) The whole plant

    (B) The leaf

    (C) The hook

5 min R on the whole plant

+ 10 min FR on

    (A) The hook (+ leaf)

    (B) The hook (- leaf)

    (C) The leaf

5 min R on hook + 10 min FR on leaf

5 min R on leaf + 10 min FR on hook

Figure 19.6 *Studies on hook opening by selective light treatments of different organs compared with the classical type of experiments on this subject. Solid rectangles represent a relative measure of the degree of hook opening in dark controls. Open rectangles show the additional effect of hook opening caused by the light treatments indicated*

defoliated plant. There was no reversibility when the leaves were selectively exposed to light. Selective pre-irradiations of the hook with red followed by far-red light caused a partial reversal, unbending being reduced to the same degree as if the leaves were pre-illuminated only with red light. The same result was obtained when intact plants were fully exposed to red light followed by selective far-red irradiation of the hook. Also, red light pre-irradiation of the whole plant could not be obviated by subsequent and selective treatment of the leaves with far-red light. Hook opening induced by selective treatment of the hook with red light was not reversed by selective far-red irradiation of the leaves and *vice versa*. Cutting the leaves immediately after their photo-induction did not change the hook response. This result indicates that the transfer of the light stimulus is very fast. It supports our concept that the primary action of phytochrome should be localised on a biophysical level. By means of a transmission system (primitive nervous system), phytochrome directs the flow of information of perceived light signals throughout the whole plant body.

In contrast to chlorophyll synthesis in the primary leaves, which is reversibly controlled by red/far-red action on the embryonic axis, opening of the hook by stimulation of the leaves with red light is not reversed by far-red light. A significant difference between both transmission phenomena is found in the opposite direction of the stimulus transfer with regard to the plant axis (root–stem–leaves). Since apical meristems develop in one direction, a polarised disposition of the light trigger system is reasonably conceivable. The red light stimulus is effectively transmitted to the embryonic axis functioning as a co-ordination centre. The far-red signal only acts *in situ* at the level of the regulatory mechanisms involved. According to our concept of phytochrome-mediated inter-organ co-operation, hook opening can then be explained by a kind of relay system (*Figure 19.7*).

LEAF EXPANSION

The most convincing evidence for the co-ordinating influence of the embryonic axis in photomorphogenic reactions of the etiolated bean seedling was found in experiments on leaf expansion. Parker *et al.* (1949) reported that the action spectrum for leaf expansion is the same as for other photomorphogenic responses under phytochrome control. Working with intact plants, Downs (1955) demonstrated that brief irradiation with red light promotes leaf expansion while the effect potentiated by red light is reversed by subsequent exposure to far-red light.

Figure 19.8 gives the results of an experiment as planned by Downs (1955). Eight–day–old etiolated seedlings were illuminated with 5 min of red light daily for seven consecutive days. Another series was treated with 5 min of red + 10 min of far-red light per day during the same period. At the end of the experiment, the plants treated

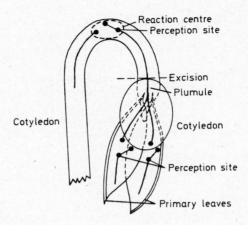

Figure 19.7 *Scheme of relay system to explain phytochrome-mediated inter-organ co-operativity in hook opening. We accept an explicit distinction between 'reaction centre' (= hook for hook opening) and 'perception site' (where the signal is perceived in the organ). The red light signal is transmitted in both directions between different perception sites, but far-red light is effective only when its signal can be integrated in the system*

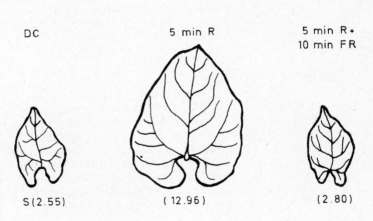

Figure 19.8 *Downs' experiment of leaf expansion. Eight-day-old etiolated seedlings were illuminated daily with 5 min of red light (5 min R) for 7 days. Another series of plants was irradiated for the same period with 5 min of red followed by 10 min of far-red light (5 min R + 10 min FR). Leaf expansion was then compared with dark controls (DC). S = average leaf size (cm²)*

with red light developed large expanded laminae while those treated with far-red light showed the same leaf size as the dark controls. There was no leaf expansion when a selective 5-min red irradiation was given daily either to the leaves or to the embryonic axis. When primary leaves were exposed to continuous white light and the remainder of the plants shielded in a black box, the leaves greened normally but did not expand. If, in addition, the embryonic axis was exposed to 5 min of red light daily in the black box (*see Figure 19.3*), leaf expansion occurred as in white light controls. Far-red following red light prevented leaf expansion. All of these results are summarised in *Figure 19.9*. Parallel to these experiments, we examined the number of mesophyll cells per unit area of leaf surface. This ratio is constant in relation to the reciprocal of the increase in leaf blade size, which means that both leaf expansion and increase in cell volume proceed together at this stage of seedling development. These macroscopic and light microscopic observations were also reflected at the ultracytological level. Leaf plastids remained underdeveloped in size and membrane structure when the embryonic axis was shielded from the light. We can conclude that normal leaf development can take place only if Pfr is present in both leaves and the embryonic axis.

Our results are contradictory to the experiments on expansion of etiolated bean leaf discs (Miller, 1952). When we used excised leaves of etiolated plants and incubated them in a Hoagland sucrose medium under continuous white light, we observed leaf growth only to the level of full-grown dark controls. After adventitious roots were formed at the base of the petiole (regeneration of new meristematic tissue), the leaves started to expand slowly. In order to achieve a better understanding of Miller's experiment, we examined the anatomical stages of leaf growth and the effect of light on these stages. It is known that the expansion in area of the blade is brought about by cell division and more especially by cell expansion. It is difficult to separate these two processes and to indicate precisely the stage where cell division stops since the decline in cell number per unit area may be due to both decreasing rates of cell division and cellular enlargement (Maksymowych, 1973). Probably the two processes overlap during leaf development under natural conditions.

With this knowledge as a background, we designed an experiment to discriminate between the two processes as far as they are modulated by light. Very young etiolated seedlings and much older ones were subjected to continuous white light for 9 days. With the former the leaves are still growing and cell division is predominant, while with the latter the leaves are fully grown so that cell division must have stopped. At the end of the experiment the leaf blades were about the same size, but the cell number per unit area was about 50 per cent higher in the leaves that were exposed to light at an early age (*Figure 19.10*). The effect of light on cell division was not fully reversible by far-red light.

When discs of incompletely grown etiolated leaves are used to demonstrate red/far-red light effects on leaf expansion, we are probably dealing

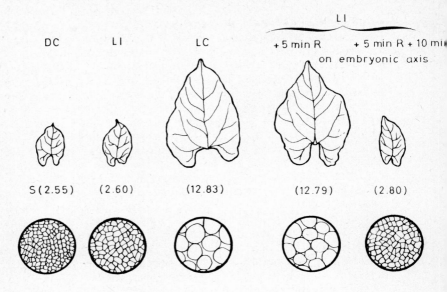

*Figure 19.9   Influence of illumination of the embryonic axis on leaf expansion.
Average leaf size (S, cm²) and number of mesophyll cells per unit area to the
leaf surface are illustrated.   DC and LC, dark and light controls, respectively;
LI, leaves selectively illuminated*

with a light effect on cell division, an inherent feature of the leaf
itself.   Cell expansion contributing to an increase of the blade in area
after full dark growth is dependent on the light sensitivity of the
embryonic axis and fully controlled by phytochrome.

CONCLUSION

From the results obtained so far, we assume that the initial photochemical transformation of phytochrome is strongly related to two
immediate consequences:
   (1) A very fast transmission of the light stimulus between different
       organs and co-ordinated by the embryonic axis.
   (2) A slow macrophysiological effect of the photomorphogenic
       phenomenon caused by changed metabolic activities which are
       controlled through the embryonic axis (long-term effects).

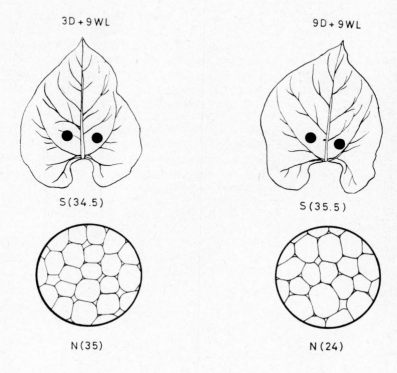

3D + 9WL

S(34.5)

N(35)

9D + 9WL

S(35.5)

N(24)

Figure 19.10 Anatomical differences in leaf expansion of etiolated bean seedlings depending on time of illumination.   Three-day-old (3D) and nine-day-old (9D) etiolated seedlings were continuously illuminated with white light for 9 days. Thereafter the average leaf area (S, cm^2) and average cell number per unit surface area (N) were determined

Biochemical Significance of the Dark Incubation in Photomorphogenic Processes

After the photoinduction of greening, hook opening and leaf expansion there is a lag period of several hours before an effect becomes visible. A maximum effect was obtained after 12–16 h. The exact nature of the intervening dark reactions is not yet understood. Since leaf greening and leaf expansion require a high activity of metabolic reactions, we followed respiration in leaf discs of light-treated etiolated plants and of dark controls as a function of time. *Figure 19.11* gives the results of these experiments. Oxygen uptake in leaves is increased when whole plants are illuminated, the maximum effect being reached after 16 h of white light (50 per cent more than the dark controls). However, the respiratory activity of leaf discs taken from plants of which the leaves were selectively exposed to continuous white light did not differ from dark control values. If an additional irradiation of 5 min of red light per day was given to the embryonic axis, the same results were obtained as when the plants were fully exposed to white light. From these results, it is obvious that the light-induced changes in oxygen uptake by the leaves of intact plants are directly related to the light activation of the embryonic axis.

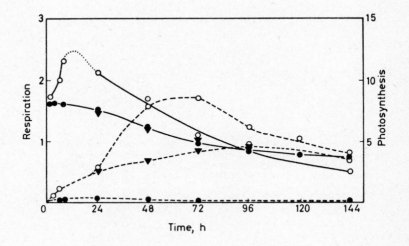

Figure 19.11   Kinetics of respiratory (solid lines) and photosynthetic (dashed lines) activities in leaf tissue. Eight-day-old etiolated bean seedlings were brought under continuous white light (○) or selectively illuminated on their leaves (▼) and compared with dark controls of the same physiological age (●)

Gordon (1964) suggested that phytochrome conversion might function by regulating energy transfer through oxidative phosphorylation. However, studies with 2,4-dinitrophenol (DNP) yielded conflicting results (Halldal, 1958). In our non-photosynthetic system, at the time of pre-irradiating the seedlings, the dependence of greening capacity on oxidative phosphorylation could easily be tested by using an uncoupler such as DNP. We infiltrated the leaves on intact plants with 0.25 mM DNP either before or after the red light pre-treatment. When the infiltration was carried out 4 h before the pre-irradiation with red light, chlorophyll formation under white light after the dark incubation was inhibited as much as in DNP-treated dark controls. When DNP was administered 4 h after the treatment with red light, chlorophyll synthesis was much higher than in DNP-treated dark controls. Plants infiltrated with DNP immediately before or after pre-treatment with red light still displayed a net effect of the red-light stimulus, but they showed 50 per cent reduced chlorophyll values compared with the red light-treated plants. Washing out the uncoupler during the dark incubation by increased transpiration completely restored the chlorophyll synthesising capacity in all DNP-treated series.

We draw the following conclusions from these experiments:

(1) The photomorphogenic expression of phytochrome action is intimately related to energy metabolism.
(2) Immediately after the photochemical transformation the active form of phytochrome is involved in some energy-dependent processes.
(3) The active form of phytochrome can be maintained for longer periods of uncoupling.

Fast Reactions Occurring at the Onset of Illumination

CHANGES IN ATP AND P_i LEVELS

On the basis of the above results, we decided to examine the kinetics of energetic metabolites in very short time ranges. These experiments were carried out with 9-day-old etiolated seedlings illuminated with red or white light for periods ranging from 30 s to 5 min. After irradiation, the plant tops were immediately frozen by immersion in liquid nitrogen and then dissected in order to collect the different organs. These organs were homogenised in Tris at pH 7.4 and samples were taken from the homogenates for P_i level determinations and for ATP measurements. An aliquot of each homogenate was incubated at 27 °C for 30 min in an ATP solution of optimal concentration. The P_i levels were determined in the supernatant of the reaction mixture. The results are given in *Figure 19.12*. Specificity for ATP was proved with substitution of ATP by ADP and β-glycerine phosphate.

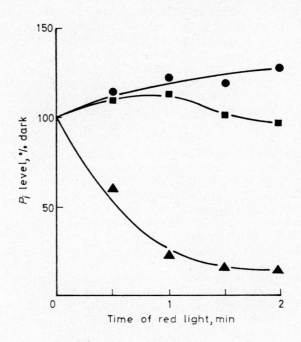

Figure 19.12   Effects of irradiation with red light (in minutes) on P_i levels after incubation of homogenates of different organs in extra ATP.   Dark control values were normalised to 100 per cent.   ●, Leaves;   ■, cotyledons;   ▲, hook region

In cotyledons the P_i values oscillated around the dark value, while in primary leaves they showed a slight increase.   In the epicotylar hook region, however, the P_i level decreased markedly within 30 s after the onset of the red light irradiation and decreased smoothly under further illumination to a constant minimum value.   Comparing the amounts of ATP that remained in dark controls after the dark incubation period with the kinetics of P_i present after various illumination periods, P_i levels in leaves and the hook region were much too low for the amounts of external ATP that had disappeared.   For the cotyledons the P_i values can be explained only by accepting the total hydrolysis of disappeared ATP to AMP.

Therefore, in leaves and hooks the light-modulated changes in P_i levels indicate a fast and intensive consumption of the P_i liberated. It could be used in light-stimulated phosphorylating activities.   In agreement with this assumption, we found in preliminary experiments that the formation of endogenous ATP was considerably stimulated in leaves within 1 min of red light irradiation and even more in the epicotylar hook after 2 min of red light. Afterwards the ATP content decreased in both cases.

Similar results were found in mesocotyl tissue by Hodges *et al.* (1972) and in etiolated bean buds by White and Pike (1974). The latter demonstrated that the promotion of ATP content is red/far-red reversible. In the cotyledons we noticed only slight oscillations in ATP values. When homogenates of dark controls, prepared from hook tissue and cotyledons, were centrifuged before incubation in extra ATP, the highest enzyme activity was found in the supernatant, containing mainly membrane vesicles. This result is in agreement with the concept that ATPase is associated with membrane fractions (Kielley, 1961; Sandmeier and Ivart, 1972; Lai and Thompson, 1972). ATPases play an important role in many aspects of cell metabolism. It is a coupling factor in oxidative phosphorylation (Penefsky *et al.*, 1960) and it is possibly involved in the transport of Na^+ and K^+ across cell membranes (Albers, 1967). Our results support the view that the membrane properties are changed at the onset of illumination. Enzyme complexes of energy metabolism at the membrane level seem to be involved directly in these primary events.

BIOPHYSICAL EFFECTS AND SITE OF PRIMARY ACTION

Exposure of the embryonic axis of etiolated seedlings to white light resulted in fast changes in its surface potential within the same time range as the ATP and P_i values were modulated by light. A temporary depolarisation of 1-1.5 mV can be measured during 100-120 s after a lag phase of 15-20 s. Dark values of the rest potential are subsequently recovered (*Figure 19.13*). A light microscopic study of dark-grown plants revealed that ATPase activity in the hook region was predominant in epidermal and hypodermal cell layers of the epicotylar ribs. With the electron microscope we noticed that this enzymatic activity is obviously limited to the plasmalemma of the epidermal and hypodermal collenchyma cells. The presence of the enzyme complex(es) in these surface tissues can be related to a possible site for light-sensitive reactions.

These data are consistent with the concept that the primary effects of Pfr in short-term phenomena are associated with membranes (Hendricks and Borthwick, 1967). Recently, much evidence has accumulated that active phytochrome is a membrane-bound protein (Smith, 1970; Haupt, 1973, Marmé, Boisard and Briggs, 1973). Membrane-bound Pfr could then regulate membrane permeability and thus the electrical potential of cell membranes (Tanada, 1968a,b; Jaffe, 1968; Newman and Briggs, 1972), depending on the activity of ATP-driven ion pumps (Satter, Marinoff and Galston, 1970; Satter, Sabnis and Galston, 1970; Yunghans and Jaffe, 1970, 1972). In other rapid

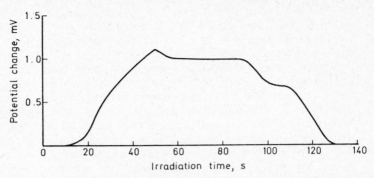

Figure 19.13   Electrical response of the embryonic axis to white light illumination starting at time zero

photoresponses the cell membrane itself seems not to be involved directly, although the importance of high-energy compounds is evident. In this respect, Yunghans and Jaffe (1972) reported that red light irradiation of etiolated mung bean root tips caused rapid changes in P_i levels and the rate of oxygen uptake. Manabe and Furuya (1973) found that within 2 min of red light illumination there was a large increase in the content of NADPH in a particulate fraction obtained from etiolated bean hypocotyls.

General Conclusions

Studying the transition from heterotrophic to autotrophic growth in the *Phaseolus* system, we found that the embryonic axis is functioning as a co-ordinating centre for phytochrome-mediated morphogenic reactions taking place in the primary leaves and in the hook region. The transmission of the light stimuli between these organs was very fast. Particular attention was paid to short-term effects induced by light. Rapid changes in P_i levels and ATP content of hook and leaf tissue were observed within the first few minutes of irradiation. The activity of ATPase(s) was very high in both dark controls and light-stimulated plants. By means of a histochemical method and electron microscopy, the enzyme complex was localised at the level of the cytoplasmic membrane, predominantly in the outer-stelar cell layers of the embryonic axis.

Within the same time range of illumination, fast changes in the surface potential of the embryonic axis were recorded. This could be the result of a changed ionic permeability of the cells belonging to the surface tissues. If the active form of phytochrome is part of the cell membrane (or at least associated with it), Pfr may modulate the activity of ATP-dependent ion pump systems and phosphorylating activities at the membrane level in many rapid energy-dependent photomorphogenic processes.

References

ALBERS, R.W. (1967). *Annu. Rev. Biochem.,* **36,** 727

DE GREEF, J.A. and CAUBERGS, R. (1972). *Physiol. Plant.,* **26,** 157

DE GREEF, J.A. and CAUBERGS, R. (1973). *Physiol. Plant.,* **28,** 71

DOWNS, R.J. (1955). *Plant Physiol.,* **30,** 468

FREDERICQ, H. and DE GREEF, J.A. (1968). *Physiol. Plant.,* **21,** 346

GARNER, W.W. and ALLARD, H.A. (1920). *J. Agr. Res.,* **18,** 553

GORDON, S.A. (1964). *Q. Rev. Biol.,* **39,** 19

HALLDAL, P. (1958). *Physiol. Plant.,* **11,** 118

HAMNER, K.C. and BONNER, J. (1938). *Bot. Gaz.,* **100,** 388

HAUPT, W. (1973). *BioScience,* **23,** 289

HEINZE, P.H., PARKER, M.W. and BORTHWICK, H.A. (1942). *Bot. Gaz.,* **103,** 517

HENDRICKS, S.B. and BORTHWICK, H.A. (1967). *Proc. Nat. Acad. Sci. U.S.A.,* **58,** 2125

HODGES, T.K., LEONARD, R.T., BRACKER, C.E. and KEENAN, T.W. (1972). *Proc. Nat. Acad. Sci. U.S.A.,* **69,** 3307

JAFFE, M.J. (1968). *Science, N.Y.,* **162,** 1016

KIELLEY, W.W. (1961). In *The Enzymes,* 2nd Ed., Vol.5, p.149. Ed. Boyer, P.D., Lardy, H. and Myzbäck, K. Academic Press, New York

KLEIN, W.H., WITHROW, R.B. and ELSTAD, V.B. (1956). *Plant Physiol.,* **31,** 289

KNOTT, J.E. (1934). *Proc. Am. Soc. Hort. Sci.,* **31,** 152

LAI, Y.F. and THOMPSON, J.E. (1972). *Plant Physiol.,* **50,** 452

LIRO, J.I. (1909). *Ann. Acad. Sci. Fenn.,* Ser. A1, 1

LONG, E.M. (1939). *Bot. Gaz.,* **101,** 168

MAKSYMOWYCH, R. (1973). In *Analysis of Leaf Development,* Ed. Abercrombie, M., Newth, D.R. and Torrey, J.G. Developmental and Cell Biology Series, Cambridge University Press, Cambridge

MANABE, K. and FURUYA, M. (1973). *Plant Physiol.,* **51,** 982

MARMÉ, D., BOISARD, J. and BRIGGS, W.R. (1973). *Proc. Nat. Acad. Sci. U.S.A.,* **70,** 3861

MILLER, C.O. (1952). *Plant Physiol.,* **27,** 408

NAYLOR, H.W. (1953). In *Growth and Development in Plants,* Ed. Loomis, W. University of Iowa Press, Ames, Iowa

NEWMAN, I.A. and BRIGGS, W.R. (1972). *Plant Physiol.,* **50,** 687

PARKER, M.W., HENDRICKS, S.B., BORTHWICK, H. and WENT, F.W. (1949). *Am. J. Bot.,* **36,** 104

PENEFSKY, H.S., PULLMAN, M.E., DATTA, A. and RACKER, E. (1960). *J. Biol. Chem.,* **235,** 3330

POWELL, R.D. and MORGAN, P.W. (1970). *Plant Physiol.,* **45,** 548

PRICE, L. and KLEIN, W.H. (1961). *Plant Physiol.,* **36,** 733

SANDMEIER, H. and IVART, J. (1972). *Photochem. Photobiol.,* **16,** 51

SATTER, R.L., SABNIS, D. and GALSTON, A.W. (1970). *Am. J. Bot.,* **57,** 374

SATTER, R.L., MARINOFF, P. and GALSTON, A.W. (1970). *Am. J. Bot.,* **57,** 916

SMITH, H. (1970). *Nature, Lond.,* **227,** 665

TANADA, T. (1968a). *Proc. Nat. Acad. Sci. U.S.A.,* **59,** 376

TANADA, T. (1968b). *Plant Physiol.,* **43,** 2070

THIMANN, K.V. and SKOOG, F. (1933). *Proc. Nat. Acad. Sci. U.S.A.,* **19,** 714

THIMANN, K.V. (1937). *Am. J. Bot.,* **24,** 407

VIRGIN, H.I. (1961). *Physiol. Plant.,* **14,** 439

WHITE, J.M. and PIKE, C.S. (1974). *Plant Physiol.,* **53,** 76

WITHROW, R.B., WOLFF, J. and PRICE, L. (1956). *Plant Physiol.,* **31,** xiii

WITHROW, R.B., KLEIN, W.H. and ELSTAD, V.B. (1957). *Plant Physiol.,* **32,** 453

WOOLF, C.M. (1968). In *Principles of Biometry,* pp.163–179. Van Nostrand, Princeton

YUNGHANS, H. and JAFFE, M.J. (1970). *Physiol. Plant.,* **23,** 1004

YUNGHANS, H. and JAFFE, M.J. (1972). *Plant Physiol.,* **49,** 1

20

INTER-ORGAN EFFECTS IN THE PHOTOCONTROL OF GROWTH

M. BLACK
JANET SHUTTLEWORTH
Department of Biology, Queen Elizabeth College, University of London, London, U.K.

Developmental changes in certain parts of the plant are well known to be controlled by specific morphogenic influences coming from distant organs. Many of these morphogenic controls are themselves instituted by the photoactivation of amply described pigment systems such as phytochrome. Two common examples are the photoperiodic control of flowering and of bud dormancy. In the former, the inception of floral organs at the apex is a response to signals emanating from leaves which have experienced inductive photoperiodic conditions. Similarly, the formation of dormant terminal buds is, in many woody species, determined by transmissible factors made by those leaves which are exposed to suitable photoinductive regimes. In both cases, of course, phytochrome is part of the photoperiodic timing mechanism.

The control of extension growth by transmission from distant organs is less well documented. An effect on internode elongation caused by irradiation at a site removed from the growing region was reported by Borthwick, Hendricks and Parker (1951). They observed that inhibition of extension of the second internode of dark-grown barley seedlings occurred even when only the tips of extended leaves were illuminated; but it is not clear what wavelengths were used in this particular experiment. Long-distance photocontrol of elongation is, perhaps, best seen in phototropism. In coleoptiles, the tip transmits the effect of uni-lateral blue light to the elongating cells further down the organ (e.g. Curry, 1969) and in dicotyledonous seedlings the cotyledons are thought to be similarly involved in the generation of transmissible influences which travel to the hypocotyl (Shibaoka and Yamaki, 1959; Lam and Leopold, 1966).

This aspect of transmissible photocontrol, from cotyledons to hypo-cotyl, is the subject of this chapter. The photoinhibition of hypocotyl elongation has, of course, already been extensively studied (*see* e.g. Vince, 1964; Mohr, 1972). It exhibits the characteristics of 'classical' phytochrome responses involving short-term irradiation with red light as well as those of the high-energy or high-irradiance reactions which

317

require long-term exposure to blue or far-red radiation.   However, most, if not all, studies of this aspect of photomorphogenesis have been concerned with dark-grown seedlings, and none, as far as we know, has paid much attention to the possible role of the cotyledons.

In light-grown (i.e. de-etiolated) seedlings of several species, on the other hand, it seems that the cotyledons may play an extremely important part in the photocontrol of hypocotyl growth.   An indication of this can initially be obtained by wrapping one cotyledon of the de-etiolated seedling with opaque material, as described by Lam and Leopold (1966) in their studies on so-called phototropism in *Helianthus annuus*.   In several species which we have examined this treatment causes bending of the hypocotyl because of greater elongation on the side beneath the darkened cotyledon.   But, as we might expect, when both cotyledons are covered, both sides of the hypocotyl extend more than when the cotyledons are fully illuminated, i.e. the hypocotyl shows a greater growth increment (*Table 20.1*).

Table 20.1   Growth of de-etiolated seedlings with covered and fully exposed cotyledons in white fluorescent light (1500 μW cm^{-2})

Species	Extension (%)	
	Covered cotyledons	Uncovered cotyledons
Lycopersicon esculentum	30.0	15.0
Brassica oleracea var. capitata	42.0	16.0
Cosmos bipinnatus	70.0	47.0
Helianthus annuus	53.0	36.0
Cucumis sativus cv. Ridge Greenline	90.0	20.0
Lactuca sativa cv. Grand Rapids	39.0	11.6

This effect is well exhibited by *Cucumis* seedlings.   These seedlings are sturdy, with relatively large cotyledons, and are therefore suitable for further investigation of the phenomenon (Black and Shuttleworth, 1974).   The stimulatory effect of covering the cotyledons on extension growth of the hypocotyl is maintained over at least 3 days (*Figure 20.1*) and at this time the inhibition by white fluorescent light is reduced by 45 per cent.   Shading of the hypocotyl or interference with gas exchange by the cotyledons does not account for the observed growth promotions (*see below*).   These findings strongly suggest that an appreciable fraction of the total inhibitory effect of white fluorescent light on hypocotyl elongation of de-etiolated seedlings is perceived by and transmitted from the cotyledons.

Inhibition of growth in etiolated seedlings of *Cucumis sativus* is due to the red, blue and far-red regions of the spectrum (Engelsma and Meijer, 1965a; Meijer, 1968).   In de-etiolated seedlings, however, only blue and red light reduce the growth rate; sensitivity to far-red

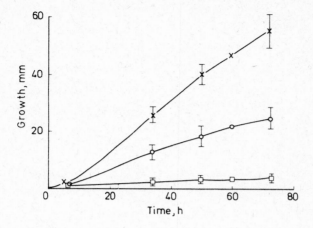

Figure 20.1 Participation of the cotyledons in the response to white fluorescent light. Growth was recorded of the 1-cm apical region of de-etiolated 6¼-day-old seedlings of Cucumis sativus *L. cv. Ridge Greenline. Seedlings were placed in the light with cotyledons uncovered (□) or covered with foil (○). Dark controls were de-etiolated seedlings returned to darkness. Intensity: 950 µW cm⁻². (Vertical bars = standard deviation)*

light seems to be lost (*Figure 20.2*). These results were obtained with continuous illumination but intermittent red light irradiation (5 min every 2 h) is almost equally effective and is reversible by 15 min of far-red light. Operational criteria for the involvement of phytochrome are therefore satisfied.

If blue light (presumably a high-irradiance reaction) and red light (phytochrome) can both inhibit hypocotyl extension, the question now arises as to which of these might act through the cotyledons. The effectiveness of blue light is unaltered by covering the cotyledons but the action of red light is strikingly reduced (*Figure 20.3*). Enclosing the cotyledons with transparent polythene has no distinguishable effect on the photoinhibition of hypocotyl growth, nor is there a strong effect resulting from covering mainly the lower surface of the cotyledons with aluminium foil (*Table 20.2*). Hindrance to gas exchange and the shading of the hypocotyl can therefore be ruled out as interfering factors.

A large proportion (in some experiments as much as 80 per cent) of the inhibitory action of red light on hypocotyl elongation in de-etiolated *Cucumis* seedlings is therefore transmitted from the cotyledons. The different contributions to the total photoinhibition made by cotyledons and hypocotyl is shown in *Figure 20.4*. The hypocotyl itself evidently has some direct perception of red light but its major sensitivity is to blue light, which acts directly upon it and whose action is not mediated by the cotyledons.

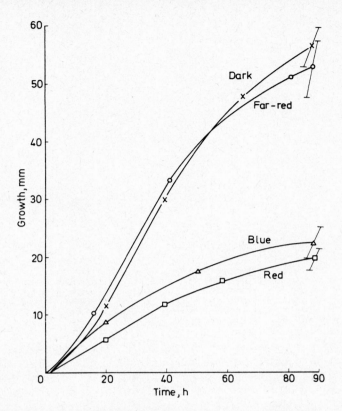

Figure 20.2   Spectral sensitivity of 6¼-day-old de-etiolated seedlings.   Illumination was continuous.   Growth is of the 1-cm apical region.   Intensities: blue = 300 µW cm⁻²; far-red = 500 µW cm⁻²; red = 350 µW cm⁻²

The energy levels of red light required for this 'cotyledon effect' in de-etiolated seedlings are much higher than those normally involved in phytochrome-controlled responses; in fact, most of our experiments have been carried out at irradiances which, although high, have nevertheless been below saturation.   This requirement for high irradiances probably results from the chlorophyll content of the cotyledons.   When dark-grown seedlings are de-etiolated by far-red light, the apical hooks straighten and the cotyledons expand but chlorophyll production is poor.   In the case of such pale-green seedlings, much lower red light intensities are effective upon the cotyledons and saturation is achieved at energy levels that are barely active upon fully green seedlings (*Figure 20.5*).

The photomorphogenic participation of the cotyledons in hypocotyl growth is a feature of de-etiolated seedlings; what is the situation in etiolated material?   The spectral sensitivity of etiolated *Cucumis* seedlings includes the far-red region, although its effect diminishes with

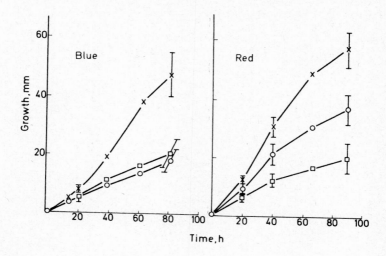

Figure 20.3 Participation of the cotyledons in the spectral responses of de-etiolated seedlings. 6¼-day-old de-etiolated seedlings were kept in continuous light with cotyledons uncovered (□) or covered (○). x, Dark controls. Intensities as in Figure 20.2. *Growth is of the 1-cm apical region*

Table 20.2 Effect of covering the cotyledons on hypocotyl growth of 6¼-day-old de-etiolated seedlings

Cotyledons were darkened by covering them with aluminium foil. Gas exchange through illuminated cotyledons was impeded by covering them with polythene 'envelopes'. To determine if shading of the hypocotyl can account for the ptomoting effect of darkened cotyledons, the undersides only were covered with foil. Continuous red light (350 μW cm^{-2}). Growth of the 1-cm apical region was recorded after 40 h. Ten seedlings were used for each treatment

Treatment	Growth (mm)
Whole seedling in darkness	62.9 ± 8.8
Whole seedling in light	19.3 ± 7.5
Cotyledons fully darkened	42.4 ± 7.6
Underside of cotyledons covered	32.2 ± 6.3
Cotyledons in polythene cover	17.4 ± 6.3

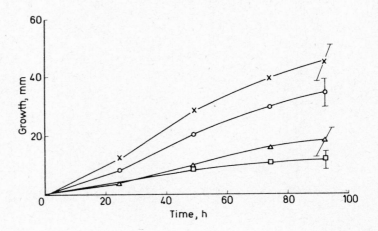

Figure 20.4   Sensitivities of hypocotyls and cotyledons to red light. Growth is of the 1-cm apical region of 6¼-day-old de-etiolated seedlings in continuous light with cotyledons covered (○) or hypocotyls covered (△).   □, Fully exposed seedlings;   x, dark controls. Intensities as in Figure 20.2

age (*Figure 20.6*); blue and red light are also inhibitory. As expected, the action of blue light is not modified by covering the cotyledons and, surprisingly, nor is that of far-red light. Moreover, protecting the cotyledons from illumination achieves only a slight relief of red light inhibition (*Figure 20.7*). The photophysiology of dark-grown seedlings with respect to the inhibition of hypocotyl growth is therefore rather different from that of de-etiolated plants.

We shall now consider the phenomenon of phototropism in green seedlings and the part of the cotyledons therein. As mentioned above, it is thought that the cotyledons are analogous to the coleoptile tip which perceives blue light and transmits the effect to the elongating zone. Experiments designed to simulate the effect of unilateral white light have been carried out with green seedlings (e.g. *Helianthus annuus*) by covering one cotyledon (Lam and Leopold, 1966). Under these circumstances, the hypocotyl grows more on the side below the darkened cotyledon, so producing a growth curvature. Such a curvature can also be induced in *Cucumis* by the same method. However, as shown above, inhibition of growth by irradiated cotyledons is due to red light only. This is also true, of course, even when only one cotyledon is illuminated. In fact, the simulation of a 'phototropic' curvature in *Cucumis* by darkening one cotyledon is possible in red light but no curvatures occur in blue light. The phenomenon is

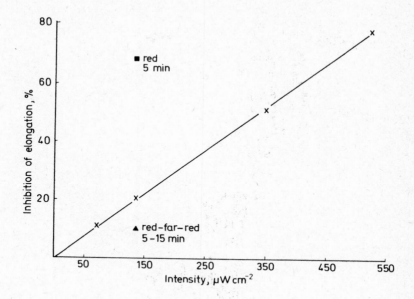

Figure 20.5   Intensity/response curve for the 'cotyledon effect' in de-etiolated seedlings.   Seedlings were de-etiolated by exposure to white fluorescent light for 30 h and were then exposed to continuous red light at different intensities. The isolated points, separate from the curve, represent the response of seedlings de-etiolated with far-red light.   They were treated with 5 min of red light every 2 h at the intensity shown.   Reversibility by far-red light is also shown.   ■ and ▲ refer to pale green seedlings expanded in far-red light

therefore not phototropism as generally accepted.   Moreover, true phototropic curvatures are induced by unilateral blue light (but not by red light) even when both cotyledons are covered with aluminium foil (*Table 20.3*).   In *Cucumis*, then, the hypocotyl itself perceives the unilateral blue light which provokes phototropic bending.   Experiments with one covered cotyledon, performed in white light, lead to incorrect conclusions concerning the role of cotyledons in so-called phototropism.

De-etiolated *Cucumis* seedlings therefore provide a good example of organ interaction in the photocontrol of extension growth.   The question which now arises is how do cotyledons irradiated with red light inhibit the growth of the hypocotyl?   Answers can be given in hormonal terms, invoking two kinds of chemical communication between the two organs.   Firstly, phytochrome activity might interfere with the supply of essential growth substances to the hypocotyl, either by reducing their transport or by lowering their concentrations in the cotyledons.   Secondly, red light might cause an increase in transmission

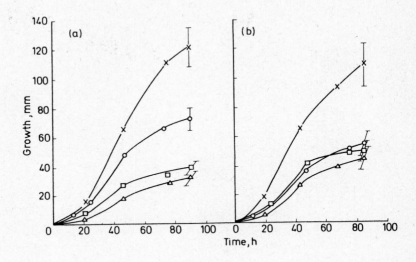

Figure 20.6 Effect of age upon the spectral response of etiolated seedlings. (a) 5 days old; (b) 4 days old. Growth of the 1-cm apical region is shown. Continuous light at intensities as in Figure 20.2. Δ, Blue; □, red; ○, far-red; x, dark

Table 20.3 Participation of cotyledons in inducible growth curvatures in *Cucumis sativus* hypocotyls

Treatment	Angle of curvature (°)	
	Red light	*Blue light*
Non-unilateral light: one cotyledon covered	+39.1 ± 7.2[1]	0
Unilateral light: both cotyledons covered	0	+34.6 ± 6.5
Unilateral light: cotyledons uncovered	0	35.6 ± 7.6

[1] Curvature is away from the covered cotyledon

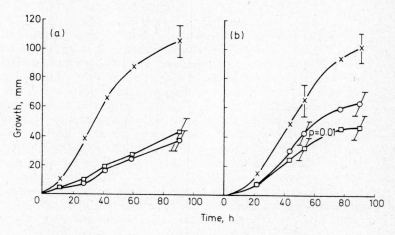

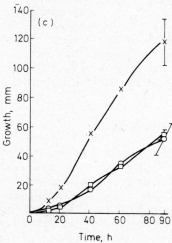

Figure 20.7 Participation of the cotyledons in photocontrol of hypocotyl elongation in 4-day-old etiolated seedlings. Growth of the 1-cm apical region is shown. □, Cotyledons uncovered; ○, cotyledons covered; x, dark control. Continuous light intensities as in Figure 20.2. (a) Blue light; (b) red light; (c) far-red light

of inhibitors from the cotyledons by promoting their transport and/or their biosynthesis.

Our studies so far lead us to favour the second possibility and to suggest that this may at least partly explain the role of the cotyledons.

'Diffusates' were collected in water from cotyledons (excised just below the cotyledonary node) which were held in darkness or in red light for 40 h. These were then fractionated as shown in *Figure 20.8*. Assayed with wheat coleoptiles and *Cucumis* seedlings (in darkness), the neutral fractions of the diffusates from red light-treated cotyledons showed considerably more inhibitory activity than those from cotyledons kept in darkness (*Figure 20.9*). Thin-layer chromatography of these neutral fractions reveals two inhibitory zones active against wheat coleoptiles (*Figure 20.10*). Both zones are more inhibitory in the neutral fraction of diffusates from red light-treated cotyledons. The slower running inhibitor has the same R_F value as xanthoxin in this solvent system (Zeevaart, 1974).

This region of the chromatogram shows no inhibition when tested with *Cucumis* seedlings but the faster running inhibitor is detectable, again with more in the diffusate from illuminated cotyledons. Thus, the evidence from these preliminary studies suggests that two inhibitors, one of which may be xanthoxin, are present in greater abundance in the material diffusing from red light-treated cotyledons. We have so far been unable to find any difference in the yield of abscisic acid following illumination of cotyledons. Further, no changes in the levels of gibberellins and auxins have yet been detected following exposure to light, at least over the time period of our experiments.

General Discussion and Conclusions

De-etiolation of *Cucumis* seedlings changes their photophysiology with respect to hypocotyl elongation. Firstly, their spectral sensitivity alters so that only red and blue light inhibit extension growth. Secondly, there develops to a large extent a morphological separation of the actions of these two wavelength regions. Red light now activates phytochrome in the expanded cotyledons so that these organs consequently exert an inhibitory influence upon elongation of the hypocotyl. The cotyledons are insensitive to blue light, however, which continues to act directly upon the hypocotyl. These findings explain why the inhibition by white light of *Cucumis* seedling growth is partially relieved when the cotyledons are covered. They also account for the curvature which develops in white light when one cotyledon is covered. The latter behaviour is exhibited by de-etiolated seedlings of many species, as is also the beneficial effect on hypocotyl growth of covering both cotyledons. It seems possible that the characteristics of the 'cotyledon effect' which we have described for *Cucumis* might be a widespread feature in de-etiolated seedlings. More knowledge is needed about this and we are therefore beginning to analyse the photophysiology of seedlings of several species.

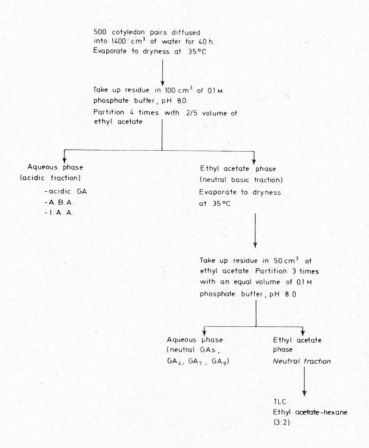

Figure 20.8   Scheme for fractionation of diffusates from cotyledons

We might place this 'cotyledon effect' in an environmental context and relate it to the behaviour of seedlings in nature.   When the cotyledons appear above ground and begin to expand because of their exposure to light, it is likely that they commence to regulate the growth of the hypocotyl even though this itself is in darkness.   Activation of phytochrome in the expanded cotyledons generates the 'signals' which decrease the growth rate of the underground hypocotyl.

The control exerted by the cotyledons in de-etiolated seedlings raises some questions concerning our understanding of photomorphogenesis.   For plausible reasons, analytical studies on the photophysiology

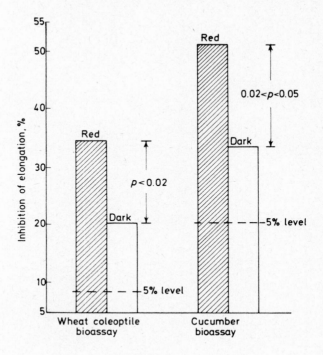

Figure 20.9   Inhibitory activity of 'diffusates' from dark and red light-treated cotyledons.   Diffusates were collected over 40 h, after which they were fractionated as indicated in Figure 20.8.   *The neutral fraction was transferred to water and assayed against wheat coleoptiles and de-etiolated* Cucumis *seedlings which were returned to darkness.   (The diffusates equivalent to that from 13 and 150 cotyledon pairs were used for the* Cucumis *and wheat assay, respectively)*

of hypocotyl growth have been performed with dark-grown material (e.g. Vince, 1964; Meijer, 1968; Turner and Vince, 1969; Mohr, 1972). Such studies have laid the foundation of our knowledge of the photocontrol of growth and their contribution has been extremely valuable. In nature, however, the dark-grown (i.e. etiolated) condition is only transient and for most of its life history the control of growth by light is exerted over the plant in its de-etiolated condition.   Leading from our observations on *Cucumis* seedlings, we might wonder how far the photophysiology of more mature plants differs from that of dark-grown seedlings.   At least one possibility is that the true leaves may make an important photomorphogenic contribution to the control of stem growth.   This clearly needs to be investigated.

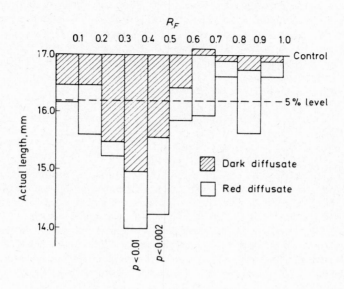

Figure 20.10 Inhibitory activity of chromatographed neutral fractions of diffusates from cotyledons. Neutral fractions of diffusates (collected over 40 h) from 300 red light-treated and 300 dark-treated cotyledon pairs were subjected to thin-layer chromatography (Eastman 13179 silica gel plates. Solvent system: hexane–ethyl acetate, 2:3). Chromatograms were bioassayed with wheat coleoptiles

The morphological separation of action of blue and red light in de-etiolated seedlings suggests that these two spectral regions might operate through different physiological mechanisms. There is still some uncertainty as to the pigments involved in the high-energy blue and far-red reactions. There is good evidence to support the view that high-energy far-red light acts through the phytochrome system but the situation regarding blue light is unclear. Our observations that a clear-cut phytochrome-mediated response occurs in one organ (i.e. the cotyledons) whereas a blue light response occurs in another might be a contribution to this debate.

Transmission of the action of red light from cotyledons to hypocotyl which controls growth in *Cucumis* is only one of the transmissible phytochrome effects in plants which have been reported (*see* e.g. Grill and Vince, 1964; De Greef and Caubergs, 1972a,b,c; Oelze-Karow and Mohr, 1973). In *Cucumis* itself (in gherkin), the synthesis of phenolic acids in the hypocotyl apparently depends upon exposure of the cotyledons to red light (Engelsma and Meijer, 1965a,b). As we have discussed above, one explanation for the organ interaction affecting

330 Inter-organ effects in the photocontrol of growth

hypocotyl elongation can be given in hormonal terms. A promising approach follows from the observation that diffusates from red light-illuminated cotyledons have more inhibitory activity than those from cotyledons held in darkness. Interestingly, the evidence suggests the presence of xanthoxin in the inhibitory diffusate. Several workers have found with other species that the level of this inhibitor increases upon exposure of plants to red light (Burden *et al.*, 1971; Anstis, Friend and Gardner, 1975). The movement of xanthoxin from cotyledons to the hypocotyl would seem to confer upon it a hormonal status.

Xanthoxin in plant tissues may originate from carotenoids such as violaxanthin (Taylor and Smith, 1967; Firn and Friend, 1972). In darkness, the biosynthesis of carotenoids remains at a relatively low level but accumulation begins after illumination with, for example, far-red light (Schnarrenberger and Mohr, 1970). Thus, light is needed in order to provide the precursors for xanthoxin production. We might speculate that this is why only expanded de-etiolated cotyledons operate in the photoregulation of hypocotyl growth, since cotyledons of dark-grown seedlings would have low carotenoid levels.

The function of the cotyledons might, of course, turn out to be more complicated, but whatever is the final explanation, the finding of a transmitted photocontrol of hypocotyl growth in de-etiolated plants would seem to be a suitable starting point for a hormonal theory of the photoregulation of growth.

References

ANSTIS, P.J.P., FRIEND, J. and GARDNER, D.C.J. (1975). *Phytochemistry*, **14**, 31

BLACK, M. and SHUTTLEWORTH, J.E. (1974). *Planta*, **117**, 57

BORTHWICK, H.A., HENDRICKS, S.B. and PARKER, M.W. (1951). *Bot. Gaz.*, **113**, 95

BURDEN, R.S., FIRN, R.D., HIRON, R.W.P., TAYLOR, H.E. and WRIGHT, S.T.C. (1971). *Nature New Biol.*, **234**, 95

CURRY, G.M. (1969). In *Physiology of Plant Growth and Development*. Ed. Wilkins, M. McGraw-Hill, London

DE GREEF, J.A. and CAUBERGS, R. (1972a). *Physiol. Plant.*, **26**, 157

DE GREEF, J.A. and CAUBERGS, R. (1972b). *Arch. Int. Physiol. Biochim.*, **80**, 961

DE GREEF, J.A. and CAUBERGS, R. (1972c). *Arch. Int. Physiol. Biochim.*, **80**, 959

ENGELSMA, G. and MEIJER, G. (1965a). *Acta. Bot. Neerl.*, **14**, 54

ENGELSMA, G. and MEIJER, G. (1965b). *Acta Bot. Neerl.*, **14**, 73

FIRN, R.D. and FRIEND, J. (1972). *Planta*, **103**, 263

GRILL, R. and VINCE, D. (1964). *Planta*, **63**, 1

LAM, S.L. and LEOPOLD, A.C. (1966). *Plant Physiol.*, **41**, 847

MEIJER, G. (1968). *Acta Bot. Neerl.*, **17**, 9

MOHR, H. (1972). *Lectures on Photomorphogenesis.* Springer, Berlin, Heidelberg, New York

OELZE-KAROW, H. and MOHR, H. (1973). *Photochem. Photobiol.,* **18,** 319

SCHNARRENBERGER, C. and MOHR, H. (1970). *Planta,* **94,** 296

SHIBAOKA, H. and YAMAKI, T. (1959). *Sci. Pap. Coll. Gen. Educ. Univ. Tokyo,* **9,** 105

TAYLOR, H.F. and SMITH, T.A. (1967). *Nature, Lond.,* **215,** 1513

TURNER, M.R. and VINCE, D. (1969). *Planta,* **84,** 368

VINCE, D. (1964). *Biol. Rev.,* **39,** 506

ZEEVAART, J.A.D. (1974). *Plant Physiol.,* **53,** 644

21

PHYTOCHROME-CONTROLLED ACETYLCHOLINE SYNTHESIS AT THE ENDOPLASMIC RETICULUM

M.J. JAFFE
Botany Department, Ohio University, Athens, Ohio, U.S.A.

Introduction

In order to know how phytochrome mediates development and growth, questions must be answered concerning its intracellular location and how it is linked to pertinent cellular processes. In the past few years, several laboratories have been seeking, by various means, to answer the first question. Thus Marmé *et al.* (1974) and Boisard, Marmé and Briggs (1974) have reported that phytochrome can be shown to be associated with vesicles derived from cell membranes, although they could not identify the origin of those membranes. In a preliminary report, Williamson and Morré (1974) reported centrifuging crude membrane vesicle preparations through a sucrose gradient and, by using previously determined recognition criteria, they were able tentatively to identify the rough endoplasmic reticulum and possibly the plasmalemma as intracellular locations for native phytochrome. This work, which was carried out in collaboration with my laboratory, will be summarised in this chapter and published in detail elsewhere.

The answer to the question of the function of phytochrome must be dependent on a prior knowledge of its location in the cell. If, for the sake of argument, we assume that physiologically active phytochrome must be associated with some membrane or membranes, this still does not tell us how it acts. One possibility is that it may act through the mediation of changes in the energetics of the cells or tissues in which it resides. Evidence for such a possibility in bean roots has been provided by Yunghans and Jaffe (1972), who showed that red light induced a rapid increase in oxygen uptake which was reversed by far-red light, and that ATP was used up during red light irradiation. Such an effect was also seen by White and Pike (1974) in etiolated bean buds, but they found an increase in ATP levels which followed the initial decrease.

If aerobic respiration is involved in the transduction of light energy to a form usable for developmental phenomena, phytochrome might be expected to be present in mitochondria. Manabe and Furuya (1974) reported that effect and further demonstrated that irradiated mitochondria were capable of phytochrome-dependent reduction of nicotinamide nucleotide. Another suggestion for a possible mode of action for phytochrome was made by Jaffe (1970). He demonstrated that acetylcholine (ACh), the neurotransmitter of the animal central nervous system, was present in bean roots, and that its titre was increased by irradiation with red light and decreased by far-red light. *Table 21.1* gives a summary of the differences in endogenous ACh between red- and far-red light-irradiated tissues of various plants, in the only three cases where such experiments have been reported in the literature. In both etiolated mung bean root tips (assayed by bioassay; Jaffe, 1970) and a callus derived from moss, assayed by gas–liquid chromatography (Hartmann and Kilbinger, 1974), the ACh titre of red light-irradiated tissue is much higher than that of tissue irradiated with far-red light or kept in the dark. When the small tertiary pulvinules of *Albizzia julibrissin* were irradiated and the extracts examined for ACh by bioassay (Satter, Applewhite and Galston, 1972), no differences were found between red- and far-red light-irradiated tissue. The authors pointed out, however, that it is only one side of the pulvinule which contracts due to phytochrome mediation, and putative ACh changes in this side might be masked by opposite changes in the other side. Although we have presented evidence that ACh is photomimetic for red light in inducing respiratory and ATP changes in bean roots (Yunghans and Jaffe, 1972) and that cholinergic drugs can modify phytochrome-mediated changes in the uptake of ^{14}C-labelled sodium acetate, in ways that would be predicted if the acetylcholine system that exists in animals were also present in the plant tissue (Jaffe and Thoma, 1973), it is not known if ACh mediates phytochrome control of growth and development, or if the changes in its titre are merely reflections of other, more pertinent, cellular events.

In order to explain the phytochrome-mediated changes in the endogenous level of ACh, it is necessary to know if the synthetic and metabolic apparatus for ACh exists in plants. Such a system is known to exist in neurons of the central nervous system of animals, and is

Table 21.1 Effect of red and far-red light on the endogenous levels of acetylcholine in different plant systems

Experiment	*Effect of red and far-red light on the ACh titre of several plants (percentage of red value)*		
	Dark	*Red*	*Far-red*
Mung bean root tips, ACh detected by bioassay	42	100	20
Tertiary pulvinules of *Albizzia julibrissin*, ACh detected by bioassay	–	100	95
Callus derived from moss, ACh detected by gas–liquid chromatography	0	100	2

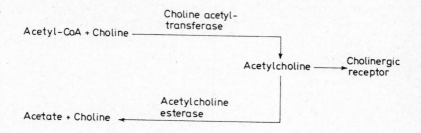

Figure 21.1   Schematic representation of the components of the acetylcholine system in animal central nervous systems

illustrated in *Figure 21.1*.   ACh is synthesised in animals by the enzyme cholineacetyl transferase (ChAT) from acetyl coenzyme A plus choline.   In nerves, it is packaged in small cholinergic vesicles in the presynaptic region of the synapse or motor end-plate.   When the propagating nerve impulse reaches the synapse, the vesicles, which are up against the plasma membrane, merge with it and release their contents to the synaptic cleft.   On the other side, at the post-synaptic membrane, they are 'grasped' by the acetylcholine receptor protein (AChR), which sets off the nerve impulse in that new neuron.   Immediately thereafter, the ACh is removed by hydrolysis from the receptor by the enzyme acetylcholineesterase (AChE).   Thus, by analogy, the increase of ACh in red light might be due to an increase in ChAT activity or a decrease in the activity of AChE.   Therefore, an effort was indicated for the identification of these components, and their ability to regulate the ACh titre during different irradiation regimes.   The available literature up to 1974 has recently been reviewed by Fluck and Jaffe (1974d).

Results and Discussion

ACETYLCHOLINE

As reviewed elsewhere (Fluck and Jaffe, 1974d), there are many reports in the literature of the presence of acetylcholine in plants, but most of these reports suffer from a lack of critical diagnostic observations.   However, recent work confirms that acetylcholine is indeed present in plants.   Jaffe (1970) chromatographed ACh with plant extracts, and found that the principle active in the clam heart bioassay (Florey, 1967) co-chromatographed with authentic ACh in two solvent systems.   He further found that when the extract was allowed to incubate with AChE of animal origin, it lost its ability to retard the beat of the clam's ventricle. Hartmann and Kilbinger (1974) detected ACh in extracts of plant tissue by means of gas–liquid chromatography. Thus it is probable that ACh is a native component of plants.

ACETYLCHOLINESTERASE

There have been a number of previous reports in the literature that acetylcholine-hydrolysing enzymic ability is present in plants (Fluck and Jaffe, 1974b). Almost all of these reports, however, present little or no evidence that such activity is due to an acetylcholinesterase or even a cholinesterase. The only previous adequate work in this area was that of Schwartz (1967), who made a preliminary characterisation of an AChE derived from etiolated pea seedlings. Riov was able to demonstrate true acetylcholinesterase activity in green and etiolated mung bean plants (Riov and Jaffe, 1973; *Table 21.2*). The AChE which has been found in members of the Leguminoseae (Fluck and Jaffe, 1974b) meets the three important criteria of a true acetylcholinesterase. Firstly, it has a greater affinity for acetylcholine as a substrate than for any other substrate tested (Riov and Jaffe, 1973; Fluck and Jaffe, 1974d); secondly, it is inhibited by high concentrations of substrate ACh (Riov and Jaffe, 1973), indicating it is probably a regulator or allosteric protein (Oosterbaan and Jansz, 1965); and thirdly, it is inhibited by low concentrations of quaternary ammonium and organophosphate molecules (Riov and Jaffe, 1973). In addition to the above criteria, plant AChE has also been found to resemble AChE of animal origin in other respects. Both have pH optima in the same region and both are of similar molecular weight and have similar subunits (Riov and Jaffe, 1973). The plant AChE is found in the polysaccharide region surrounding the protoplast (Fluck and Jaffe, 1974a) and the animal enzyme is also believed to be associated with polysaccharides on the cell surface. In addition, the basic group of the esteratic subsite of the animal AChE catalytic site has been shown to contain an imidazole group (probably on histidine) as an active component (Oosterbaan and Jansz, 1965).

By the use of reagents that are known specifically and irreversibly to bind to certain chemical groups (Baker, 1967), *Table 21.3* shows that only molecules which modify the imidazole group also inhibit plant AChE. Although AChE must assuredly be synthesised inside the cell, Fluck found that over 95 per cent of it is to be found in the cell walls of bean roots (Fluck and Jaffe, 1974a). A summary of the evidence for this is presented in *Table 21.4*. High salt concentrations, but not detergents, are capable of solubilising the enzyme from crude homogenates, indicating that it is not membrane bound. Differential centrifugation results in 95 per cent of the activity remaining in the pellet of the first, low-speed centrifugate, which contains most of the

Table 21.2 Comparison of the properties of acetylcholinesterases from bean plants with those of animal origin

Characteristic	Animal	Plant
Hydrolysis of choline esters[1]	A > P > B	A > P > B
Hydrolysis of non-choline esters	+	+
pH optimum	8.0–8.3	8.5–8.7
Curve shape of dosage response to substrate	Bell-shaped	Bell-shaped
K_m	200 μM	78 μM
Molecular weight of largest form	260 000	>200 000
Molecular weight of monomer	65 000	80 000
Inhibition by tertiary ammonium (eserine)	+++	+
Inhibition by quaternary ammonium (neostigmine)	+++	+++
Inhibition by organophosphates (paraoxon)	+++	+++
Inhibition by carbamo-ammonium (AMO-1618)	++	++
Effect of choline	Inhibits	Stimulates

[1] A = acetylcholine; B = butyrylcholine; P = propionylcholine

Table 21.3 Effect of group-specific modifying reagents on the activity of acetylcholinesterase from beans
   The AChE was pre-incubated for 30 min with the additive before making the determination.

Chemical modifier	Groups modified[1]	I_{50}[2]
Bromoacetic acid	Imidizole (++) NH$_2$ (+) SH (+)	3.7 × 10^{-4} M
N-Ethylmaleimide	SH (++)	None
2,4,6-Trinitrobenzenesulphonic acid	SH (++)	None
p-Nitrobenzene diazonium fluoroborate	Phenolic OH (++)	None
Iodine at pH 9.0	Imidizole (++) SH (+)	5 × 10^{-8} M

[1] (+) = Weak affinity; (++) = strong affinity
[2] Concentration giving 50 per cent inhibition of AChE

Table 21.4 Evidence for the subcellular localisation of acetylcholinesterase in the cell walls of bean roots

Method	Results
Treatment with detergents (Triton X-100, sodium deoxycholate)	Solubilises 3.1–5.9 per cent of AChE activity
Treatment with high salt concentration [KCl or (NH$_4$)$_2$SO$_4$]	Solubilises 29.0–36.7 per cent of AChE activity
Incubate whole roots in reaction mixture	Neostigmine-sensitive hydrolysis of ACh occurs
Differential centrifugation	92.4–96.6 per cent of activity remains in the residue
Electron micrographic cytochemical localisation of AChE activity	Found only in cell walls, primarily in the inner cell wall, adjacent to the plasma membrane

cell wall material. Cytochemical procedures coupled with electron micrography clearly indicate the AChE activity to be in the inner region of the cell walls. Finally, when whole roots are incubated in the reaction mixture, AChE activity is readily demonstrated, a result which might not be possible if both the substrate and the product have to pass the plasmalemma in order for an activity to be observed.

Plant AChE has been shown not to be pectinase (Fluck and Jaffe, 1974c). Because of its location in the cell, AChE activity could be measured in three different kinds of preparations: whole root tips, crude particulate fractions containing masses of cell wall material, and enzyme that had been extracted and solubilised (Fluck and Jaffe, 1974a).

Using the preparations mentioned above, we looked for an effect of red and far-red light on bean root AChE activity (*Table 21.5*). When intact roots were irradiated, and the subsequently extracted soluble enzyme activity was measured, it was found that there was no difference between the treatments with red and far-red light. Because this might have been due to intracellular compartmentalisation of the enzyme and an inhibitor of some part of the reaction, which we have, in fact, isolated (Fluck and Jaffe, 1974b), we tried irradiating particulate root fractions and also found no difference between the effects of red and far-red light. Finally, when whole roots were irradiated directly in the reaction mixture, the AChE activities in both red and far-red light were almost identical. On the basis of these experiments, it was concluded that the native acetylcholinesterase activity had nothing to do with the phytochrome-mediated changes in endogenous ACh.

CHOLINE ACETYL TRANSFERASE

Emmelin and Feldberg (1947) demonstrated that a mixture of histamine and acetylcholine is the toxic principle of the leaves of the stinging nettle (*Urtica urens*). Although they could find no AChE in *U. dioica* leaves, Barlow and Dixon (1973) were able to show that there was synthesising activity present. Using the acetone powder method of Fonnum (1966), Riov and Jaffe (1972) were unable to demonstrate ChAT activity due to the presence of a thioltransacetylase which acetylated the sulphydryl protectors in the reaction mixture, instead of allowing the choline to become acetylated. It was therefore necessary to find a means of removing this enzyme, which was effected as follows.

Intracellular location of phytochrome

In a collaborative effort with Morré and Williamson, an effort was made to determine the kind of cellular membranes with which

Table 21.5   Lack of effect of red (R) and far-red (FR) light on the activity of bean root acetylcholinesterase

Treatment	AChE activity (nmol min^{-1} g fresh wt.$^{-1}$)	
	4 min R	*4 min FR*
Assay on intact roots irradiated with R or FR	44.8	44.0
Intact roots irradiated with R or FR, extracted for soluble AChE, and the extracts assayed	1.14	1.20
Particulate root extract prepared, and aliquots irradiated with R or FR	0.36	0.38

phytochrome was associated. Using methods developed by Morré and co-workers, we homogenised etiolated soybean hypocotyls using very low shear forces in the presence of de-vesiculated coconut milk, SH protectors and 0.5 M sucrose as an osmoticum (Morré, Roland and Lembi, 1969). Under these conditions, the cell membranes were fragmented and spontaneously vesiculated. After a preliminary low-speed centrifugation to remove whole cells and nuclei, the remaining supernatant fluid was centrifuged at high speed through a step sucrose density gradient. The material resting at each interface was collected and re-centrifuged on a 2 M sucrose cushion for concentration.

A portion of each of these pellets was processed for electron microscopy and identification was made of the origin of the membrane vesicles (Morré, Roland and Lembi, 1969). *Figure 21.2* shows the appearance of various fractions under the electron microscope and indicates the ease with which the origin of the membranes can be identified. In A, the pellet obtained from material processed in the presence of magnesium is shown. Under this condition, phytochrome (Quail, Marmé and Schafer, 1973) is known to be more highly associated with vesicles and ribosomes are known to be associated with vesicles derived from endoplasmic reticular membrane (Sabatini, Tashiro and Palade, 1966). The remainder of the pellet, mixed with calcium carbonate as a scattering agent, was assayed for phytochrome content using a Ratiospect.

Table 21.6 shows the results of experiments designed to compare the presence of phytochrome with the presence of vesicles derived from various cellular membranes. In the absence of 9.5 mM Mg^{2+}, soluble phytochrome (non-membrane bound) is found in the supernatant fluid, and membrane-associated phytochrome is found in the fractions containing about 50 per cent plasma membrane. It may be, however, that there is little, if any, phytochrome associated with the plasma membrane vesicles, because of the results of the same experiment carried out in the presence of magnesium. In this case, in addition to the soluble phytochrome, only about 10 per cent of the particulate phytochrome is found in the fractions that contain about

50 per cent plasma membrane, whereas 48 per cent is found in the lowest pellet, where 78 per cent of the vesicles are derived from rough endoplasmic reticulum.   At no time have we found any phytochrome associated with the nuclei, or with fractions that were rich with mitochondria.   In the latter case, the present observations are at variance with those of Manabe and Furuya (1974).   This may be due to differences in plant material and preparatory procedures, but there is also the possibility that their data may be explicable in terms of connections which have been demonstrated *in vivo* between endoplasmic reticulum and mitochondria (Morré, Merritt and Lembi, 1971).

In the present work, the best correlations (r = +0.9491) have been found between rough endoplasmic reticulum membranes and phytochrome, although neither microsomes nor some sedimentable, non-membraneous cell component such as microfilaments, microtubules or lipoprotein micelles can be ruled out.   Plasma membrane also cannot be entirely ruled out, but the correlation coefficient in this case is very small (r = +0.2171) in the presence of Mg^{2+}, and the higher correlation in its absence may be due to unidentifiable smooth endoplasmic reticulum.   There is also the slight possibility that the vesicular binding of the ribosomes in the presence of Mg^{2+} is artifactual, occurring after their separation from the endoplasmic reticulum during homogenisation.   For this reason, it would be desirable to have another marker for endoplasmic reticulum membrane, and such a marker has been reported in the endoplasmic reticulum of castor bean endosperm. Lord *et al.* (1973) have shown that phosphorylcholine-glyceride transferase, the last enzyme needed for the synthesis of lecithin (the major lipid component of plant plasma membranes), is to be found only in the endoplasmic reticulum.   We hope to be able to use this as an ER marker in future work.

Phytochrome-mediated control of ChAT

Since it had been impossible to demonstrate ChAT activity in bean extracts prepared by the method used for animal material, and since the phytochrome itself had been found to be associated with rough ER vesicles, it seemed worth looking for ChAT activity in the same fraction.   For this experiment, the vesicles were prepared entirely in the dark, combined with the rest of the reaction mixture, irradiated for 4 min and incubated in the dark for 40 min, as indicated in *Table 21.7*.   The finished reaction mixture was then passed through a small Dowex 1-X8 (Cl⁻) anion-exchange column.   Radioactive acetyl-CoA and vesicles were retained on the column, while choline and newly synthesised radioactive ACh were eluted.   *Table 21.7* shows that there is a small increase in apparent ACh synthesising activity when the vesicles are irradiated with red light, over those which were irradiated with far-red light or left in the dark.

Table 21.6 Distribution of phytochrome in the sucrose density gradient in the presence or absence of Mg^{2+}

Fraction	Sucrose (M)	Prepared in absence of Mg^{2+}		Prepared in presence of Mg^{2+}	
		Components of fraction[1]	$\Delta(\Delta A)$ per mg protein	Components of fraction[1]	$\Delta(\Delta A$ per mg protein)
500 g pellet		Nuclei and starch	0.0	Nuclei and starch	0.0
7 000-90 000 g pellet					
90 000 g supernatant		Microsomes	0.009	Microsomes	0.038
A	0.8	Soluble protein 82% unknown and other	0.076	Soluble protein 88% unknown	0.059
B		27% PM, 11% Golgi, 59% unknown + some ribosomes	0.0	52% Golgi, 48% unknown	0.0
C (top)	1.0	39% PM, 61% other	0.0	55% PM, 45% unknown	0.014
C (bottom)		91% mitochondria	0.0	—	0.013
D (top)	1.2	49% PM, 51% other	0.0	47% PM, 53% other	0.022
D (bottom)		77% mitochondria, 17% other	0.058	80% mitochondria	0.001
Pellet (top)	1.4	49% PM, 50% other	0.043	78% rough ER	0.137
Pellet (bottom)		19% PM, 32% mitos, 49% other	0.0	78% rough ER	0.137

[1] PM = plasma membranes; mitos = mitochondria; ER = endoplasmic reticulum

Table 21.7   Phytochrome-mediated control *in vitro* of the synthesis of acetylcholine by membrane vesicles

Reaction mixture: 100 μl PO_4 buffer + choline, plus 50 μl [^{14}C] acetyl-CoA, plus 50 μl vesicle prep., or water.   Irradiation: 0 min (dark), or 4 min red, or 4 min far-red.   Incubation: 40 min at 26 °C in the dark.   Add 20 μl of water, or Triton X-100 to make 0.5 per cent.   ACh determination: add the 0.22 ml to a 5-cm anion-exchange column, then elute with two successive 0.6-ml aliquots of distilled water directly into a scintillation vial.

Condition	d.p.m. per 25 g fresh weight		
	Dark	*Red*	*Far-red*
Minus Triton X-100	199	358	286
Plus Triton X-100	111	2014	20

Such a small increase was unsatisfactory in the light of the 5-fold greater ACh titre in red light-irradiated roots over far-red light-irradiated roots (*Table 21.1*).   It seemed possible that, as it was synthesised, the ACh became trapped in the membrane of the vesicles, or in their lumens, so a membrane-disruptive treatment with Triton X-100 was included immediately following incubation and prior to layering on the anion-exchange column.   When this was done, the results were more consistent with those in *Table 21.1*.   Vesicles irradiated with red light synthesised 100 times more apparent ACh than did those irradiated with far-red light.   Critical characterisation of the product of the reaction has not yet been carried out, but the conditions are those which result in only labelled ACh when preparations of animal origin are used as the source of the enzyme.   It is assumed that the detergent effect of Triton X-100 is to break up the ER vesicle membrane, releasing the radioactive ACh and allowing it to elute from the column. If the ACh is preferentially secreted into the lumen of the vesicle as it is synthesised, there is an implication of 'sidedness' to the membrane, whereas if the newly synthesised ACh is incorporated into the membrane itself, 'sidedness' is not indicated.   The nature of the physiological role of ACh may be dependent on the answer to that question.

Since the lecithin-synthesising enzyme is also known to be in the ER membrane, we shall consider if phytochrome can also mediate its activity.   If it can, such mediation may be either direct or indirect. There is reason to believe that such a mediation may occur, since Hokin and Hokin (1953) and Hokin-Neaverson (1974) have shown an effect of ACh on phospholipid metabolism in pigeon pancreas.   When slices of pancreas were incubated with $^{32}P_i$, the addition of ACh to the incubation medium produced an increased labelling of phospholipids.   This 'phospholipid effect' has been demonstrated in a large number of animal tissue and cell types (Fluck and Jaffe, 1974b), and is usually an increased turnover of the phosphate groups in phosphatidic acid and phosphatidylinositol.   The 'phospholipid effect' may represent a more general function of ACh, and since lecithin is a phosphatidyl-choline derivative, it may represent a useful approach to the question of the function of ACh in plant tissues. It may be that one way in which phytochrome mediates plant growth and development is by regulating cell membrane synthesis at the endoplasmic reticulum.

Acknowledgements

I am pleased to acknowledge the following colleagues for their stimulating collaboration in various aspects (specified in the text) of the above work: Prof. D. James Morré, Prof. Richard Fluck, Dr Joseph Riov and Dr Francis Williamson. I am grateful to Dr Carl Norris of the United States Department of Agriculture Plant Industry Station, Beltsville, Md., for the loan of the Ratiospect instrument, and the U.S. National Science Foundation for a research grant (GB33257) which provided financial support for parts of this work.

References

BAKER, B.R. (1967). *Design of Active-Site-Directed Irreversible Enzyme Inhibitors: The Organic Chemistry of the Enzymic Active Site,* 325pp. Wiley, New York

BARLOW, R.B. and DIXON, R.O.D. (1973). *Biochem. J.,* **132,** 15

BOISARD, J., MARMÉ, D. and BRIGGS, W.R. (1974). *Plant Physiol.,* **54,** 272

EMMELIN, N. and FELDBERG, W. (1947). *J. Physiol.,* **106,** 440

FLOREY, E. (1967). *Comp. Biochem. Physiol.,* **20,** 365

FONNUM, F. (1966). *Biochem. J.,* **100,** 479

FLUCK, R.A. and JAFFE, M.J. (1974a). *Plant Physiol.,* **53,** 752

FLUCK, R.A. and JAFFE, M.J. (1974b). *Phytochemistry,* **13,** 2475

FLUCK, R.A. and JAFFE, M.J. (1974c). *Plant Physiol.,* **54,** 797

FLUCK, R.A. and JAFFE, M.J. (1974d). *Curr. Adv. Plant Sci.,* **5,** 1

HARTMANN, E. and KILBINGER, H. (1974). *Biochem. J.,* **137,** 249

HOKIN, M.R. and HOKIN, L.E. (1953). *J. Biol. Chem.,* **203,** 967

HOKIN-NEAVERSON, M. (1974). *Biochem. Biophys. Res. Commun.,* **58,** 763

JAFFE, M.J. (1970). *Plant Physiol.,* **46,** 768

JAFFE, M.J. and THOMA, L. (1973). *Planta,* **113,** 283

LORD, J.M., KAGAWA, T., MOORE, T.S. and BEEVERS, H. (1973). *J. Cell Biol.,* **57,** 659

MANABE, K. and FURUYA, M. (1974). *Plant Physiol.,* **53,** 343

MARMÉ, D., MACKENZIE, J.M. Jr., BOISARD, J. and BRIGGS, W.R. (1974). *Plant Physiol.,* **54,** 263

MORRÉ, D.J., ROLAND, J.-C. and LEMBI, C.A. (1969). *Proc. Indiana Acad. Sci.,* **79,** 96

MORRÉ, D.J., MERRITT, W.D. and LEMBI, C.A. (1971). *Protoplasma,* **73,** 43

OOSTERBAAN, R.A. and JANSZ, H.S. (1965). In *Comprehensive Bio-chemistry*, Vol.16, pp.1–47. Ed. Florkin, E. and Stotz, M.

QUAIL, P.H., MARMÉ, D. and SCHAFER, E. (1973). *Nature, Lond.*, **245**, 189

RIOV, J. and JAFFE, M.J. (1972). *Phytochemistry*, **11**, 2437

RIOV, J. and JAFFE, M.J. (1973). *Plant Physiol.*, **51**, 520

ROLAND, J.-C., LEMBI, C.A. and MORRÉ, D.J. (1972). *Stain Tech.*, **47**, 195

SABATINI, D.D., TASHIRO, Y. and PALADE, G.E. (1966). *J. Mol. Biol.*, **19**, 503

SATTER, R.L., APPLEWHITE, P.B. and GALSTON, A.W. (1972). *Plant Physiol.*, **50**, 523

SCHWARZ, O.J. (1967). *Masters Thesis,* North Carolina State University, Raleigh

WHITE, J.M. and PIKE, C.S. (1974). *Plant Physiol.*, **53**, 76

WILLIAMSON, F.A. and MORRÉ, D.J. (1974). *Plant Physiol.*, Suppl., 46

YUNGHANS, H. and JAFFE, M.J. (1972). *Plant Physiol.*, **49**, 1

V

PHOTOPERIODISM, ENDOGENOUS RHYTHMS AND PHYTOCHROME

22

PHYTOCHROME AND PHOTOPERIODISM

DAPHNE VINCE-PRUE
Department of Botany, Plant Science Laboratories, University of Reading, Reading, U.K.

Seasonal adaptations are essential for the survival of plants and animals in many environments and the switch from one metabolic pattern to another is frequently controlled by the durations of light and darkness in the daily cycle. Seasonal adaptations can be effected by other environmental signals such as temperature but photoperiodism is often the overriding factor. Photoperiodic responses of many different kinds are seen in plants and animals from a variety of taxonomic groups (Vince-Prue, 1975; Altman and Dittmer, 1966; Beck, 1968; Loftes, 1970) and it may be that the formal similarities which occur among organisms that utilise the length of day and/or night as a signal for the time of year are merely reflections of a convergent evolution which obscure significant differences in underlying mechanisms (Pittendrigh, 1972). The search for a single mechanism may, therefore, prove unrewarding.

A photoperiodic response requires the ability to distinguish between light and darkness. In plants, the evidence points to phytochrome being the major and probably the only photoreceptor, but its precise role is unknown. The phenomenon of photoperiodism implies that in some sense the organism measures time, and it is the involvement with time measurement which complicates the analysis of phytochrome control of photoperiodic responses. A major debate for many years has been whether the clock that effects photoperiodic time measurement is some sort of hour-glass or a circadian oscillation of the type known to be utilised as the chronometer in instances where organisms identify phases of the daily cycle, such as in overt rhythms of leaf movement. It is still an open question whether a photoperiodic response of plants under natural conditions involves only an hour-glass component of phytochrome behaviour (Evans and King, 1969), an action of phytochrome linked to a circadian oscillation (Cumming, Hendricks and Borthwick, 1965) or is dependent on some combination of both (Hillman, 1971; Wagner and Cumming, 1970).

Even though it is abundantly clear that circadian rhythmicity is somehow involved in the photoperiodism of many plants, it is by no means clear what form this involvement takes. Pittendrigh (1972) has pointed out that there are several ways in which circadian periodicity might affect photoperiodism: it could be involved without itself effecting photoperiodic time measurement, and if it does function as a clock there is more than one way in which this might occur. Three possibilities are considered in detail by Pittendrigh and these are briefly as follows.

If the circadian clock is the timer, then photoperiodic induction might occur or fail depending on whether a particular phase of the rhythm coincides in time with a light (or dark) signal in the daily cycle. Only in certain photoperiods will the entrained steady state cause the coincidence of the inducible phase with light (or dark). This 'external coincidence' model (*see Figure 22.1*) is considered by many workers to be the operative mechanism in photoperiodism. The change in effectiveness of periodic light perturbations given during long

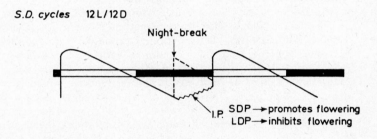

Figure 22.1   Circadian model for photoperiodic responses based on that for the Xanthium *clock.   I.P. = inducible phase.   SDP = short-day plant, LDP = long-day plant.   (From Papenfuss and Salisbury, 1967)*

daily dark periods [for example in *Glycine* (Coulter and Hamner, 1964) and *Chenopodium* (Cumming, Hendricks and Borthwick, 1965)] indicates that, at least in some plants, there is a rhythmic change in the ability of a light signal to induce flowering. This is usually interpreted as coincidence or non-coincidence of the light signal with an inducible phase which varies rhythmically. The use of short light pulses in the short-day plant *Lemna perpusilla* grown in axenic culture has revealed a rhythm of sensitivity to light which originates with a light-off signal when plants are transferred from continuous light to darkness; flowering is delayed when light pulses are given at some phases of the postulated rhythm and accelerated by light given at other times (Hillman, 1964, 1970). Again an external coincidence model can explain these results.

The second possibility for circadian clock timing in photoperiodism is an 'internal coincidence' mechanism. This model assumes a population of circadian oscillations. The phase of one oscillator is set by dawn and that of a second is set by dusk; only in some photoperiods would the two oscillations be entrained in such a way that their critical phase points coincide and effect a photoperiodic response. Such an internal coincidence model seems to explain best the date for control of diapause in a wasp, *Nasonia vitripennis* (Saunders, 1973). If photoperiodic responses are effected by the internal coincidence of a multi-oscillator system, the environmental light cycle would play no direct part in the photoperiodic process but serve only to entrain the oscillators. Rhythms set by dawn (light-on) and dusk (light-off) signals have been demonstrated in plants and have been implicated in photoperiodic responses. For example, a model for the photoperiodic control of flowering in *Xanthium* (Papenfuss and Salisbury, 1967; *see Figure 22.1*) assumes that two rhythms are involved. Transfer to light (light-on) starts a rhythm that begins to oscillate into the dark state after about 10 h: however, after about 13 h of light the clock is suspended and it begins to oscillate again only on transfer to darkness, i.e. in response to a light-off signal. A recent study of leaf movements in *Xanthium* (Hoshizaki, Brest and Hamner, 1974) clearly demonstrated that light-on and light-off signals acted to phase two different rhythms. Light-on resulted in a peak of downward movement 16 h later, while light-off gave an immediate and sudden upward movement, reaching a peak in about 4 h. In various light/dark (L/D) cycles, the two rhythms interacted. In cycles between 8L/16D and 14L/10D, the downward position of leaves anticipated the forthcoming light signal by a time which gradually decreased as the photoperiod was lengthened until, with a 16L/8D cycle, it occurred 30 min after the light-on. It is interesting that the critical photoperiod for flowering in *Xanthium* is between 14 and 16 h. The same workers observed that in another short-day plant, *Glycine max,* a downward position of the leaves at the beginning of the photoperiod also occurred only in cycles favourable for flowering (*Table 22.1*).

Both light-on and light-off rhythms have also been demonstrated in the control of flowering in the short-day plant *Pharbitis nil,* but it was

Table 22.1 Effects of cycle length on flowering and on the direction of leaf movement at the beginning of the light period for *Glycine max* cv. Biloxi (From Brest, Hoshizaki and Hamner, 1971)

Cycle length (h)	Leaf position prior to beginning of light period	Number of flowering nodes
24	Down	32
32	Up	1
48	Down	34
60	Up	8
72	Down	31

concluded that the critical night length was determined by the latter (Takimoto, 1966). Flower induction in *Pharbitis* appeared to involve an hour-glass component in addition to the light-off rhythm because, with a single inductive night, the response was quantitative and increased in magnitude depending on the absolute duration of darkness.

The third possibility considered by Pittendrigh (1972) is that physiological function might be changed if the rhythm is not in resonance with the entraining cycle, and some responses commonly interpreted as being due to external coincidence might arise in this way (for example the rhythmic variations in flowering which are often obtained with different cycle lengths, as in *Table 22.1*). In such a case the circadian organisation is not necessarily involved in time measurement.

The physiological responses of photoperiodism are removed in time and space from the initial photo-act. It is not surprising, therefore, that the precise function of phytochrome is poorly understood and, despite much phenomenology, no really satisfying model for phytochrome action in photoperiodism has yet been developed. If time measurement is effected by a circadian clock, then phytochrome could interact with an inducible light-sensitive phase of a rhythm generated by the light to dark transitions at dawn and/or dusk. However, phytochrome is known to act as a synchroniser or *zeitgeber* in some overt rhythms (Hillman, 1971; Wilkins and Harris, Chapter 25), and Papenfuss and Salisbury (1967) have proposed that the only action of light in the photoperiodic control of flowering in *Xanthium* is on the operation of the clock. If an internal coincidence mechanism model operates, phytochrome would be expected to play no direct role in the inductive process but would only affect the phases of the endogenous circadian oscillators. Finally, phytochrome might interact with an hour-glass component of timing or even (perhaps by reversion) generate such a component.

A more detailed consideration of some experiments concerning effects of phytochrome in photoperiodic responses will now be presented. An important fact about the photoperiodic mechanism is that it operates only in de-etiolated plants and dark-grown plants do not appear to show photoperiodic responses. Seedlings of the short-day plant *Pharbitis nil* do not flower in darkness but can be made sensitive to an inductive dark period by giving two exposures to red light each of 1 min spaced about 24 h apart; far-red reversibility demonstrated the involvement of phytochrome (Friend, 1975). The action of Pfr could be to initiate and phase a circadian oscillation, but other possibilities are not excluded. Binding of Pfr to membranes (Quail, Marmé and Schäfer, 1973), filling of reaction centre sites (Raven and Spruit, 1973) or re-distribution of phytochrome in the cell (McKenzie *et al.,* 1975) might be a necessary prerequisite for the induction of the photoperiodic mechanism in leaves or cotyledons. An interesting recent observation with *Pharbitis* is that plants raised in a constant environment in continuous light and given a single 12-h inductive dark period showed a circadian variation in the amount of flowering depending on the time from sowing when the dark period began (Spector and Paraska, 1973). The rhythm seemed to be synchronised by the time of emergence from the soil (Paraska and Spector, 1974) and the results indicated that a light-on signal resulting from the first exposure of the seedling to light persisted for several days. Plants did not, however, become photoperiodically sensitive until after about 3 days of continuous light (cf. Friend, 1975).

In photoperiodism, plants respond to the duration and timing of light and dark periods in the daily cycle. A sharply defined 'critical daylength' often marks the transition between long-day and short-day responses. A critical daylength of this kind can be observed in the transition between vegetative growth and flowering in both short-day plants (e.g. *Xanthium strumarium*) and long-day plants (e.g. *Hyoscyamus niger*). It is also seen in the transition between growth and dormancy in some woody plants (e.g. *Populus tacamahaca*; Nitsch, 1957), the onset of bulbing in onion (Magruder and Allard, 1937) and tuber formation in *Begonia evansiana* (Esashi, 1961). When such a sharp transition in critical daylength occurs, it has been shown that the measurement of time is primarily concerned with the length of the dark period. When long days are coupled with long nights in non-24-h cycles, a short-day type of response occurs, while short days with short nights give a long-day response. The discovery that the effect of a long dark period may be annulled when it is interrupted near the middle with a brief exposure to light emphasised the importance of the reactions which occur in darkness. This has sometimes led to the concept that the events which occur in darkness are those which determine photoperiodic responses, light functioning merely to prevent these dark reactions from taking place. More recent results, however, point to the conclusion that the photoperiodic control of physiological processes involves at least two types of reaction, one occurring during darkness, leading to a 'long night' effect, and one occurring during long periods of light, leading to a 'long day' effect. These two processes probably operate to different extents in different plants and in different physiological processes.

Dark Reactions

The onset of darkness initiates a process which, after a sufficiently long period, results in a long night response. If the dark period is shorter than the critical, no response occurs. Similarly, when a night break is given at the appropriate time, which is usually several hours from the beginning of darkness and is constant for any one species, the dark response is prevented. Time measurement thus occurs during darkness and it appears to be linked with its beginning. Two questions then arise: how is the transition to darkness detected, and what is the nature of the phytochrome changes which occur? In simple terms, the most commonly accepted model is some variant of that shown in *Figure 22.2*. At the close of the daily photoperiod, part of the phytochrome in the leaf is present as Pfr. A brief saturating exposure to red light has little effect at the end of the day or during the first few hours of darkness, but several hours later red light inhibits the long night effect, preventing flowering in short-day plants and promoting flowering in long-day plants. The action

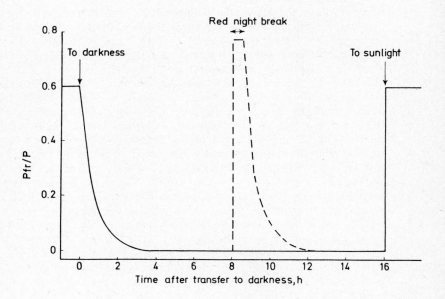

Figure 22.2   Possible changes in Pfr/P ratio in leaves during the course of a 16-h night.   The dashed line shows the effect of a 30-min night break with red light which re-establishes a high level of Pfr; this prevents flowering in short-day plants and promotes flowering in long-day plants.   It is assumed that Pfr decays at a constant rate in darkness with a half-life of 0.5 h

spectrum for this night break effect is similar to that for other phytochrome-mediated responses, although with a rather broader peak (Parker *et al.,* 1946; Parker, Hendricks and Borthwick, 1950). Reversibility by far-red light has been demonstrated in many plants, confirming that phytochrome is the controlling pigment. Examples of red/far-red reversibility are found in the inhibition of flowering in short-day plants (in *Xanthium strumarium;* Downs, 1956), the pro-motion of flowering in long-day plants (in *Hyoscyamus niger;* Downs, 1956), the control of dormancy (in *Weigela florida;* Vince-Prue, 1975), the prevention of cold acclimation (in *Cornus stolonifera;* McKenzie, Weiser and Burke, 1974) and the inhibition of tuber formation (in *Begonia evansiana;* Esashi, 1966). Reversibility cannot, however, be demonstrated in some cases. The response to red light in the night-break effect indicates the presence of the red-absorbing form of phyto-chrome, Pr, and it is usually concluded that this is formed from Pfr by reversion during the intervening hours of darkness. Implicit in this model is the idea that the Pfr level falls on transfer to darkness, and that this is coupled to a light-off signal. One of the earliest proposals for the involvement of phytochrome in photoperiodism was that the critical night length might represent the time taken for the concentra-tion of Pfr in the leaves to fall below a threshold value which no longer prevented a dark or 'low Pfr' reaction from taking place. If dark reversion of Pfr to some threshold value takes an appreciable time, then by-passing this process by photoconverting Pfr to Pr at the beginning of darkness would be expected to shorten the critical night length, or at least to increase the effectiveness of a given duration of dark. Some experiments of this kind have been partly successful. Far-red light at the end of the day induced flowering with a night length shorter than the critical length in the short-day plants *Xanthium* (Borthwick, Hendricks and Parker, 1952), *Setaria italica* (Downs, 1959) and *Sorghum vulgare* (Lane, 1963). Recently, another short-day induced phenomenon, that of cold acclimation in *Cornus stolonifera,* was found to occur with long photoperiods of 15 h when these were immediately followed by 30 min of far-red light, although it did not occur when 15-h photoperiods were terminated with 30 min of red light (McKenzie, Weiser and Burke, 1974). In a long-day plant, *Lolium temulentum,* a physiologically significant reduction of Pfr appeared to occur within 30–40 min of transfer to darkness, as measured by comparing the effectiveness of varying durations of darkness preceded, or not preceded, by a brief exposure to far-red light (Holland and Vince, 1971). As the decrease in critical night length in short-day plants did not usually exceed about 60 min, these results are usually interpreted as indicating that Pfr disappears from the leaf tissues, or falls to some threshold value within this length of time, coupling then to some dark process which is involved in time measurement. After the critical night length a 'low Pfr' reaction leads to the observed long night response. Such a model is shown schematically below:

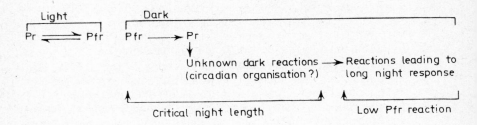

In this scheme, only Pfr reversion is considered. The photoperiodic mechanism operates in mature light-grown leaves and any decay process would almost certainly have gone to completion during the previous exposure to light. For example, an inductive dark period in *Pharbitis* can immediately follow exposure to continuous fluorescent light for many days. Contrary to the suggestion of Lisansky and Galston (1974), therefore, the lowering of Pfr in darkness by a decay process of the kind observed in etiolated seedlings is unlikely. The possibility of new synthesis of Pr during several hours of darkness cannot be excluded but there is no evidence that it occurs in mature leaves.

If the loss of Pfr is involved in timing, a knowledge of the dark reversion process in green leaves is of great importance. Unfortunately, the low concentration of spectrophotometrically photoreversible phytochrome in light-grown tissues and the effect of chlorophyll in reducing measurable $\Delta(\Delta A)$ (Grill, 1972) mean that direct measurements *in vivo* are still not possible. However, results from *in vitro* measurements and *in vivo* studies of etiolated tissues are instructive. No reversion has been detected in etiolated seedlings of several Gramineae (e.g. *Zea;* Pratt and Briggs, 1966) or in the Centrospermae (Kendrick and Hillman, 1971) despite the fact that the mature green plant may respond to an inductive night. Many cultivars of *Zea,* for example, are short-day plants. In several dark-grown dicotyledons, reversion of Pfr to Pr begins immediately on transfer to darkness after a saturating exposure to red light. In *Raphanus sativus* (a long-day plant), *Glycine max* (a short-day plant) and *Phaseolus vulgaris* (a day-neutral plant), reversion was essentially complete within about 60 min (Hopkins and Hillman, 1965). Similarly, in seedlings of *Sinapis alba* (a long-day plant) and Alaska pea, reversion began immediately and occupied 30–60 min; dark destruction of Pfr continued for a longer period (Kendrick and Hillman, 1970; McArthur and Briggs, 1971). Only part of the Pfr formed by red light underwent reversion in each case; this was about 13 per cent in *Sinapis.* Recent experiments on *Cucurbita* showed that the rapid loss of photoreversibility following exposure to red light was related directly to the fraction of phytochrome which was membrane bound;

during the first 30 min the total soluble phytochrome remained unchanged (Boisard, Marmé and Briggs, 1974). *In vivo* reversion in etiolated seedlings may, therefore, be a property shown only by soluble phytochrome.

The physiological significance of reversion in photoperiodism still remains obscure even though night-break studies have demonstrated that a low Pfr level in tissues is necessary during part of an inductive long night. An important result from recent studies of phytochrome in solution and *in vivo* is that its properties are considerably modified by the molecular environment. *In vivo* reversion rates can vary with the kind of tissue, with tissue age and with irradiation pre-treatment. Reversion rates also vary *in vitro*. Perhaps of greatest interest is the marked acceleration in the presence of reducing compounds such as ferredoxin, NADH and dithionite (Mumford and Jenner, 1971; Pike and Briggs, 1972; Kendrick and Spruit, 1973a). Phytochrome has been shown to be associated with etioplasts (Wellburn and Wellburn, 1973; Evans, Chapter 10), and if it is also associated with chloroplasts it would be in a strongly reducing environment at the end of the light period; reversion might, therefore, occur extremely rapidly. The exact location of the photoperiodic perception mechanism has yet to be determined, however.

In some environments, reversion can be extremely slow. In partly hydrated seeds, complete far-red reversibility persisted for 24 h or more in darkness, after a short exposure to red light (Loercher, 1974, Hsiao and Vidaver, 1973), while at a water content of 7 per cent seed could be stored for more than 1 year without any functionally apparent dark reversion (Vidaver and Hsiao, 1972). Localised cell environments might also result in slow or negligible rates of Pfr reversion in fully hydrated cells. For example, there was no evidence for loss of Pfr in darkness in experiments where phytochrome control of stem elongation in *Fuchsia* and *Phaseolus* was studied (Downs, Hendricks and Borthwick, 1957; Vince-Prue, 1973, 1975). In *Fuchsia* a 1-h exposure to far-red light given at any time during a 16-h dark period increased stem elongation compared with the short-day controls, and reversibility experiments confirmed that the effect was due to a lowering of Pfr (*Figure 22.3*). The amount of stem elongation appeared to be a linear function of the time during which Pfr was absent from the tissue, and the results showed that ground-state Pfr was present and active in inhibiting stem elongation throughout the whole of the 16-h night. Experiments with end-of-day mixtures of red and far-red light also indicated that no reversion occurred and the results did not fit a threshold model of Pfr action (Vince-Prue, 1973). There is evidence that this effect of light is a direct action of Pfr in the responding internode and may not reflect the situation in photoperiodically sensitive leaves. Studies of phytochrome-controlled anthocyanin synthesis (*Figure 22.3*) seemed to indicate that ground-state Pfr also persisted in the leaves throughout most of a 16-h dark period, but the anthocyanin-forming sites are likely to be different from those active in photoperiodism.

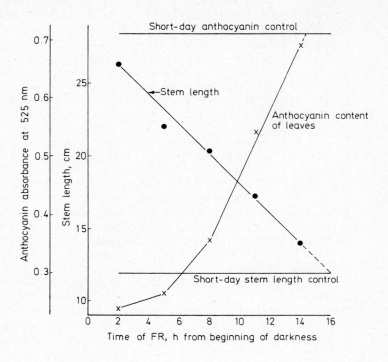

Figure 22.3   Effect of a 1-h night break with far-red light on stem elongation and anthocyanin content of leaves of Fuchsia hybrida *cv. Lord Byron.   Plants were grown in 8-h days in daylight and received a night break for 1 h with far-red light at the time indicated on the abscissa*

Flowering occurred in *Fuchsia* only when a night break was given (Holland and Vince, 1968), indicating that the Pfr level had fallen at this time and was no longer adequate for photoperiodic induction.

Few attempts have been made to study reversion *in vivo* during photoperiodic induction.   The most detailed are still those of Evans and King (1969) and Cumming, Hendricks and Borthwick (1965), who attempted to follow phytochrome changes using a null method.   Various mixtures of red and far-red light were applied at different times during an inductive night, the rationale being that if the Pfr/P value present in the tissue at any time is exactly the same as that established by a particular ratio of red to far-red light, then no physiological change will occur.   Consequently, the Pfr/P values at any time can be calculated from the null, or no-effect, mixture at that time.   There are several objections to this rationale, one of the most important being that phytochrome is cycled during the light perturbation so that the ground-state system cannot be left undisturbed. However, the null results indicated that, in *Pharbitis,* Pfr remained more or less unchanged at the value established by end-of-day red light for 6 h before disappearing fairly rapidly during the next 2 h of darkness. Similar studies with *Chenopodium rubrum* indicated the presence of ground-state Pfr even after 30 h of darkness. In both of these experiments, transfer to darkness did not appear to initiate an immediate decrease in Pfr. Indeed, Evans and King's results indicated that, if the photoperiod was terminated by light with a high far-red to red ratio, Pfr gradually reappeared during the early hours of darkness. We should, therefore, perhaps question the generally accepted concept that a lowering of Pfr is a necessary part of time measurement during a critical night. A low Pfr level is, however, thought to be required for completion of the long-night effect because this is prevented by a red night break, which is reversed by far-red light.

One way in which phytochrome could give an immediate light-off signal appears generally not to have been considered. In light the pigment cycles from one form to the other, and at any time part of the total is present as weakly absorbing intermediates. The rate of cycling depends on the irradiance and wavelength of the light (Kendrick and Spruit, 1973b). Rapid cycling occurs in mixed red and far-red wavelengths such as in sunlight and, because of overlapping absorption bands, cycling also occurs in red light (Briggs and Fork, 1969). On transfer to darkness, cycling stops immediately and there is an increase in Pfr from the weakly absorbing Pbl intermediate (Kendrick and Spruit, 1973b). The rate of cycling in sunlight depends on irradiance and would gradually decrease during twilight. A characteristic of the natural light environment is, therefore, that phytochrome cycles from one form to the other; this cycling stops in darkness and could give a light-off signal which might be coupled to a dark timing process. Such a model implies an ecological significance for the photoreversible reaction in green plants in addition to the accepted one of physiological adaptation to low red to far-red ratios under canopies (Vince-Prue, 1973).

Another characteristic change in metabolism following a light to dark transfer is the cessation of photosynthesis. The fact that photoperiodic mechanism does not operate in completely etiolated plants and that phytochrome has been shown to be associated with etioplasts at least points to the possibility of some involvement of photosynthesis with photoperiodism.

Light Reactions

THE 'HIGH-PFR' REACTION

It has been shown that exposure to several hours of light must normally precede a dark period in order to obtain a long-night response. The effect of the duration of a previous exposure to light on the flowering response of the short-day plant *Xanthium strumarium* to a 12-h night was examined by Salisbury (1965). No response occurred with light periods shorter than 4 h, and flowering increased with increase in the duration of light up to a maximum at 12 h. Salisbury also found that red was the most effective waveband and only low irradiances were needed. Similar effects were found for the inhibitory action of darkness on flowering in the long-day plant *Hyoscyamus niger*; dark inhibition was reduced when the preceding light period was very short (Joustra, 1970). Evidence that this requirement for light involves Pfr is indicated by far-red/red reversibility experiments and by action spectra. The need for a Pfr-requiring process as a necessary prerequisite for the effects of long periods of darkness was first demonstrated for the induction of flowering in the short-day plant *Pharbitis nil.* Flowering in young seedlings was depressed by giving a brief exposure to far-red light at the end of the photoperiod and was restored by giving a subsequent brief exposure to red light. The action spectra for inhibition by far-red light and for re-induction by red light were characteristic of phytochrome (Nakayama, Borthwick and Hendricks, 1960). Thus, at that time in the daily cycle, Pfr promoted flowering, whereas after several hours of darkness a red night break inhibited flowering (*Figure 22.4*). Inhibition of flowering by far-red light applied at the end of the photoperiod or early in the night has been demonstrated in many short-day plants and the response seems to be general (*see* Vince, 1972; Evans, 1971). Most investigators have concluded that phytochrome has a dual action in flowering in short-day plants; floral induction first requires a reaction that depends on the presence of Pfr phytochrome, whereas at a later time in the cycle flowering is inhibited when Pfr is formed by red light. In at least three long-day plants, *Hyoscyamus niger, Anethum graveolens* (Borthwick, 1959) and *Lemna gibba* (Ishiguri and Oda, 1974), an end-of-day treatment with far-red light has been shown to induce flowering in short day/long night cycles and it is reasonable to conclude that here the inhibitory effect of the long dark period was prevented. It is possible to generalise, therefore, that a 'long night' effect, whether it is inhibitory or inductive, occurs only when preceded by some reaction which requires the presence of Pfr phytochrome. A characteristic feature of this phenomenon is that under conditions where a terminal exposure to far-red light reduces the effect of a subsequent long night (often only following a very short day) there is also a loss of reversibility of the night-break effect because far-red light alone is inhibitory at this time. Thus, under these conditions there is a marked

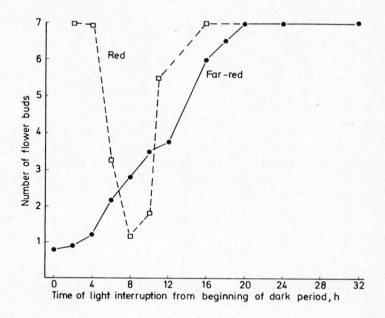

Figure 22.4 Flowering response of Pharbitis nil *exposed to a single interruption with 5 min of red or far-red light given at different times during a 48-h dark period at 19 °C. (Data of Takimoto and Hamner, 1965)*

change in the effect of red light from promoting at the end of the photoperiod to inhibitory 8-10 h later, while far-red light is inhibitory during the whole of this time (*Figure 22.4*). Such results have led to the suggestion that the inhibitory effect of far-red light on the dark reactions is always due to removal of ground-state Pfr, while the inhibitory action of red light during a night break might be due to some other reaction.

An exactly similar phenomenon has been described for the long-night induction of tuber formation in *Begonia evansiana*. With cycles of 12L/12D, terminal far-red light did not prevent the dark effect and a night break showed red/far-red reversibility. With 8L/16D cycles, however, on exposure to far-red light at the end of the light period prevented tuber formation and both red and far-red light were inhibitory in the middle of the 16 h night (Esashi, 1966). Recently, the dark inhibition of flowering in the long-day plant *Lemna gibba* has been shown to exhibit a comparable response. Certain short-day sequences of red followed by far-red light (e.g. 7 h red, 3 h far-red) allowed flowering, presumably by preventing the inhibitory long-night effect.   At the end of such a sequence, red light restores the inhibitory effect of the long night and this was reversed by far-red light.   After several hours of darkness, flowering was promoted by red light (i.e. a red night break counteracted the inhibitory effect of darkness) and at this time no far-red reversal occurred (Ishiguri and Oda, 1974).

These experiments show that a long-night effect is dependent on Pfr-requiring events which occur at the end of the preceding photoperiod. Few studies of the photoperiod itself have been made.   Recently, King (1974) showed that the events of darkness can be affected by giving far-red light at different times before transferring plants from continuous light to a 12-h inductive night, even though Pfr was re-established with red light before entry to darkness.   When 1.5 h of far-red light was applied 3 h before the end of the photoperiod, flowering in the short-day plant *Pharbitis nil* was increased.   Flowering was inhibited, however, when far-red light was applied 9 h before entry to darkness.   There were some indications of a rhythmic change in response to the timing of far-red light, even though plants were in continuous light (cf. the results of Paraska and Spector, 1974, and Spector and Paraska, 1973, cited earlier).   The use of intermittent exposures and the demonstration of at least partial red/far-red reversibility indicated that these effects were due to phytochrome.   Exposure to far-red light during the photoperiod was found to affect the timing responses of the ensuing dark period.   A treatment which promoted flowering (far-red light beginning 3 h before the end of the day) shortened the critical night length by about 50 min and advanced the time of maximum response to a night break; a treatment which inhibited flowering (far-red light 9 h before the end of the day) lengthened the critical night and delayed night-break timing.   These results emphasise that the operation of the phytochrome system during the photoperiod affects the events of the dark period, and they are not inconsistent with the general conclusion that a 'high Pfr' reaction (the presence of Pfr for several hours during the photoperiod or early part of the night) is a prerequisite for the consummation of the 'low Pfr' or dark reaction.   The relative timing of both reactions in the daily cycle is clearly important.

THE 'LONG-DAY' EFFECT

In many long-day responses, light appears not only to counteract the effect of darkness but also to have a positive effect which requires long daily exposures.   In *Lolium temulentum*, for example, the inhibitory effect of a long night can partly be counteracted by giving a relatively short night break with red light at the appropriate time (Vince, 1965) or by nitrogen gas (Evans, 1962); a much greater promotion of flowering can, however, be obtained by exposing plants to a single long photoperiod.   Flowering in many other long-day plants is also accelerated much more by long daily exposures to light than by a night break of 15–30 min (Vince, 1970).   Flowering in *Brassica campestris* not only requires long daily light periods, but also has a high threshold irradiance value of about 1 000 lx (Friend, 1969). The spectral characteristics of such 'long-day' responses are now well documented (Vince, 1969, 1972; Vince-Prue, 1975; Evans, 1971) and need be referred to only briefly here. In contrast to the greater effect of red light and the typical Pfr action spectrum obtained with short night breaks, long night breaks of several hours are more effective when mixtures of red and far-red light are used, and they show a different action spectrum with a peak at about 710 nm (Schneider, Borthwick and Hendricks, 1967). A possible implication of phytochrome intermediates in this reaction has been postulated but the action spectrum for intermediate production shows a peak between 685 and 700 nm. The inductive effect of a long night break for *Hyoscyamus* was sharply reduced at 675 nm (Schneider, Borthwick and Hendricks, 1967), where intermediates are still near their maximum (Kendrick and Spruit, 1973b). Red/far-red reversibility often cannot be demonstrated and, therefore, despite the effectiveness of the red and far-red regions of the spectrum, the evidence for phytochrome as the only photoreceptor in such long-day responses is not conclusive.

Photoperiodic responses in plants appear to depend both on 'long-night' and 'long-day' processes. In order to counteract the long-night effect, light of a sufficiently high red content is needed. For long-day responses, however, the maximum effect of a day extension is achieved with an approximately equal ratio of red to far-red light. Another characteristic feature of long-day responses is that red light is inhibitory in the second part of a 16 h photoperiod but is promoting in the first part. The sequence 8 h sunlight/8 h red light/8 h dark had little promoting effect on flowering in *Lolium temulentum*, despite the fact that 8 h is less than the critical night length. The sequence 8 h red light/8 h sunlight/8 h dark, on the other hand, strongly promoted flowering (Vince, 1965). Such sequences (*Table 22.2*) can be used to distinguish between those responses which are controlled primarily by a long-night process and those which also depend on a long-day process. The difference can be seen clearly if a comparison is made between petiole growth and flowering in the short-day plant strawberry and elongation growth and dormancy in *Weigela florida* (*Table 22.2*). In strawberry, the promotion of petiole elongation and the inhibition of flowering depend on a long-day process (Vince-Prue and Guttridge, 1973), and they occurred in a 16–17-h day only when an 8–9-h period in red light preceded the 8-h day in sunlight. In *Weigela*, stem elongation and the onset of dormancy depend on a long-night process; this was prevented when the dark period was shortened to 8 h, irrespective of whether the 8-h period in red light preceded or followed the short day in sunlight.

Examples of both processes are found in short- and long-day plants. Flowering in many short-day plants appears to depend primarily on a long-night process which promotes flowering, but in strawberry (Vince-Prue and Guttridge, 1973) and probably also in *Portulaca oleracea*

Table 22.2 Effect of different light sequences on photoperiodic responses

Type of response	Plant[1]	Response	Treatment		
			Red before short day	Red after short day	Short day
Long-day response[2]	Lolium temulentum (LDP)	Flowering (%)	100	0	0
		Spike length (mm)	3.4	0.91	0.80
	Fragaria × Ananassa (SDP)	Flowering (%)	20	100	100
		Petiole length (mm)	211	108	77
	Portulaca oleracea (SDP)	No. of leaves to flower (data of Gutterman, 1974)	15.8	6.8	7.9
	Picea abies	Dormant plants (%)	41	100	100
		Δ stem length (mm)	26.7	8.4	5.3
Long-night response[3]	Fuchsia hybrida cv. Lord Byron (LDP)	Days to anthesis	30.4	31.7	Vegetative
	Zygocactus truncatus (SDP)	Flowering (%)	0	0	50
	Kalanchoe blossfeldiana (SDP)	Flowering (%)	0	0	100
	Weigela florida	Dormant plants (%)	0	0	100
		Δ stem length (mm)	182	204	0

[1] LDP = long-day plant; SDP = short-day plant.

[2] In long-day responses, the effect of an 8–9-h day extension with red light given immediately after a short day in sunlight resembles that of a short day, whereas when the day extension with red light precedes the short day, a different response is obtained. The prevention of a long-night response may contribute to the effect of long photoperiods in some cases but this cannot be determined from these results.

[3] In long-night responses, both types of long-day sequence produce similar results, which differ from those in a short day.

(Gutterman, 1974) flowering is controlled mainly by an inhibitory process which occurs in long days. The short-day plant *Glycine max* is particularly interesting. Some cultivars flower only slowly under long (20-h) photoperiods in fluorescent light; in others the inhibition of flowering requires the addition of far-red light (tungsten-filament lamps) during the long photoperiod (Buzzell, 1971). In the former, flowering appears to be determined primarily by a promoting effect of the long night, which is prevented by light with a sufficiently high red to far-red ratio. In the latter, a positive inhibitory effect of long-light periods (requiring a mixture of red and far-red light) is needed in order to prevent flowering. The two types differ by only a single gene pair and crosses segregate in the ratio 3:1.

Flowering in most long-day plants seems to depend primarily on the long-day process but a long-night effect is also seen. *Lolium temulentum* has already been cited as an example of a long-day plant where flowering is both inhibited by long dark periods and accelerated by long daily exposures to light; removal of the inhibitory effect of darkness results in some flowering but this is greatly increased by giving a long period of light. *Hyoscyamus niger* almost certainly behaves in a similar way. Flowering occurs when the inhibitory effect is removed by defoliation in short days (Lang, 1965) but the presence of a leaf in long days gives a greater promotion of flowering. Some promotion of flowering also occurs when the inhibiting effect of a short day/long night cycle is prevented by applying far-red light at the end of the day (Borthwick, 1959), or by a short night break in a 12-h dark period (Parker, Hendricks and Borthwick, 1950). However, there is a greater acceleration of flowering with long daily photoperiods and the spectral characteristics are those of a long-day type of response (Schneider, Borthwick and Hendricks, 1967).

Fuchsia hybrida appears to be an example of a long-day plant where the control of flowering is primarily via an inhibitory effect of short day/long night cycles. Flowering is accelerated almost as much by a night break as by a long photoperiod and a mixture of red and far-red light promotes flowering little more than red light alone (*Table 22.3*). As expected, a comparison of the effects of applying red light in the first or second half of a 16 h photoperiod reveals little difference (*Table 22.2*)

Dormancy phenomena in woody plants also exhibit both processes. In *Weigela florida,* a short red night break prevents dormancy as effectively as a long day and the sequence in which red and red + far-red occurs during a long photoperiod is unimportant (*Table 22.2*). These results indicate that the induction of dormancy in this plant is primarily the result of a dormancy-promoting reaction which occurs during long nights. In *Picea abies,* however, an 8 h period in red light had little effect when it was given immediately after a short day in sunlight, whereas dormancy was largely prevented when the 8 h of red light immediately preceded the short day (*Table 22.2*). Dormancy in this plant thus appears to be primarily controlled by a dormancy-inhibiting process which occurs during the long daily light period.

Table 22.3   Effect of night breaks on flowering in *Fuchsia hybrida* cv. Lord Byron.   Results given are the number of days to anthesis

Night break	Duration of night break (h)					
	½	1	2	4	8	16
Red night break	43.4	43.2	42.7	42.1	41.5	39.2
Red plus far-red night break	44.5	43.7	42.4	42.0	38.9	37.5

It is possible that the 'long-day' and 'long-night' responses are related to rhythms or modulations that are out of phase with one another. There is evidence that in both processes there are inhibitory and promoting effects of red light.   In 'long-day' responses, red light is inhibitory (i.e. the response is increased by adding or substituting far-red light, or even by darkness) from about the 6th hour of the daily photoperiod; after about the 16–18th hour, red light is strongly promoting (Vince-Prue, 1975).   For the induction of a 'long-night' response, red light is required during the first 8–12 h of the daily photoperiod; after this time red light begins to inhibit the response.   These changes are shown schematically in *Figure 22.5*.

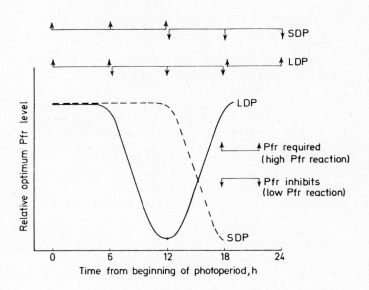

Figure 22.5   Schematic representation of 'high Pfr' and 'low Pfr' reactions in the flowering of long- and short-day plants.   In short-day plants the Pfr-requiring process occurs early in the photoperiod and is followed by a period when Pfr inhibits flowering.   In long-day plants similar reactions are seen but their timing is markedly different.   (From Vince-Prue, 1975.   Copyright © 1975, McGraw-Hill Book Company (UK) Limited.   D. Vince-Prue, Photoperiodism in Plants. *Reproduced by permission)*

The Action of Phytochrome in Photoperiodism

Until details of the action of phytochrome in simpler systems are
resolved, it is premature to expect a complete understanding of its
role in photoperiodism. The overt physiological responses such as
flowering and dormancy are slow and probably arise as a consequence
of a multiplicity of transmissible factors generated in the leaf during
photoperiodic induction. Among other substances, rapid daylength-
induced changes have been observed in cytokinins (Henson and Wareing,
1974), gibberellins (Cleland and Zeevaart, 1970) and abscisic acid
(Zeevaart, 1974).

The results of experiments with specific inhibitors and anti-metabolites
have yielded little evidence that RNA and protein synthesis in the leaf
are essential for photoperiodic induction (*see* Vince-Prue, 1975). Consequently, phytochrome action via gene activation or repression seems
unlikely. However, it is important to emphasise that photoperiodic
responses are inductive in the sense that, after a few favourable cycles,
the leaves become 'induced' and more or less permanent changes in
metabolism occur. The longer term changes may involve some effect
on the expression of genetic information. In this context, it is interesting to note the experiments of Queiroz and Morel (1974). In
Xanthium, PEP carboxylase activity increases in short days, partly as
a result of a circadian oscillation control which operates immediately
on transfer to short-day cycles, and gives a peak of activity early in
the dark period. There is also a cumulative effect of continued short-
day cycles which increases enzyme capacity to a maximum after 35
short days. With increasing numbers of short days there are changes
in the circadian variations, namely an increase in amplitude and a
delay in peak time to near midnight by day 60. Similar changes
were found in other enzymes. Progressive changes in leaf metabolism
during photoperiodic exposures are strongly indicated by these results.

Despite the possibility of such long-term alterations in metabolism,
there are several plants where flowering can be evoked by a single
appropriate photoperiodic cycle and it is in such plants that most
studies of phytochrome action have been made. If an internal coincidence model for the involvement of circadian organisation with photoperiodism proves to be correct, the role of phytochrome must be to
phase the multi-oscillator system. Studies of the mechanism of action
of phytochrome in overt rhythms would then be immediately relevant
to photoperiodism. Satter and Galston (1971) have shown that the
mechanism which operates leaflet movement in *Albizzia* is the same
whether leaflets close under the control of phytochrome or by a circadian
oscillation. Evidence is increasing that circadian systems involve membrane changes and it has been suggested that oscillatory changes in
cell membranes are closely associated with the clock mechanism (Scott
and Gulline, 1975; Njus, Sulzman and Hastings, 1974; Bünning and
Moser, 1972; Sweeney, 1974). The phase or period of oscillation can
be changed by treatment with ethanol (Sweeney, 1974). As the

ethanol effect was prevented by valinomycin, which is known to complex with K^+, enhancing its transport through biological membranes, it was concluded that ethanol probably affected the rhythm by affecting membrane properties.   In a circadian rhythm of stimulated bioluminescence in *Gonyaulux,* ethanol mimicked the light effect fairly closely, producing phase shifts of the same direction and magnitude as those produced by bright light pulses.   Similarities in the responses to ethanol and light have been observed in other systems (Bünning and Moser, 1973), and it has been suggested that the phase-shifting effects of light may be membrane mediated.   The membrane-regulating properties of phytochrome are, therefore, of immediate relevance in circadian systems and probably also in photoperiodism.

There is still little direct evidence that phytochrome-dependent membrane changes control photoperiodic responses.   In long days, lithium chloride at 10^{-3} M inhibited flowering in the long-day plant *Lemna gibba* and promoted flowering in the short-day plant *Lemna perpusilla.* ADP, on the other hand, appeared to enhance phytochrome action, increasing flowering in *L. gibba* and inhibiting that of *L. perpusilla* (Kandeler, 1970).   Lithium is an antagonist of K^+, and ADP and K^+ are co-factors for the phytochrome-mediated regulation of membrane properties in root tip adhesion.   Thus, although the additives were not shown to be affecting the process of induction in the leaves, the results could be interpreted as evidence for the involvement of phytochrome-mediated membrane changes in photoperiodic induction.   Lithium also slows the *Kalanchoe* clock (Engelmann, 1972).

Daily transfer to distilled water was found to inhibit the flowering response of *Lemna perpusilla,* and the inhibitory effect of such a transfer was greatest when plants were most sensitive to the inhibitory action of night-break light (Halaban and Hillman, 1971, 1974).   The distilled water effect was overcome by adding calcium ions, by lowering the temperature of the distilled water to 20 °C, and by supplementing it with a flower-promoting material obtained after incubating plants in water in the dark.   The promoting substance could be obtained by transferring plants to water at any time during darkness and thus its rate of production could not be significant for time measurement. Leakage of the flower-promoting material could be prevented by light but high irradiances were needed and phytochrome did not appear to be involved.   It was suggested that the action of phytochrome in flowering was to control the directional transport of ions through cells and that distilled water may simulate the effect of membrane changes. This hypothesis is supported by the fact that the distilled water effect is prevented by calcium ions, which are known to be essential for the maintenance of membrane integrity.

In conclusion, it is instructive to look at other types of experiment with membranes.   Using electron spin resonance techniques, temperatures below about 10 °C have been shown to cause changes from a liquid crystalline state to a solid gel state in the membranes of gloxysomes, mitochondria and proplastids (Wade *et al.,* 1974).   A marked increase in the leakage of glucose through an artificial liposomal system occurred at 25 °C and the addition of GA$_3$ lowered this transition temperature by 5–10 °C (Wood and Paleg, 1974). Changes in membrane properties can thus be effected by low temperature and by GA$_3$ as well as by phytochrome (Roux and Yguerabide, 1973). It is perhaps significant that photoperiodic processes often interact with low-temperature effects (e.g. in vernalisation) and that the application of GA$_3$ can substitute for a photoperiodic requirement in some plants.

References

ALTMAN, P.L. and DITTMER, D.S. (1966). *Environmental Biology.* Federation of American Societies for Experimental Biology, Washington

BECK, S.D. (1968). *Insect Photoperiodism.* Academic Press, New York

BOISARD, J., MARMÉ, D. and BRIGGS, W.R. (1974). *Plant Physiol.,* **54**, 272

BORTHWICK, H.A. (1959). In *Photoperiodism and Related Phenomena in Plants and Animals,* p.275. Ed. Withrow, R.B. American Association for the Advancement of Science, Washington

BORTHWICK, H.A., HENDRICKS, S.B. and PARKER, M.W. (1952). *Proc. Nat. Acad. Sci. U.S.A.,* **38**, 929

BREST, D.E., HOSHIZAKI, T. and HAMNER, K.C. (1971). *Plant Physiol.,* **47**, 676

BRIGGS, W.T. and FORK, D.C. (1969). *Plant Physiol.,* **44**, 1089

BÜNNING, E. and MOSER, I. (1972). *Proc. Nat. Acad. Sci. U.S.A.,* **69**, 2732

BÜNNING, E. and MOSER, I. (1973). *Proc. Nat. Acad. Sci. U.S.A.,* **70**, 3387

BUZZELL, R.J. (1971). *Can. J. Genet. Cytol.,* **13**, 703

CLELAND, C.F. and ZEEVAART, J.A.D. (1970). *Plant Physiol.,* **46**, 392

COULTER, M.W. and HAMNER, K.C. (1964). *Plant Physiol.,* **39**, 848

CUMMING, B.G., HENDRICKS, S.B. and BORTHWICK, H.A. (1965). *Can. J. Bot.,* **43**, 825

DOWNS, R.J. (1956). *Plant Physiol.,* **31**, 279

DOWNS, J. (1959). In *Photoperiodism and Related Phenomena in Plants and Animals,* p.129. Ed. Withrow, R.B. American Association for the Advancement of Science, Washington

DOWNS, R.J., HENDRICKS, S.B. and BORTHWICK, H.A. (1957). *Bot. Gaz.,* **118**, 199

ENGELMANN, W. (1972). *Z. Naturforsch.,* **27B**, 477

ESASHI, Y. (1961). *Sci. Rep. Tohoku Univ.,* **27**, 101

ESASHI, Y. (1966). *Plant Cell Physiol.,* **7**, 405

EVANS, L.T. (1962). *Aust. J. Biol. Sci.,* **15**, 281

EVANS, L.T. (1971). *Annu. Rev. Plant Physiol.,* **22**, 365

EVANS, L.T. and KING, R.W. (1969). *Z. PflPhysiol.,* **60**, 277

FRIEND, D.J.C. (1969). In *The Induction of Flowering,* p.344. Ed. Evans, L.T. Macmillan of Australia, South Melbourne

FRIEND, D.J.C. (1975). *Physiol. Plant.*, **35**, 286

GRILL, R. (1972). *Planta*, **108**, 185

GUTTERMAN, Y. (1974). *Oecologia*, **17**, 27

HALABAN, R. and HILLMAN, W.S. (1971). *Plant Physiol.*, **48**, 760

HALABAN, R. and HILLMAN, W.S. (1974). In *Chronobiology*, p.666. Scheving, L.E., Halberg, F. and Pauly, J.E. Igaku Shoin Ltd., Tokyo,

HENSON, I.E. and WAREING, P.F. (1974). *Physiol. Plant.*, **32**, 185

HILLMAN, W.S. (1964). *Am. Nat.*, **98**, 323

HILLMAN, W.S. (1970). *Plant Physiol.*, **45**, 273

HILLMAN, W.S. (1971). *Plant Physiol.*, **48**, 770

HOLLAND, R.W.K. and VINCE, D. (1968). *Nature, Lond.*, **219**, 511

HOLLAND, R.W.K. and VINCE, D. (1971). *Planta*, **98**, 232

HOPKINS, W.G. and HILLMAN, W.S. (1965). *Am. J. Bot.*, **52**, 427

HOSHIZAKI, T., BREST, D.E. and HAMNER, K.C. (1974). *Plant Physiol.*, **53**, 176

HSIAO, A.I. and VIDAVER, W. (1973). *Plant Physiol.*, **51**, 459

ISHIGURI, Y. and ODA, Y. (1974). *Plant Cell Physiol.*, **15**, 287

JOUSTRA, M.K. (1970). *Meded. Landbouwhogesch. Wageningen*, **70**, 19

KANDELER, R. (1970). *Planta*, **90**, 203

KENDRICK, R.E. and HILLMAN, W.S. (1970). *Plant Physiol.*, **46**, 596

KENDRICK, R.E. and HILLMAN, W.S. (1971). *Am. J. Bot.*, **58**, 424

KENDRICK, R.E. and SPRUIT, C.J.P. (1973a). *Plant Physiol.*, **52**, 327

KENDRICK, R.E. and SPRUIT, C.J.P. (1973b). *Photochem. Photobiol.*, **18**, 139

KING, R.W. (1974). *Aust. J. Plant Physiol.*, **1**, 445

LANE, H.C. (1963). *Crop Sci.*, **3**, 496

LANG, A. (1965). In *Encyclopedia of Plant Physiology*, Vol.XV/1, p.1380. Ed. Ruhland, W. Springer-Verlag, Berlin

LISANSKY, S.G. and GALSTON, A.W. (1974). *Plant Physiol.*, **53**, 352

LOERCHER, L. (1974). *Plant Physiol.*, **53**, 503

LOFTES, B. (1970). *Animal Photoperiodism*. Edward Arnold, London

MAGRUDER, R. and ALLARD, H.A. (1937). *J. Agr. Res.*, **54**, 715

McARTHUR, J.A. and BRIGGS, W.R. (1971). *Plant Physiol.*, **48**, 46

McKENZIE, J.M. Jr., COLEMAN, R.A., BRIGGS, W.R. and PRATT, L.H. (1975). *Proc. Nat. Acad. Sci. U.S.A.*, **72**, 799

McKENZIE, J.S., WEISER, C.J. and BURKE, M.J. (1974). *Plant Physiol.*, **53**, 783

MUMFORD, F.E. and JENNER, E.L. (1971). *Biochemistry, N.Y.*, **10**, 98

NAKAYAMA, S., BORTHWICK, H.A. and HENDRICKS, S.B. (1960). *Bot. Gaz.*, **121**, 237

NITSCH, J.P. (1957). *Proc. Am. Soc. Hort. Sci.*, **70**, 512

NJUS, D., SULZMAN, F.M. and HASTINGS, J.W. (1974). *Nature, Lond.*, **248**, 116

PARASKA, J.R. and SPECTOR, C. (1974). *Physiol. Plant.*, **32**, 62

PAPENFUSS, H.D. and SALISBURY, F.B. (1967). *Plant Physiol.*, **42**, 1562

PARKER, M.W., HENDRICKS, S.B. and BORTHWICK, H.A. (1950). *Bot. Gaz.*, **111**, 242

PARKER, M.W., HENDRICKS, S.B., BORTHWICK, H.A. and SCULLY, N.J. (1946). *Bot. Gaz.,* **108,** 1

PIKE, C.S. and BRIGGS, W.R. (1972). *Plant Physiol.,* **49,** 514

PITTENDRIGH, C.S. (1972). *Proc. Nat. Acad. Sci. U.S.A.,* **69,** 2734

PRATT, L.H. and BRIGGS, W.R. (1966). *Plant Physiol.,* **41,** 467

QUAIL, P.H., MARMÉ, D. and SCHÄFER, E. (1973). *Nature New Biol.,* **245,** 189

QUEIROZ, O. and MOREL, C. (1974). *Plant Physiol.,* **53,** 596

RAVEN, C.W. and SPRUIT, C.J.P. (1973). *Acta Bot. Neerl.,* **22,** 135

ROUX, S.J. and YGUERABIDE, J. (1973). *Proc. Nat. Acad. Sci. U.S.A.,* **70,** 762

SALISBURY, F.B. (1965). *Planta,* **66,** 1

SATTER, R.L. and GALSTON, A.W. (1971). *Science, N.Y.,* **174,** 518

SAUNDERS, D.S. (1973). *Science, N.Y.,* **181,** 358

SCHNEIDER, M.J., BORTHWICK, H.A. and HENDRICKS, S.B. (1967). *Am. J. Bot.,* **54,** 1241

SCOTT, B.I.H. and GULLINE, H.F. (1975). *Nature, Lond.,* **254,** 69

SPECTOR, C. and PARASKA, J.R. (1973). *Physiol. Plant.,* **29,** 402

SWEENEY, B.M. (1974). *Plant Physiol.,* **53,** 337

TAKIMOTO, A. (1966). *Bot. Mag., Tokyo,* **79,** 474

TAKIMOTO, A. and HAMNER, K.C. (1965). *Plant Physiol.,* **40,** 859

VIDAVER, W. and HSIAO, A.I. (1972). *Can. J. Bot.,* **50,** 687

VINCE, D. (1965). *Physiol. Plant.,* **18,** 474

VINCE, D. (1969). *Acta Hort.,* **14,** 91

VINCE, D. (1970). *Proc. 18th Int. Hort. Congr.,* **5,** 169

VINCE, D. (1972). In *Phytochrome,* p.257. Ed. Mitrakos, K. and Shropshire, W. Academic Press, London, New York

VINCE-PRUE, D. (1973). *An. Acad. Bras. Ciênc.,* **45,** Suppl., 93

VINCE-PRUE, D. (1975). *Photoperiodism in Plants.* McGraw-Hill, Maidenhead

VINCE-PRUE, D. and GUTTRIDGE, C.G. (1973). *Planta,* **110,** 165

WADE, N.L., BREIDENBACH, R.W., LYONS, J.L. and KEITH, A.D. (1974). *Plant Physiol.,* **54,** 320

WAGNER, E. and CUMMING, B.G. (1970). *Can. J. Bot.,* **48,** 1

WELLBURN, F.A.M. and WELLBURN, A.R. (1973). *New Phytol.,* **72,** 55

WOOD, A. and PALEG, L.G. (1974). *Aust. J. Plant Physiol.,* **1,** 31

ZEEVAART, J.A.D. (1974). *Plant Physiol.,* **53,** 644

23

PHOTOPERIODISM IN LIVERWORTS

W.W. SCHWABE
Wye College, University of London, Wye, Ashford, Kent, U.K.

Introduction

Light effects on liverworts have been known for many years and studies include work on all the attributes of light action: direction, intensity, duration and spectral composition. Surprisingly, the number of species used experimentally has been small.

Effects of light duration or photoperiod have been recorded for both vegetative growth and reproduction, but most investigations have concentrated on aspects of vegetative growth. Moreover, it is not certain whether some of the quantitative differences recorded in vegetative reproduction by means of gemmae were specific effects or resulted from overall growth effects on the plant (Voth and Hamner, 1940). Observations by Benson-Evans and Hughes (1955) suggested that in the British strain of *Lunularia cruciata* sexual reproductive development could be promoted by the change from short to long daylength when preceded by a low-temperature period. Seasonal changes could thus be due to several factors of the external environment and perhaps also the previous history of the plants under natural conditions, making it difficult to pinpoint specific factors. Some of the effects of light on growth or morphogenesis of liverworts are listed in *Table 23.1*.

Effects of Photoperiod and Spectrum

The most detailed studies on etiolation, senescence and tropic effects of photoperiod and light spectra are those of Fredericq (1964) and De Greef and co-workers (e.g. Fredericq and De Greef, 1968). These interesting investigations were largely centred on the behaviour of the meristems of thalli, their chlorophyll content, etc., and clearly established the involvement of phytochrome responses in these phenomena, with red/far-red reversibility of etiolation and chlorophyll destruction.

Some time previously, Schwabe and Nachmony-Bascomb (1963) and Nachmony-Bascomb and Schwabe (1963) studied the photoperiodic responses in *Lunularia cruciata* L. In this species, overall growth and

Table 23.1 Some effects of light on liverworts

Effect	Species	Response[1]	Reference
Vegetative growth and dormancy	Marchantia polymorpha Lunularia cruciata	Growth promoted by LD Vegetative growth promoted by SD Dormancy induced by LD	Voth and Hamner, 1940 Schwabe and Nachmony-Bascomb, 1963
Etiolation	M. polymorpha	Etiolation induced by darkness or low-intensity light Etiolation induced by FR	Förster, 1927; Stephan, 1928 Fredericq, 1964; Ninnemann and Halbsguth, 1965; Fredericq and De Greef, 1968
Drought resistance	L. cruciata	Promoted by LD (in dormant state)	Schwabe and Nachmony-Bascomb, 1963
Senescence and loss of chlorophyll	M. polymorpha	Promoted by white light (1 h) by FR. Inhibited by R (5 min)	De Greef et al., 1971
Vegetative reproduction by gemmae (gemma cups)	M. polymorpha M. nepalensis	More cups in SD Cup production inhibited by very low light intensity	Voth and Hamner, 1940 Chopra and Sood, 1970
'Germination' of gemmae by rhizoid production	M. polymorpha L. cruciata	Promotion by R and reversal by FR Promotion by R and reversal by FR	Ninnemann and Halbsguth, 1965 Valio and Schwabe, 1969
Dorsi-ventral polarity of gemmae	M. polymorpha M. polymorpha	Iso-bilateral development of rhizoids with bilateral illumination Bilateral rhizoid development in continuous unilateral light in presence of phalloidin	Fitting, 1939 Halbsguth, 1958
Sexual reproduction	M. polymorpha L. cruciata	Promoted by LD Promoted by LD	Wann, 1925; Voth and Hamner, 1940 Benson-Evans and Hughes, 1955

[1] SD = short days; LD = long days; R = red light; FR = far-red light

dormancy are controlled by daylength, particularly in the Mediterranean strains (strains from Italy, Corfu and especially Israel have been tested). The Israel strain used is highly sensitive to photoperiodic conditions and in warm conditions becomes completely dormant after exposure to less than 10 long days. While there is no critical daylength for the induction of dormancy, the longer the day, the more rapid is the cessation of growth. This response is highly sensitive to temperature and (over the range 12–24 °C) the higher the temperature, the more rapid is the onset of dormancy. The reverse of dormancy induction is achieved more rapidly, and as little as 4 short days (8 h) are needed to break dormancy.

Perhaps the most striking result of dormancy induction is the drought hardening effect. Thus thalli exposed to 14 or more long days can be air dried and stored in this condition for some years without losing viability, while similar drying of short-day grown thalli leads to their immediate death. These responses have been confirmed to be directly photoperiodic, being brought about as effectively by light-break treatments as prolonged illumination. Measured in terms of gemmaling growth in area after 18 days (i.e. 10 days of treatment followed by 8 short days) the following values were obtained:

Short-day control:	4.50 mm²
Continuous light control	2.85 ± 0.29 mm²
Nightly light break (10 min red, 630-680 nm):	3.21 mm²

It has also been shown that the photoperiodic response itself is phytochrome mediated (Wilson and Schwabe, 1964). These experiments on growth and subsequent studies on growth and 'germination' of gemmae by Valio and Schwabe (1969) and Schwabe and Valio (1970) have shown that light breaks consisting of red light (wavelength about 650 nm) are effective in bringing about the long-day response and that these effects are reversible by far-red irradiation (735 nm) (*Table 23.2*).

Table 23.2 Growth inhibition in *Lunularia* by treatment with continuous light or red light breaks and reversal by brief far-red irradiation for 10 days followed by 2 weeks in short day

Treatment	Area growth (mm²)
Continuous light	2.3
Short day control	5.2
Red/far-red	3.7
Red	1.8
Far-red/red	1.8
Far-red	3.5
LSD	0.68

The sensitivity of *Lunularia* to such irradiation is, in fact, considerable and a long-day response can be brought about by a nightly exposure of as little as 3 min of red light, whereas reversal is at least partially effective with as little as 5 s of far-red light, although far-red light itself is somewhat inhibitory compared with total darkness.

Valio and Schwabe also tested the response to different parts of the spectrum and it was found that not only the red and far-red parts of the spectrum were involved, but also a band in the blue region in which phytochrome is also known to absorb. The effects on thallus and gemmaling growth of irradiation treatments with specified narrow wavebands and their reversal were tested in a number of experiments. The results shown in *Table 23.3* are an example of such experiments, in which the timing of light treatments during the 24 h was also varied.

Table 23.3 Effect of daily red (R), far-red (FR) and blue (B) radiation treatment (5 min), and the times of irradiation, on area growth of *Lunularia* gemmalings after 10 days

All plants were grown in short day (8 h light + 16 h dark). Results are mean areas (mm^2) of four replicates of three gemmalings

Treatment	Controls		Irradiation time	
		Middle of main light period	Beginning of dark period	Middle of dark period
SD	2.90			
CL[1]	1.42			
R		3.17	2.97	2.39
FR		2.68	2.04	2.90
R/FR		2.67	2.37	2.93
B		2.91	2.97	2.67
	LSD (5%)	0.29		

[1] CL = continuous light

Further, it was found that the initial growth made by dormant gemmae taken from the cup is also subject to phytochrome control. Gemmae do not 'germinate' in total darkness, if the term germination may be specifically applied to the start of growth which, in this instance, is signalled by the formation of rhizoids (Fitting, 1939). Gemmae may stay alive in total darkness and not produce rhizoids, without losing the capacity to grow, for as long as 6 months or more. At a temperature of 20 °C, a period of about 2 h of light is required before rhizoid production is subsequently induced in the dark in approximately 50 per cent of the gemmae (Valio and Schwabe, 1969). A large number of experiments on the induction of rhizoids by means of exposure to irradiation from different spectral wavebands have been carried out and a clearcut situation of promotion, reversal (inhibition) and re-promotion is found, red light promoting and far-red and blue light preventing germination. Results from two representative experiments are given in *Table 23.4.*

Table 23.4 Reversibility of red or far-red irradiation effects on rhizoid production

Gemmae taken from cup of mother thallus were given some hours of white light and then exposed for 1 min to red (R), far-red (FR) and blue (B) light, before being placed into complete darkness. The time interval between two irradiations was 1 min.

Treatment	Gemmae with rhizoids (%)	Treatment	Gemmae with rhizoids (%)
R	100	B	0
FR	10	B/R	90
FR/R	95	R/B	0
R/FR	0	R/B/R	100
R/FR/R	95	R/B/R/B	0
R/FR/R/FR	0		

Strikingly, not all of these effects are identical with those on dormancy induction of thalli, as will be seen from a comparison of *Tables 23.3* and *23.4,* in this response blue light having the same effect as red and not far-red light.

Lunularic Acid

Further efforts have centred on the biochemical identification of the factors involved in the inhibition of growth. The inhibitor predicted by Schwabe and Nachmony-Bascomb (1963) was ultimately verified when a new growth inhibitor, since given the trivial name of 'lunularic acid', was identified and characterised by Valio, Burden and Schwabe (1969). Lunularic acid is almost certainly identical with the inhibitory substance described by Fries (1964).

Much of the more recent work has concentrated on the role of this substance in the metabolism of liverworts. Thus it has been shown that lunularic acid (LNA) is present in some 70 species of liverworts so far tested (Gorham, unpublished work) (*see also* Huneck and Pryce, 1971; Pryce, 1971a). This list comprises all major groups of thallose and foliose liverworts, although interestingly *Anthoceros laevis* has so far given negative results, which agrees with other unusual characteristics of the Anthocerotales. If supplied in the external medium even in fairly low concentrations (approximately 10–20 p.p.m.), LNA is inhibitory to growth. When LNA was applied at the level of 8×10^{-8} mol, growth of *Marchantia polymorpha* was reduced by 70 per cent and *Lunularia cruciata* by 40 per cent compared with controls, while high concentrations have actually proved lethal to *Lunularia*.

A claim that algae also contain endogenous LNA has recently been made (Pryce, 1971b) on the basis of analyses in which relatively small amounts of material were used. However, using much larger amounts of algal material in an attempt to repeat this result, no LNA was found even though small amounts (1 part in 10^8 of fresh material) would have been detected with the GLC and TLC techniques employed. The amounts of material from the several species used for extraction are shown in *Table 23.5* (Thomas, unpublished work).

Table 23.5 Fresh weights of algal material extracted in negative tests for LNA content

In all cases no sign of LNA could be detected. The first three species mentioned were grown in axenic culture.

Species	Weight or volume extracted
Chlorella pyranoidosa	5 g fresh wt.
Monodus subterraneum	5 g fresh wt.
Anabaena inaequalis	5 g fresh wt.
Ulva lactuca	20 g fresh wt.
Sargassum muticum	10 g fresh wt.
(Mixed) pond algae	50 g dry wt.
Algal culture filtrate	20 l

Regrettably, so far we have no clearcut evidence of the mode of action of LNA, but this is a position not very different from that of all other major plant hormones. Work on the metabolism of LNA has suggested that its synthesis is via the phenyl–propanoid–polymalonate pathway (Pryce, 1971b).

The photoperiodic effect on the amounts of LNA found in *Lunularia* suggests that possibly light activation of phenylalanine ammonia-lyase (PAL) may be involved in this control but so far conclusive evidence is still lacking. However, it has now been proved (Gorham, unpublished work) that PAL is the enzyme involved in the synthesis of LNA and it is suggested that the synthesis proceeds via cinnamic acid involving cinnamic acid 4-hydroxylyase, demonstrated in *Conocephalum*. LNA itself is readily decarboxylated by a specific aromatic decarboxylase to lunularin (Pryce, 1972; Pryce and Linton, 1974). Gorham (unpublished work) has now characterised the enzyme, its activity in relation to temperature, pH, etc.

Ecological Aspects

From the ecological point of view, it has been shown in *Lunularia* that LNA and its decarboxylation product lunularin (LN) are responsible for the inhibition of growth of gemmae while in the cup, LN being more inhibitory than LNA, especially in *Marchantia*. The inhibitor needs to be diffused away, even from the actively growing thallus, in order to permit optimum growth (Schwabe and Valio, 1970). At the same time it may have an apparently allelopathic function. Thus, at relatively high concentrations the growth of fungi (fungal spore germination) is inhibited or partly inhibited (*Table 23.6*). The germination of seeds and elongation of seedling hypocotyl and root are partly suppressed (*Table 23.7*). Similar results were also found by Fries (1964), especially with *Avena*. So far, experimental attempts to detect

Table 23.6 Fungal spore germination as a percentage of control

Species	Concentration of lunularic acid (p.p.m.)			
	500	250	100	50
Botrytis cinerea	0	65	100	100
Alternaria brassicicola	0	0	71	100
Glommerella cingulata	0	0	23	100
Sclerotinium fructigena	0	0	35	100

Table 23.7 Seed germination tests
Values given are germination as a percentage of control

Expt. no.	Cress seed (Lepidium) on filter paper soaked in purified extract of Lunularia	Lettuce seed on 'Lunularia soil'
1	73	89
2	58	71
3	60	62
4	–	92
Mean	63.7	78.5

effective inhibition of seed germination, etc., in soils on which liver-worts have been grown have given variable results; failure to find consistent inhibition may perhaps be attributed to bacterial breakdown in the soil of LNA, since it has been found to serve as a substrate for bacterial growth. In particular, algal growth is severely curtailed or inhibited (*Table 23.8, Figures 23.1* and *23.2*). However, in the plant itself, high concentrations of LNA can be found and in *Conocephalum conicum* concentrations of over 600 p.p.m. have been recorded while the concentration of LN was maintained roughly at 5 per cent of the LNA (Gorham, unpublished work).

Table 23.8 Growth inhibition of algae by lunularic acid after 2 weeks' treatment

Species	Parameter	Control	Lunularic acid concentration (p.p.m.)	
			1	10
Chlamydomonas	Cell no. increase as percentage of starting value	141	101	56
Tribonema	Increase in filament length as percentage of starting value	216	63	30
Anabena	Mean no. of cells per filament	40.3	19.3	19.8

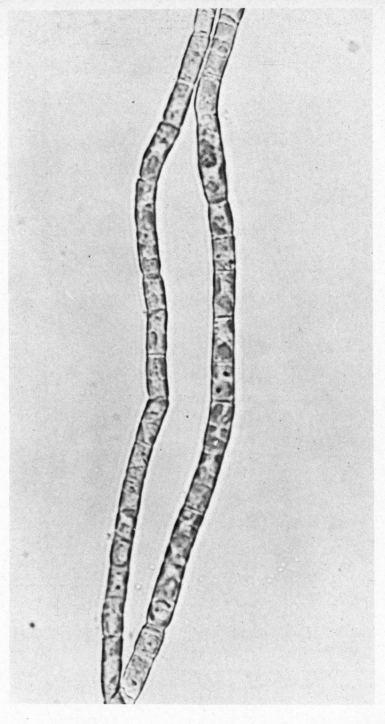

Figure 23.1 Tribonema *sp. grown in axenic culture for 2 weeks from inoculation*
(control)

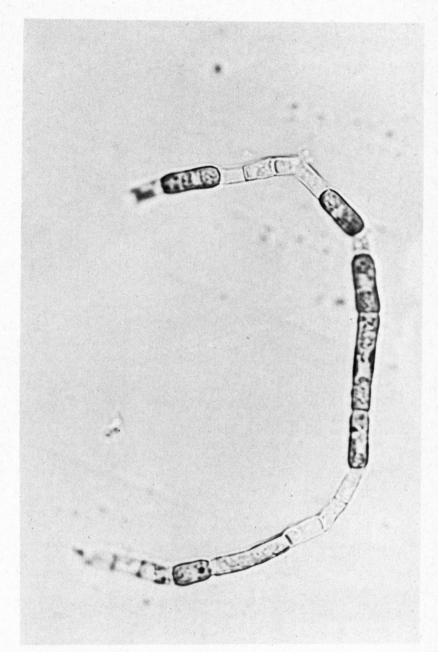

Figure 23.2 Tribonema *sp. grown in axenic culture in the presence of 100 p.p.m. of LNA for 2 weeks from inoculation. N.B., dead cells in short fractured filament and individual swollen cells*

LNA appears to fulfil the role of abscisic acid (ABA) in liverworts. Incidentally, ABA also inhibits growth in *Lunularia*. While no ABA has so far been detected in any liverwort, LNA has not been found in higher plants. A test of LNA effects on stomatal opening in *Commelina* also proved negative (Thomas, unpublished work).

It is still uncertain whether inhibitory action of LNA or LN on the growth of the thallus is also directly responsible for the induction of drought resistance in *Lunularia*, but the results in *Table 23.9* suggest that there is some direct or indirect action.

Table 23.9 Effect of lunularic acid on drought resistance of 1-cm thallus tips (50 mm^2) of *Lunularia cruciata*

Treatment[1]	Survival (%)	Final area per thallus 1 month after drying (mm^2)
SD control	14	11
3 wk SD + LNA	60	78
1 wk LD	38	105
1 wk LD + LNA	30	57
2 wk LD	60	48
2 wk LD + LNA	82	56
3 wk LD	92	290
3 wk LD + LNA	84	140

[1]SD = short day; LD = long day; LNA = 220 p.p.m. lunularic acid

In previous experiments (Schwabe and Valio, 1970), a striking effect of chelating agents (EDTA) promoting growth in *Lunularia cruciata* was found. When gemmae were grown on a nylon cloth substrate, area growth was virtually doubled in the presence of 100 p.p.m. of EDTA. More recently, Gorham (unpublished work) found a similar favourable effect in *Marchantia*, which, in this instance, could be related to the probable presence of minute amounts of heavy metal residues in unwashed filter paper. However, he also found that the lunularate decarboxylase is largely inhibited by the presence of EDTA while the presence of copper or mercury(II) ions was less inhibitory (*Table 23.10*). This might suggest the possibility of the growth inhibition being due to LN rather than LNA. However, at least with external application of LN and LNA, this has not yet been demonstrated.

In conclusion, it may be said that LNA production, and its functioning in relation to phytochrome activity, seem to control growth and reproduction of the liverwort in a manner that allows maximum adaptation to the environmental fluctuations requiring dormancy or growth. The same substance also seems to function, at least to some extent, allelopathically, reducing germination of seeds, fungal attack and competition from algae. Under natural conditions, however, growth of liverworts may be dependent on their being able to remove LNA into the surrounding environment or substrate, probably by liquid diffusion. If this does not occur, growth may be halted. Thus, substrates capable of absorbing LNA may be beneficial to growth, and something similar

Table 23.10 Effect of chelating agent (EDTA) and heavy metal ions on lunularate decarboxylase activity isolated from *Conocephalum*

Treatment	Activity as percentage of control
Control	100
10^{-3} M EDTA	25.3
10^{-3} M Na$_2$EDTA	39.0
10^{-3} M copper(II) sulphate	78.2
10^{-3} M mercury(II) chloride	50.3

may be involved in the growth promotion found in *Funaria hygrometrica* cultured on activated charcoal (Klein and Bopp, 1971). Perhaps this also is the explanation for the markedly improved growth of a liverwort described by NASA, although I am not sure it was *Lunularia,* on lunar soil, which, with its glassy structure, may have a particularly good absorptive capacity.

References

BENSON-EVANS, K. and HUGHES, J.G. (1955). *Trans. Br. Biol. Soc.,* **2,** 513

CHOPRA, R.N. and SOOD, S. (1970). *Bryologist,* **73,** 592

DE GREEF, J. BUTLER, W.L., ROTH, T.F. and FREDERICQ, H. (1971). *Plant Physiol.,* **48,** 407

FITTING, H. (1939). *Jahrb. Wiss. Bot.,* **88,** 633

FÖRSTER, K. (1927). *Planta,* **3,** 325

FREDERICQ, H. (1964). *Bull. Soc. R. Bot. Belg.,* **98,** 67

FREDERICQ, H. and DE GREEF, J. (1968). *Physiol. Plant.,* **21,** 346

FRIES, K. (1964). *Beitr. Biol. Pflanz.,* **40,** 177

HALBSGUTH, W. (1958). *Ber. Dtsch. Bot. Ges.,* **71,** 21

HUNECK, S. and PRYCE, R.J. (1971). *Z. Naturforsch.,* **26B,** 738

KLEIN, B. and BOPP, H. (1971). *Nature, Lond.,* **230,** 474

NACHMONY-BASCOMB, S. and SCHWABE, W.W. (1963). *J. Exp. Bot.,* **14,** 153

NINNEMANN, H. and HALBSGUTH, W. (1965). *Naturwissenschaften,* **52,** 110

PRYCE, R.J. (1971a). *Planta,* **97,** 354

PRYCE, R.J. (1971b). *Phytochemistry,* **10,** 2679

PRYCE, R.J. (1972). *Phytochemistry,* **11,** 1355

PRYCE, R.J. and LINTON, L. (1974). *Phytochemistry,* **13,** 2497

SCHWABE, W.W. and NACHMONY-BASCOMB, S. (1963). *J. Exp. Bot.,* **14,** 353

SCHWABE, W.W. and VALIO, I.F.M. (1970). *J. Exp. Bot.,* **21,** 122

STEPHAN, J. (1928). *Planta,* **6,** 510

VALIO, I.M.F. and SCHWABE, W.W. (1969). *J. Exp. Bot.,* **20,** 615

VALIO, I.M.F., BURDEN, R.S. and SCHWABE, W.W. (1969). *Nature, Lond.,* **223,** 1176

VOTH, P.D. and HAMNER, K.C. (1940). *Bot. Gaz.,* **102**, 169
WANN, F.B. (1925). *Am. J. Bot.,* **12**, 307
WILSON, J.R. and SCHWABE, W.W. (1964). *J. Exp. Bot.,* **15**, 368

24

LIGHT/TIMER INTERACTIONS IN PHOTOPERIODISM AND CARBON DIOXIDE OUTPUT PATTERNS: TOWARDS A REAL-TIME ANALYSIS OF PHOTOPERIODISM

W.S. HILLMAN

Biology Department, Brookhaven National Laboratory, Upton, New York 11973, U.S.A.

Introduction

An understanding of photoperiodism can only be achieved by establishing which of the many processes that occur during inductive light/dark cycles are decisive and how decisive they are. Unfortunately, most work has involved waiting until the effects of photoperiodic treatments have become apparent, long after the controlling events themselves. What is needed is some means of observing those events, or at least of obtaining information on their course, while they occur.

An important approach has emerged from work on the relation between circadian rhythmicity and photoperiodism, studies in which indicators of the phase of the rhythm have been correlated with results such as flowering. Such work originated in Bünning's observations on rhythms, leaf movements and photoperiodism; probably the most successful single investigation relating leaf movements to the photoperiodic response is that of Halaban (1968) on *Coleus*. However, it should be clear that, in general terms, the search for and use of 'real-time' indicators of photoperiodic effects need not depend on assumptions concerning circadian rhythmicity or any other form of timing. Indeed, to be most useful, such an indicator should reflect not only photoperiodic timing, but also effects of factors such as light quality.

Some studies of leaf movements (e.g. Salisbury and Denney, 1974) gave results that are ambiguous with respect to photoperiodism, suggesting what might, *a priori*, have been suspected anyway: the degree to which a given process is indicative of, or coupled to, mechanisms that underlie photoperiodism may vary from species to species or with environmental conditions. The corollary to this, however, is that if the coupling can be controlled in a defined manner, that control itself may provide information on the mechanisms in question. However, it would be useful to have as an indicator some process that can be more easily defined in biochemical terms than are leaf movements, since these movements, in the instances mentioned, are probably complex resultants of changes in both turgor and growth.

Carbon dioxide output, known since the work of Wilkins (1959) to reflect a circadian rhythm in at least one plant, is the net result of several respiratory and related reactions that are largely understood, at least in principle. In *Lemna perpusilla* under appropriate conditions, flowering and CO_2 output respond to ambiguous skeleton photoperiods in a manner suggesting that both processes are affected by the same circadian timer (Hillman, 1970, 1972). In addition, the effects of standard light schedules on CO_2 output can be modified by the nitrogen source supplied, with different phytochrome responses and timing exhibited, for instance, on nitrate, ammonium or nitrogen-free media (Hillman, 1975). The work reported here shows that the pattern of CO_2 output in this system, on some media but not on others, gives useful correlations with the photoperiodic effects of light/dark schedules. These data thus represent a first approximation to a photoperiodic indicator that can be coupled or uncoupled at will.

Materials and Methods

All work was conducted with axenic cultures of a clone designated *Lemna perpusilla* collection number 6746 by Landolt (1957), from whom it was originally obtained. According to Kandeler and Hügel (1974), this strain conforms more closely to the criteria for *L. paucicostata,* but the older designation is retained here so as to avoid confusion. Stock cultures were maintained in continuous white light. For all details on which no other references are given, see Hillman (1975).

For both the CO_2 and flowering experiments, the basal medium was generally Z (nitrogen-free M) with 30 μM EDTA, but occasionally 0.5 × full strength of a nitrogen-free modification of Hutner's medium was also used. All media contained 1 per cent (29 mM) of sucrose. As nitrogen sources, either nitrate, ammonium (with marble chips to buffer pH changes) or aspartate was supplied as indicated, each at 15 mM. Sterilisation was effected by autoclaving except that aspartate was tested in both autoclaved and sterile filtered media.

CO_2 output was measured as described earlier, except that 50 ml of medium in 125-ml Erlenmeyer flasks were used instead of 100 ml in bottles. Times of maximal and minimal output were determined after 4–6 days under the conditions being tested. On both nitrate and ammonium media, the maximum referred to is the second of the two daily peaks observable under schedules with main light periods shorter than about 8 h and the only one with longer main light periods. By arbitrary convention, maximal or minimal plateaux lasting 3 h or less are assigned to the time of their mid-point, those lasting longer to the time of their termination (i.e. the time before the start of the descent for a maximum), and if two or more maxima or minima occur within 3 h their times are averaged.

Experiments on flowering were conducted at about 26 °C either on 50 ml of medium in 125-ml Erlenmeyer flasks when testing the effects of various nitrogen sources, or on 15 ml of the stock medium with or without ammonium in 20 × 130 mm test-tubes when determining times of maximal sensitivity to night interruptions. For the first type of experiment, the cultures were maintained under a single light/dark schedule throughout (9–13 days) and then dissected to evaluate percentage flowering (FL%) (Hillman, 1959). The values presented are based on four or five cultures per treatment. In the night interruption experiments, 3–4 days of schedules with interrupted nights were usually preceded (or, occasionally, followed) by 4–6 days of the corresponding schedule without night interruptions in order to raise the values obtained. Otherwise, most of a series of interruptions, usually at 1-h intervals, from the fifth to the twelfth hours after the start of darkness, would give FL% values of zero, making designation of the minimum impossible. Again, four or five cultures were dissected per treatment and the time of interruption giving the minimal flowering in each series was recorded. The values presented are the means and standard errors of the time obtained in 9–11 series for each condition.

In order to minimise temperature changes and direct light effects on CO_2 flux, main light periods in all experiments, except most of those to date on the time of maximal light sensitivity in flowering, were replaced with skeleton photoperiods in which 3 h or less of darkness separated successive brief light exposures. To simulate main light periods of various lengths, 0.25 h light exposures were applied starting at the indicated times, with hour 0 taken arbitrarily as the start: 4 h, 0, 1.75, 2.75, 3.75; 6 h, 0, 1.75, 3.75, 5.75; 8 h, 0, 3.25, 6.5, 7.75; 10 h, 0, 3.25, 6.5, 9.75; 12 h, as 10 h plus 11.75; 14 h, as 10 h plus 12.75 and 13.75; 16 h, as 10 h plus 12.75 and 15.75. Light in all experiments was red, at an incidence in the range 10–35 μW cm^{-2}, obtained by screening cool white fluorescent tubes with Rohm and Haas Plexiglas, red 2444 (Hillman, 1967).

Results and Discussion

FLOWERING UNDER SKELETON PHOTOPERIODS ACROSS THE CRITICAL RANGE

In previous work on chelated media under full light periods, *L. perpusilla* 6746 flowered as a short-day plant, optimally under 10 h or shorter main light periods and not at all under 16 h or more of light per day (Hillman, 1959). Experiments were thus performed to determine firstly if similar photoperiodic control occurs under skeleton main light periods, and secondly whether the nitrogen source affects the response. Typical results on nitrate, ammonium and aspartate are summarised in *Figure 24.1*. The variability is greater than expected from previous studies, which may be due in part to the use of skeleton

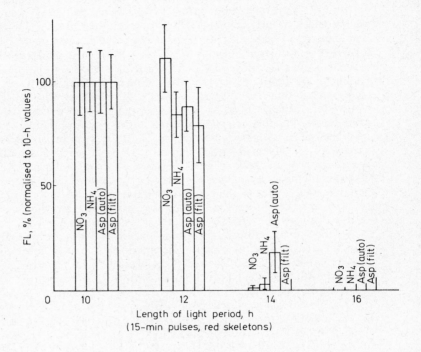

Figure 24.1 Flowering under 10, 12, 14 and 16 h skeleton main light periods on various media (see text). Means and standard errors of the 10-h values, with days' duration in parentheses, are as follows: NO_3, 28.7 ± 4.5 (11); NH_4, 41.8 ± 6.0 (10); autoclaved aspartate, 44.8 ± 6.7 (9); sterile filtered aspartate, 23.0 ± 3.0 (13)

main light periods. Unpublished observations in this laboratory, as well as recent work by Doss (1975), suggest that the effects of slight temperature fluctuations and as yet unidentified factors on *Lemna* flowering are greater under conditions of low light and skeleton photoperiods. Nevertheless, it is clear from *Figure 24.1* that photoperiodic control occurs under these conditions in about the same manner as in earlier work, and regardless of the nitrogen source. Similarly, when cultures were grown on all these media, as well as on glutamate or glutamine, under light 1 h/dark 23 h or light 0.25 h/dark 5.75 h, flowering occurred under the first schedule but not under the second.

CARBON DIOXIDE OUTPUT PATTERNS ACROSS THE CRITICAL RANGE

With the necessary basis in flowering response, the question is whether, on any of the media, the entrained patterns of CO_2 output under skeleton main light periods of 10, 12, 14 and 16 h show features that correlate informatively with the change in photoperiodic effect encountered in that range. Typical raw recorder data are given in *Figure 24.2*, while the timing of maxima and minima, representing 2–7 such experiments on each medium, is summarised in *Figures 24.3* and *24.4*, respectively.

Without *a priori* specifying criteria for an informative correlation with daylength response, inspection of *Figure 24.3* suggests that the time of maximal output on aspartate is uninformative. It occurs at a constant time after the start of each light period and falls entirely within each light period. Thus its properties are no more useful than would be the merely abstract hypothesis that something photoperiodically

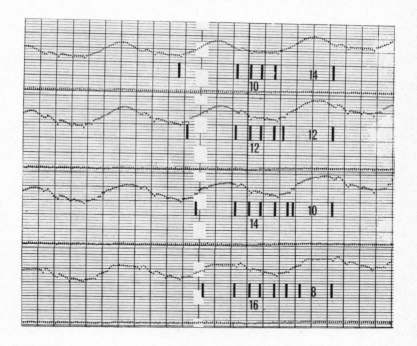

Figure 24.2 Time course of CO_2 output on NO_3 medium under 10, 12, 14 and 16 h skeleton main light periods. Light exposures (15 min each) are shown by thick black bars over one complete cycle only, but were given throughout; the figure shows the last 4 days of a 7-day experiment. Each vertical division represents 4 h, and successive points in each trace are 0.5 h apart. Each chart also contains a baseline, with successive points 0.5 h apart, representing a blank without plants. Baseline points in all charts are synchronous, although they may be imperfectly aligned in the figure. For additional results, see text and Hillman (1975)

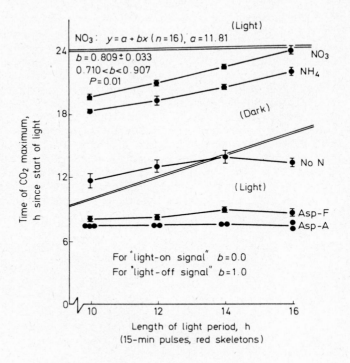

Figure 24.3   Timing of maximal CO_2 output on various media as a function of the length of the entraining skeleton main light period in the range 10–16 h. Points and brackets represent means and twice the standard errors derived from three or more experiments, while points without brackets represent values from individual experiments.   Regression for the NO_3 values calculated using four experiments for each light period length, total n = 16

significant happens at some constant time after the beginning of the light period.   The timing of maximal output in nitrogen-free medium, on the other hand, although it also remains unchanged as a function of light period length, falls across the changing time at which darkness begins in such a way that the time between it and the nearest (preceding) light exposure, as illustrated in *Figure 24.5*, decreases notably over the critical daylength range.   Yet the suggestion, from such data alone, that a crucial event in *Lemna* photoperiodism falls near the start of the dark period, would require much more evidence before it could be taken seriously.

There remain the maxima in nitrate and ammonium media, which, in contrast, appear highly informative.   Firstly, both occur well into

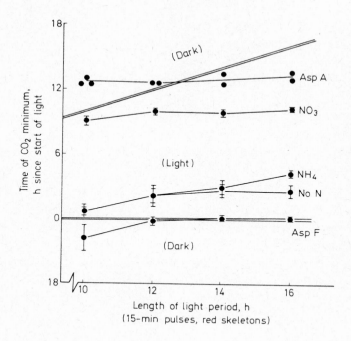

Figure 24.4 Timing of minimal CO_2 output on various media, otherwise as in Figure 24.3

the dark period, a time during which much of the literature suggests that significant photoperiodic events take place. Secondly, the time between either of these maxima and the nearest light exposure, in this case the beginning of the succeeding light period, decreases regularly as a function of daylength (*Figure 24.5*). In addition, the way in which these maxima move closer to the next light period as the length of the light period increases is reminiscent of the same effect on the time of minimum leaf position in *Coleus*, which Halaban (1968) found to occur at a constant time from the time of maximal sensitivity to a night interruption. Finally, in view of evidence that photoperiodism in many plants depends on the lengths of neither dark nor light periods, but on their interactions (e.g. Hamner and Hoshizaki, 1974), it is noteworthy that the slope of the straight line that fits the nitrate data best differs significantly from both 0.0 and 1.0. On the axes used in *Figure 24.3*, the former value would indicate simple dependence on the start of the light period, or 'light-on' signal, the latter simple dependence on the start of the dark period, or 'light-off'

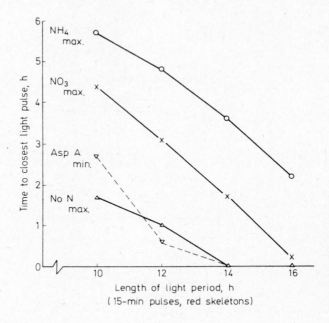

Figure 24.5   Time between the indicated feature of CO_2 output and the nearest light exposure, as a function of light period length, calculated from the data used in Figures 24.3 and 24.4

signal.   Hence, regardless of other findings, further investigation using the nitrate and ammonium maxima seemed warranted.

Turning briefly to *Figure 24.4,* all of the times of minimal CO_2 output appear to depend more or less uninformatively on the start of the light periods, except that the minimum on autoclaved aspartate can be discussed in the same terms as the maximum on nitrogen-free medium (*see also Figure 24.5*).   The timing of this feature, however, in contrast to that of all the others so far, has not proved consistently repeatable, for reasons so far undetermined.

To complete this analysis, *Figure 24.6* indicates the extent to which increasing the length of the main light period actually modifies the form of the CO_2 output patterns, in terms of the relative length of ascending and descending portions, rather than merely the phase relation of some phase to light and darkness.   On this basis, the pattern on nitrate seems substantially more responsive to increasing daylength in the critical range than any other.

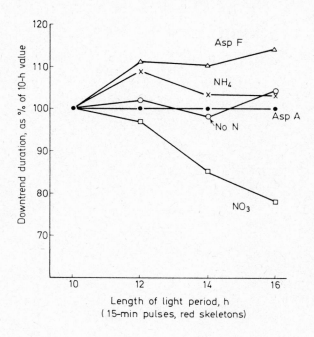

Figure 24.6   Downtrends (time elapsed from the maximal to the minimal output) of the entrained curves of CO_2 output calculated from the data used in Figures 24.3 *and* 24.4, *and normalised to the value under 10-h skeleton main light periods for each medium*

CARBON DIOXIDE OUTPUT PATTERNS AND THE TIME OF MAXIMAL SENSITIVITY TO A NIGHT INTERRUPTION

A more demanding criterion of any potential photoperiodic indicator is its relation, if any, to the way in which the time of maximal sensitivity to a night interruption, the so-called 'inducible' or 'inductive' phase (Pittendrigh, 1966; Halaban, 1968) varies with the length of the main light period.   Experiments on flowering under cycles with solid, non-skeleton red light periods of 3 or 8 h were conducted several years before the current work, but under otherwise identical conditions; typical data for one series in each group are given in *Figure 24.7*.   The overall results and a statistical treatment are summarised in *Figure 24.8*, as are data on the timing of the CO_2 output maximum on nitrate medium under 4-, 6- and 8-h skeleton main light periods.

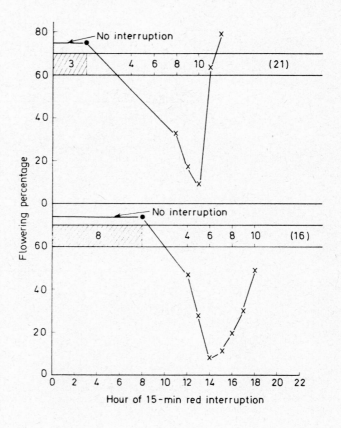

Figure 24.7 Flowering as affected by light interruptions of the dark period at various times in schedules with solid red light periods and 3 or 8 h. In each series, cultures were maintained under the schedule with uninterrupted night for 5 days, then given interruptions at the indicated times for the succeeding 5 or 4 days, respectively. Points are means of five cultures per treatment

Figure 24.8 supports two major conclusions. Firstly, there is no significant difference (t = 1.211, 0.2 < P < 0.4) between the slopes of the two functions. Thus, their timing can be regarded as expressing the same underlying process. Secondly, as the 99 per cent confidence limits show, the slopes are significantly different both from 0.0, which on the axes used would signify dependence solely on a light-off signal, and from −1.0, which would signify dependence solely on a light-on signal. An interaction of these signals is again indicated, as for the nitrate and ammonium maxima under longer main light periods.

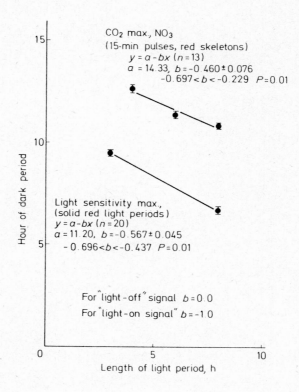

Figure 24.8 Timing of the maximal CO₂ output on NO₃ medium and of maximal sensitivity to light given in the dark period, expressed as hours of the dark period and as a function of main light period length. The successive values for CO₂ output are means, with brackets equal to twice the standard errors, of 4, 4 and 5 experiments, respectively, while those for light sensitivity are derived from 9 and 11 series, respectively, of the kind illustrated in Figure 24.7

Of course, flowering experiments with comparable skeleton light periods are required. So far, results for 10 h indicate maximal sensitivity at 5.95 ± 0.24 h of the dark period, a value that fits well with the CO_2 slope re-calculated to include its own 10-h values. Hence it seems justified to conclude at least tentatively that, in *L. perpusilla* 6746, the time of the daily CO_2 output maximum on nitrate medium is as precise an indicator of some process underlying photoperiodism as is, in *Coleus,* the minimum leaf position, and in nearly the same way. *Figure 24.9* summarises the timing of the CO_2 maxima on nitrate, ammonium and on aspartate media under daily light periods from 0.25 to 16 h long, and emphasises the difference

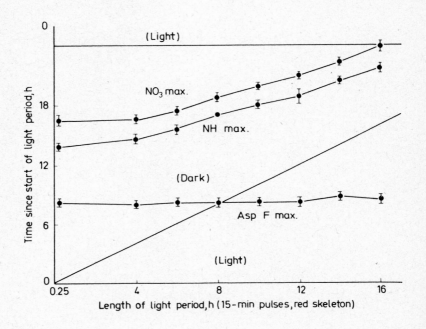

Figure 24.9 *Timing of maximal CO_2 output on various media as a function of the length of the skeleton light period in the range 0.25–16 h. Means, with brackets equal to twice the standard error, of three or more experiments in each instance*

between the simple light-on dependence of the third and the course of the others. It also suggests that the nitrate and ammonium functions are essentially parallel (*see also Figure 24.3*), so that what is true of one is true of the other, and that neither is truly linear over the entire range, although of course short segments can be approximated linearly, as in *Figures 24.3* and *24.8*.

In the *Coleus* experiments, the leaf position indicator and the 'inductive phase' both seemed dependent only on the light-on signal (Halaban, 1968) rather than on the interaction detected here; this may be a genuine species difference. However, the present results also stand in puzzling contrast to others on *L. perpusilla* 6746. Purves (1961) found that with either a 7- or 10-h main light period the maximum inhibition of flowering was given by night interruptions 9 h after the start of the dark period, indicating sole dependence on a light-off signal. This discrepancy cannot be explained at present, and suggests the value of repeating Purves' experiments as precisely as possible under the original conditions. So far, no evidence for simple dependence on a light-off signal has been observed in this laboratory in either flowering or CO_2 output data under schedules with a 24-h periodicity, on any medium, although a light-off signal can initiate oscillations in subsequent darkness (Hillman, 1975).

PROSPECTS FOR THE ANALYSIS OF PHOTOPERIODISM

Besides the work of Wilkins (1959) and previous data on *L. perpusilla* itself, results such as those of Queiroz (1974) and associates on enzymatic pathways of CO_2 metabolism in *Kalanchoë* and of Wagner, Frosch and Deitzer (1974) on energy charge and related phenomena in *Chenopodium* all place photoperiodism in a context of complex oscillations in metabolism. Thus, the fact that a feature of CO_2 output in one plane serves as a photoperiodic indicator in a way similar to, for example, leaf position in another, is not in itself particularly remarkable. Its major interest lies in the dependence on the nitrogen source: while the CO_2 maxima on nitrate or ammonium provide information relevant to photoperiodism, that on aspartate, which supports both growth and flowering equally well, provides none at all. It should thus be possible to distinguish biochemical processes which reflect photoperiodic control from those which do not by determining how the metabolism resulting in CO_2 output on aspartate medium differs from that on the other two.

In principle, the mechanism by which such effects of the nitrogen source occur seems clear enough. Since the pools of intermediates in CO_2 production are affected by the levels of related amino acids and other nitrogen compounds, changing the nitrogen source may change the proportions of the reactions contributing to CO_2 output. Only a relatively small fraction of the total output would need to be modified to bring about effects of the type observed (Hillman, 1975). However, the complexities to be expected in going from such generalities to specific differences suggest that additional physiological experiments should precede examination of individual pathways and enzymes. At least three lines of investigation are needed.

Firstly, CO_2 output patterns on nitrogen sources besides those reported here should be obtained in the test situations already studied. For example, a crude correlation derivable from the present data is that some portion of CO_2 output on inorganic nitrogen sources is coupled to photoperiodic control, while that on organic nitrogen is not. However, preliminary results indicate that, in fact, glutamine may give patterns similar to nitrate and ammonium while glutamate, depending on concentration, may even completely obscure any light-entrained pattern; these are cited only as examples of the sort of information that must be established with certainty.

Secondly, there are additional test situations in which photoperiodic response and CO_2 output should be compared, such as those provided by the interactions between light quality in main light periods and in night interruptions (Hillman, 1967). It may be that with increasing complexity only comparisons of effects on several media, not single features on one, will prove informative. At any rate, 'photoperiodic indicators' should survive the most demanding tests possible before specific biochemical hypotheses are based on them.

Thirdly, but equally important, the courses of oxygen uptake and dark CO_2 fixation under at least some of the conditions used here must be determined in order to distinguish the roles of respiration and other processes in establishing the patterns in question. Obtaining such data with the continuity and precision possible for CO_2 output poses some technical difficulties, but they are not insuperable.

In conclusion, the results presented strengthen the view that the effects of varying nitrogen source on the relationship between CO_2 output patterns and photoperiodically active light schedules constitute a powerful tool for identifying biochemical processes affected by those schedules. In spite of increasing information on phytochrome and related phenomena during the past decade, our understanding of the integrated, complex response of photoperiodism has hardly advanced to the same degree; perhaps a new approach will improve matters.

Acknowledgements

Research carried out at Brookhaven National Laboratory under the auspices of the U.S. Energy Research and Development Administration. Thanks are due to Helen J. Kelly, Neal Tempel and Rosemarie Dearing for technical assistance.

References

DOSS, R.P. (1975). *Plant Physiol.,* **55,** 108

HALABAN, R. (1968). *Plant Physiol.,* **43,** 1894

HAMNER, K.C. and HOSHIZAKI, T. (1974). *BioScience,* **24,** 407

HILLMAN, W.S. (1959). *Am. J. Bot.,* **46,** 466

HILLMAN, W.S. (1967). *Plant Cell Physiol.,* **8,** 467

HILLMAN, W.S. (1970). *Plant Physiol.,* **45,** 273

HILLMAN, W.S. (1972). *Plant Physiol.,* **49,** 907

HILLMAN, W.S. (1975). *Photochem. Photobiol.,* **21,** 39

KANDELER, R. and HÜGEL, B. (1974). *Plant System. Evol.,* **123,** 83

LANDOLT, E. (1957). *Ber. Schweiz. Bot. Ges.,* **67,** 271

PITTENDRIGH, C.S. (1966). *Z. Pflanzenphysiol.,* **54,** 275

PURVES, W.K. (1961). *Planta,* **56,** 684

QUEIROZ, O. (1974). *Annu. Rev. Plant Physiol.,* **25,** 115

SALISBURY, F.B. and DENNEY, A. (1974). In *Chronobiology,* p.679. Ed. Scheving, L.E., Halberg, F. and Pauly, J.E. Igaku Shoin Ltd., Tokyo

WAGNER, E., FROSCH, S. and DEITZER, G.F. (1974). *J. Interdiscip. Cycle Res.,* **5,** 240

WILKINS, M.B. (1959). *J. Exp. Bot.,* **10,** 377

25

PHYTOCHROME AND PHASE SETTING OF ENDOGENOUS RHYTHMS

MALCOLM B. WILKINS
PHILIP J.C. HARRIS
Department of Botany, Glasgow University, Glasgow G12 8QQ, U.K.

Introduction

The control of circadian rhythms in plants by light has been the subject of a number of investigations. A clear conclusion to be drawn from these studies is that in different groups of plants different pigments are involved in photoreception. In no case, however, has the pigment been unequivocally identified.

In the fungus *Neurospora crassa,* only the blue end of the visible spectrum is capable of suppressing the rhythm of conidiation (Sargent and Briggs, 1967) while in the photosynthetic dinoflagellate *Gonyaulax polyedra,* both red and blue light can re-set the phase of the rhythm of luminescence (Hastings and Sweeney, 1960). In the higher plants, activity seems to be associated principally, but not exclusively, with the red end of the spectrum (Bünning and Lörcher, 1957; Wilkins, 1960a; Hillman, 1971; Halaban, 1969).

The limitation of activity to the red end of the spectrum in *Bryophyllum* (Wilkins, 1960a, 1973) and the observation of reversible effects of red and far-red radiation in *Lemna perpusilla* (Hillman, 1971) and *Phaseolus multiflorus* (Bünning and Lörcher, 1957) suggest that phytochrome is involved in the photocontrol of circadian rhythms in a number of higher plants. In this chapter, we report some of our more recent studies on the possible involvement of phytochrome in the regulation of the circadian rhythm in *Bryophyllum* leaves.

Materials and Methods

The experimental plant material was *Bryophyllum (Kalanchoë) fedtschenkoi* R. Hamet et Perrier. The stock of plants had been derived vegetatively as cuttings from a single original plant, and was the same as that used in previous investigations (Wilkins, 1959, 1960a,b, 1962a,b, 1967, 1973; Warren and Wilkins, 1961). The plants were grown in a glasshouse and provided with supplementary irradiation from mercury vapour lamps to give a photoperiod of at least 16 h.   Experiments were carried out on single detached leaves of plants which had been transferred from the glasshouse to a controlled-environment room and maintained for at least 7 days in an 8-h photoperiod.   The temperature was 25 °C during the photoperiod and 15 °C during the dark period.

The method for measuring the rate of carbon dioxide output of the leaves was as described in an earlier paper (Wilkins, 1973).   Carbon dioxide-free air passed through the comparison tube of a Grubb-Parsons S.B.1 or S.B.2 infrared gas analyser and was divided into four streams with equal, monitored flow-rates of 1.55 l h^{-1}.   The gas streams were passed through separate Perspex plant chambers immersed to a depth of 150 mm in a water-bath at 15 ± 0.1 °C and directed in turn through the sample tube of the analyser for 15 min while the gas streams from the other chambers were directed to waste.   In each experiment, leaves in two of the plant chambers were irradiated while that in a third chamber remained in darkness as a control.   The fourth gas stream passed through a blank chamber and provided a check on the zero of the gas analyser.

Monochromatic radiation was obtained with Bausch and Lomb high-intensity grating monochromators, which were mounted vertically above the water-bath.   The divergent beam of radiation evenly irradiated the leaves in two of the chambers.   Spectral bands 25 nm wide were used. Two layers of Cinemoid Orange (No. 5) filter were inserted in the beam when the monochromators were set at wavelengths longer than 560 nm so as to eliminate overlapping blue light of other order spectra from the grating.   The radiant flux density was measured in J m^{-2} s^{-1} by means of a Kipp-Zonen compensated thermopile with a water screen between the thermopile and the light source.   The quantum flux density was calculated and adjusted in the different bands of the spectrum by varying the potential across the tungsten lamp of the monochromator with a rheostat.   The far-red radiation used in red/far-red reversal experiments was obtained by passing the columnated beam from a 45-W tungsten halogen lamp through a Corning 7-69 filter.

The leaves were detached from the plants, weighed, placed in the chambers and transferred to continuous darkness at 1600 h.   The experiments were continued until at least two peaks had been recorded after a light treatment, in most cases three or four peaks being recorded, in order to determine the new stable phase.   Phase shifts were calculated as the advance or delay in hours in the times of occurrence of the peaks in the treated leaves compared with those of the peaks in the untreated leaves.   The rate of carbon dioxide output of the leaves was calculated in μg CO_2 h^{-1} g (fresh weight)$^{-1}$ and plotted hourly against the time of day.

Results

PHASE CONTROL BY SINGLE LIGHT STIMULI

The relative effectiveness of a standard exposure to light in shifting the phase of the rhythm was assessed as a function of the time in the cycle at which it was applied. Leaves were exposed to white fluorescent light at a radiant flux density of 0.6 J m^{-2} s^{-1} for 4 h and 1½ cycles were scanned at intervals of 1 or 2 h beginning at 0400 h, which was 12 h after the leaves had been transferred from the growth room to continuous darkness. The results of three individual experiments are shown in *Figure 25.1* to illustrate the types of result obtained. The exposure gave rise to a phase advance (*Figure 25.1a*), no effect at all (*Figure 25.1b*) or a phase delay (*Figure 25.1c*), according to its time of application in the cycle. The results of a series of such experiments are collated into a phase-response curve shown in *Figure 25.2*. The position in the cycle at which a treatment is given clearly determines whether or not a phase shift is induced. Where a phase shift is induced, its nature (advance or delay) and its magnitude are determined by the precise time of application of the light treatment, since the next peak of the rhythm occurs a specific time after the end of the treatment.

The effectiveness of different wavelengths of light in bringing about a phase shift in the *Bryophyllum* rhythm was studied by exposing leaves to equal quantum flux densities of monochromatic radiation for 4 h from 0 h to 0400 h, the position in the cycle at which an exposure to white light induced a maximum phase shift (*Figure 25.2*).

The effectiveness of exposure for 4 h to a quantum flux density of 8.9 × 10^{-13} E cm^{-2} s^{-1} in several spectral bands 25 nm wide is shown in *Figure 25.3*, and the collated results of a series of such experiments over the spectral range 380–800 nm are shown in *Figure 25.4*. At this quantum flux density, activity is confined wholly to wavelengths of radiation between 600 and 700 nm with a peak at 640–660 nm. In another series of experiments, a higher quantum flux density of 1.9 × 10^{-11} E cm^{-2} s^{-1} was used. A slightly wider band of activity was observed, extending from 560 to 700 nm, but peak activity was again located in the 600–660-nm zone (*Figure 25.5*). At both quantum flux densities there was a sharp cut-off at 700 nm and no evidence of activity in the blue region of the spectrum.

The dependence of activity upon wavelength shown in *Figures 25.4* and *25.5* suggests the involvement of phytochrome in the reception of the light stimulus that leads to the induction of a phase shift. If this deduction is correct, then it must be possible to detect at least some degree of red/far-red reversibility. The exposure of leaves for 4 h to red light (660 nm) at a radiant flux density of 8.5 × 10^{-2} J m^{-2} s^{-1} resulted in a phase shift of about 12 h (*Figure 25.6a*), whereas exposure to far-red radiation at a radiant flux density of 7.8 J m^{-2} s^{-1} had no

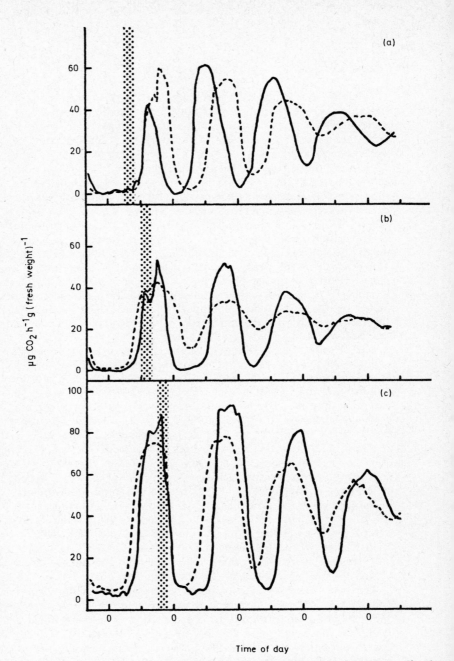

Figure 25.1   Effect of a 4-h exposure to white fluorescent light $(0.6\ J\ m^{-2}\ s^{-1})$ on the phase of the circadian rhythm of CO_2 output in Bryophyllum leaves otherwise kept in continuous darkness and at $15\,^{\circ}C$.   The positions of the light treatments in the cycle are shown by the shaded bars in (a), (b) and (c), and the rhythms in control, unirradiated leaves are shown by the broken lines. 0 = Midnight

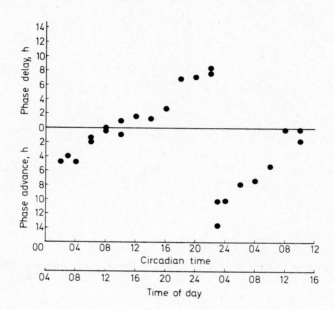

Figure 25.2 Phase-response curve for the rhythm in Bryophyllum *leaves following exposure to white fluorescent light (0.6 J m⁻² s⁻¹) for 4 h at different times in the cycle. The phase advance or delay is plotted against mid-point of the light treatment in sidereal time 0 = 24 = midnight, and circadian time where 0 = dawn, the time the photoperiod would have begun in cycles of 12 h light and 12 h darkness. Each point is the mean of two replicate leaf samples*

effect (*Figure 25.6b*). Exposure to red light for 4 h followed immediately by exposure to far-red radiation for 4 h resulted in a marked phase shift, although its magnitude was 4–5 h less than that attained with red light alone (*Figure 25.6c*). When leaves were exposed simultaneously to red and far-red radiation for 4 h, the circadian rhythm of carbon dioxide metabolism was virtually abolished and it was not possible to estimate the time of occurrence of the peaks (*Figure 25.6d*). There is no doubt that red and far-red radiation do interact in their effect on the *Bryophyllum* rhythm. Sequential exposure to red and then to far-red radiation reduces the phase shift in comparison with that achieved with red alone; their simultaneous application for 4 h has the effect of rapidly damping out the rhythm.

The lack of complete reversibility in the experiments in which 4-h exposures to radiation were used might be attributed to the fact that the exposures were sufficiently long for any effect to proceed beyond the reversible photochemical stages. Attempts to overcome this difficulty

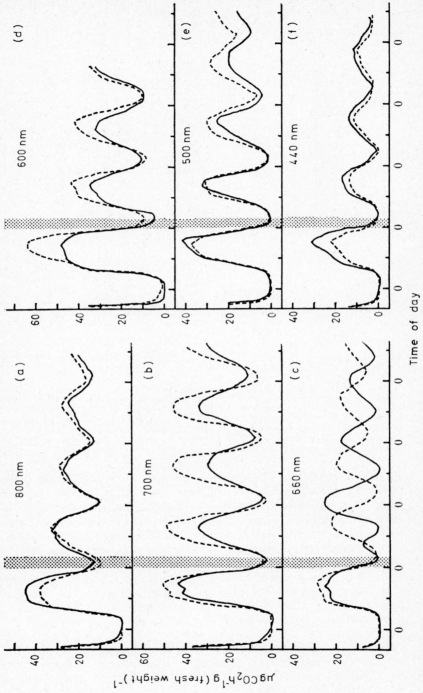

Figure 25.3 Effect of exposing Bryophyllum leaves for 4 h to an incident quantum-flux density of 8.9×10^{-13} E cm^{-2} s^{-1} at (a) 800 nm, (b) 700 nm, (c) 660 nm, (d) 600 nm, (e) 500 nm and (f) 440 nm from 0 to 0400 h between the peaks of a rhythm persisting in darkness at 15°C. Rhythms in the irradiated leaves are shown by the continuous lines and those in the unirradiated control leaves by the broken lines. Time of irradiation is shown by the shaded bar. 0 = Midnight. (From Wilkins 1973)

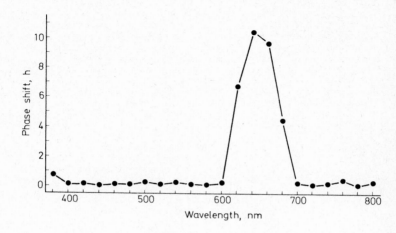

Figure 25.4 Relative effectiveness of exposing leaves for 4 h to an incident quantum-flux density of 8.9 × 10⁻¹³ E cm⁻² s⁻¹ in spectral bands at 20 nm intervals between 380 and 800 nm. Exposures were given from 0 to 0400 h between the peaks of the rhythm as indicated in Figure 25.3. *(From Wilkins, 1973)*

were made by exposing the leaves for 4 h to rapid alternation of red light and darkness. When this treatment comprised exposure to 5 min of red light alternating with 5 min of darkness for 4 h, a distinct phase shift was observed (*Figure 25.7a*). In a second experiment, leaves were exposed to 5 min of red light alternating with 5 min of far-red radiation for 4 h. The result of this treatment was not a clear reversal of the phase shift induced by red light alone, but rather an abolition of the rhythm. When the leaves were exposed to continuous far-red radiation for 4 h superimposed on the 5 min of red and 5 min of darkness experimental regime, the circadian nature of the rhythm was again lost and small peaks tended to occur at 12-h intervals. This technique thus also failed to demonstrate a clear reversal of the effectiveness of red light on exposure to far-red radiation in so far as phase shifts were concerned, but there is no doubt that the far-red radiation again interacted in some way with the red radiation, leading to the loss of a clearly defined circadian rhythm.

Some variation in the relative lengths of time for which the leaves were exposed to red light and to darkness over a period of 4 h was examined in order to determine whether or not clear evidence of red/far-red reversibility could be obtained. A sequential regime of 10 s of red light followed by 30 s of darkness led to the induction of a

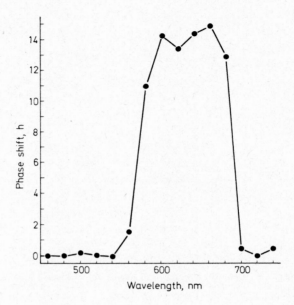

Figure 25.5   Relative effectiveness of exposing leaves for 4 h to an incident quantum-flux density of 1.9 × 10⁻¹¹ E cm⁻² s⁻¹ in spectral bands at 20-nm intervals between 460 and 740 nm. Exposures were given from 0 to 0400 h between the peaks of the rhythm.   (From Wilkins, 1973)

substantial phase shift (*Figure 25.8a*) but a regime of 10 s of red followed by 30 s of far-red radiation again led to the abolition of the rhythm (*Figure 25.8b*).   When the regime of 10 s of red light and 30 s of darkness was applied to leaves for 4 h with a simultaneous and continuous exposure to far-red radiation, the result was variable.   Either a very slight phase shift was induced in the rhythm (*Figure 25.8c*) or the rhythm lost its circadian nature and adopted a 12-h period in which it was impossible to assess whether a phase shift had occurred (*Figure 25.8d*).

The rapid alternation of either red light and darkness or of red light and far-red radiation therefore gave some evidence for the occurrence of a red/far-red reversibility.   There is no doubt, however, that red and far-red radiation interact under certain circumstances and lead to the abolition of the circadian rhythm in carbon dioxide metabolism in *Bryophyllum* leaves.

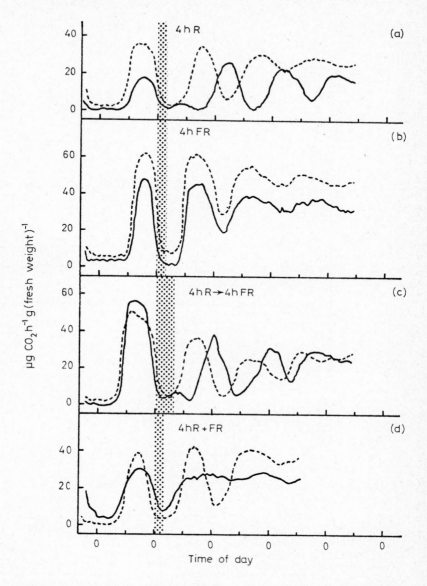

Figure 25.6 Effect of exposure (a) to red light for 4 h, (b) to far-red radiation for 4 h, (c) to 4 h of red light followed by 4 h of far-red radiation and (d) to red and far-red radiation simultaneously for 4 h on the phase of the circadian rhythm in Bryophyllum leaves. The times of the irradiations are shown by the shaded bars, and the control rhythms in total darkness are shown by the dotted lines. 0 = Midnight

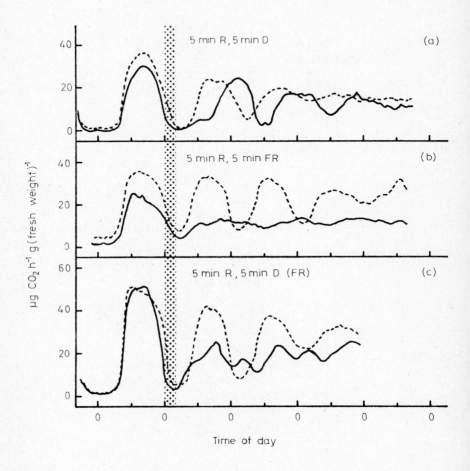

Figure 25.7 Effects of exposing leaves of Bryophyllum *for 4 h to (a) 5 min of red light alternating with 5 min of darkness, (b) 5 min of red light alternating with 5 min of far-red radiation and (c) to the regime in (a) superimposed on a 4-h continuous exposure to far-red radiation. Time of irradiation regimes is shown by the shaded bar; the broken lines show the control rhythm in unirradiated leaves. 0 = Midnight*

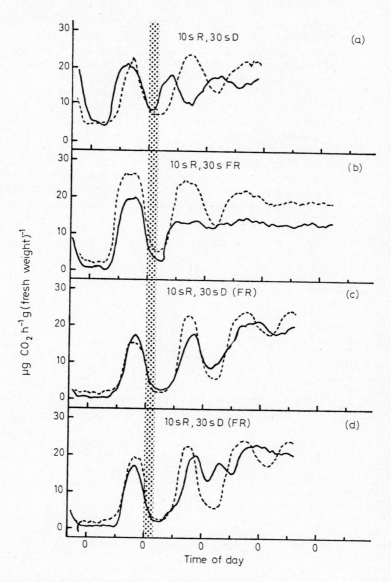

Figure 25.8 Effects of exposing leaves of Bryophyllum for 4 h to (a) 10 s of red light alternating with 30 s of darkness, (b) 10 s of red light alternating with 30 s of far-red radiation and (c) and (d) to the regime used in (a) superimposed upon a continuous exposure to far-red radiation for 4 h. Time of irradiation regimes is shown by the shaded bar; the broken lines show the control rhythm in unirradiated leaves. 0 = Midnight

ENTRAINMENT STUDIES

When plants showing circadian rhythms are subjected to cycles of light and darkness, the rhythms are usually entrained to the period of the external cycles (Wilkins, 1962) and the phase of the rhythm assumes a specific relationship to that of the entraining cycles. Entrainment of the circadian rhythm in *Bryophyllum* leaves to 12 h/12 h cycles of white light and darkness is shown in *Figure 25.9*. The phase of the entraining cycle of light and darkness in *Figure 25.9a* differs by 12 h from that in *Figure 25.9b*. In each case, the phase of the rhythm of carbon dioxide output locked on to light-dark cycles, the maximum of carbon dioxide output occurring at a specific time during the photoperiod. The regime in *Figure 25.9a* was very nearly that which the plant experienced during at least the week before the experiment began in that the end of the photoperiod occurred at 1600 h whereas the regime in *Figure 25.9b* was virtually the reverse

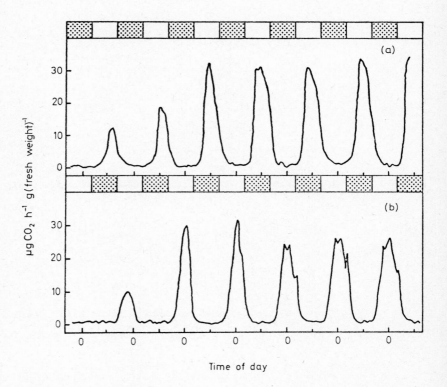

Figure 25.9   Entrainment of the rhythm to 12 h/12 h cycles of white fluorescent light and darkness.   The phase of the entraining cycles in (a) differs from that in (b) by 12 h, as shown by the bar above each figure, the shaded zone indicating darkness.   0 = Midnight

of the previous cycles, the end of the photoperiod occurring at 0400 h. In each case, entrainment to a new stable phase took place within two cycles.

More detailed analysis of the mechanism of entrainment can be achieved by exposing the plants to 24-h cycles of light and darkness in which the photoperiod is of very short duration and of a defined spectral band. Examples of regimes in which the cycle consisted of 15 min of white light and 23.75 h of darkness are shown in *Figure 25.10*, the phase of the cycles in *Figure 25.10a* being 12 h different from those in *Figure 25.10b*. In each case, the rhythm was entrained and locked on to a new phase after several cycles. In *Figure 25.10a*, the 15-min light period began at 0400 h, and a new stable phase was attained after 3 or 4 cycles, whereas in *Figure 25.10b* the light period began at 1600 h and a stable phase was achieved after one or two cycles. The rapidity with which the rhythm is entrained to a new stable phase is obviously determined by the phase of the entraining cycles and the initial phase of the rhythm which results from the light-dark cycles in the growth room where the plant was held prior to being used in the experiment. It is clear from *Figure 25.10* that a 15 min exposure to white light every 24 h is adequate to entrain the rhythm.

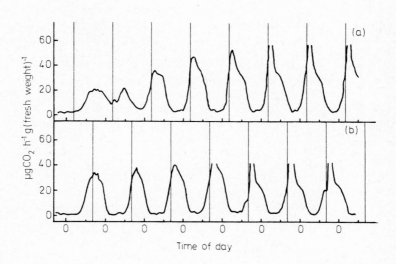

Figure 25.10 Entrainment of the rhythm in Bryophyllum *leaves to 15 min of white fluorescent light and 23.75 h of darkness every 24 h. The vertical bars indicate the times of the 15-min light treatments which differ by 12 h in (a) and (b). 0 = Midnight*

A comparison of the effect of exposing leaves of *Bryophyllum* for 15 min every 24 h to radiation of different wavelengths is shown in *Figure 25.11*. In darkness (*Figure 25.11a*), the rhythm damped out after four or five cycles. Periodic exposure to a 25-nm spectral band centred on 450 nm had no effect on either the phase or the persistence of the rhythm (*Figure 25.11b*). Periodic exposure to red light (660 nm) for 15 min caused the rhythm to persist indefinitely and to assume a specific phase in relation to the entraining cycles (*Figure 25.11c*). Far-red radiation (730 nm) seemed to act in a similar way to red light, although the amplitude of the rhythm was less and more cycles (six or seven) were required before the new stable phase was established (*Figure 25.11d*). Entrainment and continuation of the rhythm by periodic exposure to light is thus dependent upon the red end of the spectrum, and the effect shown in *Figure 25.11c,d* is one of two instances in the *Bryophyllum* rhythm in which it has been possible to obtain an essentially similar result with red and far-red radiation, although the effectiveness of the latter is always much less than the former.

Leaves were also subjected to more rapid cycles of light and darkness in which a 15-min exposure to red light was given every 12 h. Under this regime, the circadian rhythm was lost (*Figure 25.12b*) whereas in leaves that received only one 15-min irradiation with red light every 24 h the rhythm continued for at least 7 days (*Figure 25.12a*). If one of the 15-min exposures to red light in the 12-h cycles is followed immediately afterwards with a 15-min exposure to far-red radiation (*Figure 25.12c*) the rhythm is not lost, but persists in the same way and in the same phase as when the leaves received only one 15-min exposure every 24 h (*Figure 25.12a*). When the leaves are exposed to 15 min of red light every 12 h, and one of the exposures to red light is immediately preceded by a 15-min exposure to far-red radiation, the circadian rhythm is also abolished after about three cycles (*Figure 25.12d*), a result which is similar to that obtained with a 15-min exposure to red light every 12 h without any involvement of far-red radiation (*Figure 25.12b*).

There is clear evidence from the results presented in *Figure 25.12* for complete reversal of the effectiveness of red light by far-red radiation. Only in this type of experiment in which entrainment of the rhythm is studied by using 15 min of red light every 24 h or every 12 h has it been possible to observe a completely reversible red/far-red effect on the circadian system in *Bryophyllum* leaves.

Discussion

The phase-response curve for *Bryophyllum* is basically similar to those reported for other organisms (Aschoff, 1965; Pittendrigh, 1965). In most organisms, the relationship between the time in the cycle at which the light perturbation is given and the resulting phase shift is not linear

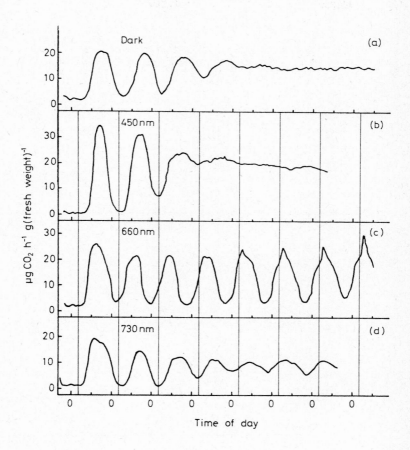

Figure 25.11 Relative effectiveness in entraining the circadian rhythm in Bryophyllum on exposing leaves to wavelengths of radiation of (b) 450 nm, (c) 660 nm and (d) 730 nm for 15 min and to darkness for 23.75 h in each 24 h. Rhythm in control leaves in darkness is shown in (a). Times of exposure to radiation are shown by the vertical bars. 0 = Midnight

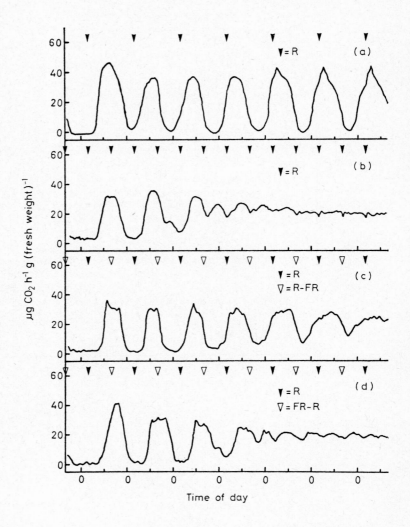

Figure 25.12 Effect on the circadian rhythm of CO_2 output in Bryophyllum leaves of exposure to 15 min of red light (a) every 23.75 h and (b) every 11.75 h. The effects of giving a 15-min exposure to far-red radiation immediately after, and immediately before, one of the exposures to red light used in (b) are shown in (c) and (d), respectively. 0 = Midnight

(e.g. *Drosophila*), whereas in *Bryophyllum* it appears to be linear, a finding which supports the previous report by Wilkins (1960a) that the magnitude of the phase shift induced by a single light perturbation is determined by the fact that the next peak occurs a specific time after the end of the perturbation. This being so, there must be a time in the cycle at which the perturbation does not give rise to a phase shift, as observed by Wilkins (1960a) and in the present phase-response curve (*Figure 25.2*).

Study of the relative effectiveness of equal quantum flux densities of radiation of different wavelengths has provided an 'action spectrum' shown in *Figure 25.4*. In a strict sense this is not an action spectrum since the Bunsen–Roscoe reciprocity law is not valid for this response; a minimum time of exposure to light appears to be required in order to achieve a maximum phase shift rather than the phase shift being determined by the quantum dose. The action spectrum for *Bryophyllum* is different from those which have been determined in *Neurospora* (Sargent and Briggs, 1967) and in *Gonyaulax* (Hastings and Sweeney, 1960) in that activity is strictly confined to wavelengths of radiation above 560 nm. This characteristic, the fact that there is maximum effectiveness at 640–660 nm and the sharp cut-off at 700 nm, suggests that phytochrome is involved in photoreception.

Several experimental approaches using sequential or simultaneous exposures to red and far-red radiation showed that these spectral bands interact strongly with each other in their effect on the rhythm. Frequently this interaction did not take the form of a clear partial or total reversal of the effect of red light by far-red radiation, but rather an abolition of the rhythm when far-red radiation was applied. The abolition of the rhythm may arise in a number of ways. For example, the far-red radiation may interact with the red light to abolish oscillation of the basic circadian system in each of the cells of the leaf. On the other hand, total reversal of the phase shift induced by red light may occur in the basic circadian system only in some cells and not in others. This would result in abolition of the overt rhythm of carbon dioxide output from the leaves because the rhythms in some cells would be 180° out of phase with those in other cells. The occurrence of such cellular desynchronisation may account for the occurrence of small peaks at 12-h intervals in some experiments.

The rhythm of carbon dioxide metabolism in *Bryophyllum* leaves can be entrained to 24-h cycles of light and darkness. As little as 15 min of light in each 24-h cycle is sufficient to entrain rapidly the rhythm and to maintain it in a precise phase relationship to the entraining cycles. Similar findings have been reported for the rhythms in *Drosophila pseudo-obscura* (Pittendrigh and Minis, 1964) and *Lemna perpusilla* (Hillman, 1971), while in other organisms, for example *Pharbitis nil*, a longer exposure to light is necessary for entrainment to be achieved (Bollig, 1974).

Entrainment of the rhythm in *Bryophyllum* to 15 min of radiation in each 24-h cycle can be achieved only with wavelengths in the red end of the spectrum. Activity was found at wavelengths above 700 nm, in the far-red region of the spectrum, although this was less than that in the 600–700-nm spectral band. This finding is in marked contrast to the sharp cut-off in activity observed at 700 nm in the action spectra for phase-shift induction by single exposures to light. We have been able to establish that this extension of activity into the far-red region of the spectrum is not attributable to the use of a higher quantum flux density in the entrainment studies than in the single exposure experiments. An explanation of the difference between the effectiveness of far-red radiation in the entrainment studies and the single-stimulus, phase-shift studies is that far-red radiation might induce only a very small phase shift when given at the time we selected for the single exposures, and this shift might have been below the limits of statistical significance. When the far-red exposure is repeated in each cycle, however, these small phase shifts would in effect be multiplied, with the result that a gradual entrainment of the rhythm would occur. Such a suggestion would also explain why entrainment to far-red radiation is slower than that to red light.

The entrainment studies have provided a clear demonstration of complete red/far-red reversibility in the *Bryophyllum* rhythm. This finding, together with the obvious interaction of red and far-red radiation in leading to the abolition of the rhythm in some of the other experiments, leave little doubt that phytochrome is the pigment involved in light absorption in the photocontrol of the circadian system in *Bryophyllum.* The characteristics of the action spectrum for phase shifting had previously indicated most strongly that phytochrome was the pigment involved. The main features of the action spectrum are very similar to those of the absorption spectrum of phytochrome with the exception of the absence of a minor peak in the blue region. This difference might be attributable to the fact that anthocyanin is present in leaf epidermal cells of *Bryophyllum fedtschenkoi,* and this red pigment might result in little blue light penetrating into inner cells of these thick and succulent leaves. However, on no occasion has an effect of blue light on the rhythm been detected, even with relatively high radiant flux densities.

We conclude, therefore, on the characteristics of the action spectrum and the occurrence of both partial and complete red/far-red reversibility, that phytochrome is the principal, and probably the only, pigment involved in the photocontrol of the circadian system in *Bryophyllum* leaves.

Phytochrome is held to be involved in entrainment of the rhythms of carbon dioxide output in *Lemna perpusilla* (Hillman, 1971) and of the leaf-movement rhythm in *Phaseolus multiflorus* (Lörcher, 1958) on the basis of the reversible effect of red and far-red radiation, although no action spectrum has been determined in these plants. Similar claims have been made for the rhythm of sensitivity to photoperiodic induction in *Xanthium* (Salisbury, 1965; Denny and Salisbury, 1970). Phytochrome is clearly not involved in the photocontrol of the circadian systems in

Neurospora crassa (Sargent and Briggs, 1967), *Gonyaulax polyedra* (Hastings and Sweeney, 1960) or *Oedogonium cardiacum* (Bühnemann, 1955), nor is it always the pigment involved in higher plants (Karvé, Engelmann and Schoser, 1961; Karvé and Jigajinni, 1966; Halaban, 1969). However, the statement of Hamner and Hoshizaki (1974) that 'phytochrome is not involved in the entrainment or rephasing of the basic circadian rhythms (clocks)' is obviously incorrect.

A change in the state of phytochrome from a low to a high level of the Pfr form either once for a few hours, or repetitively in 24-h cycles, thus modifies the operation of the rhythm in the phosphoenol-pyruvate carboxylase system which gives rise to the rhythm in carbon dioxide emission from *Bryophyllum* leaves (Warren and Wilkins, 1961; Warren, 1964). The precise mechanism by which the interconversion of phytochrome modifies the circadian rhythm in the activity of the enzyme system in the leaves is not yet understood.

References

ASCHOFF, J. (1965). In *Circadian Clocks,* pp.95-111. Ed. Aschoff, J. North-Holland, Amsterdam
BOLLIG, I. (1974). *Doctoral Dissertation,* Eberhard-Karls University, Tübingen
BÜHNEMANN, F. (1955). *Planta,* **46**, 227
BÜNNING, E. and LÖRCHER, L. (1957). *Naturwissenschaften,* **44**, 472
DENNY, A. and SALISBURY, F.B. (1970). *Plant Physiol.,* **46**, Suppl., 26
HALABAN, R. (1969). *Plant Physiol.,* **44**, 973
HAMNER, K.C. and HOSHIZAKI, T. (1974). *BioScience,* **24**, 407
HASTINGS, J.W. and SWEENEY, B.M. (1960). *J. Gen. Physiol.,* **43**, 697
HILLMAN, W.S. (1971). *Plant Physiol.,* **48**, 770
KARVÉ, A.D., ENGELMANN, W. and SCHOSER, G. (1961). *Planta,* **56**, 700
KARVÉ, A.D. and JIGAJINNI, S.G. (1966). *Naturwissenschaften,* **53**, 181
LÖRCHER, L. (1958). *Z. Bot.,* **46**, 209
PITTENDRIGH, C.S. (1965). In *Circadian Clocks,* pp.277-297 Aschoff, J. North-Holland, Amsterdam
PITTENDRIGH, C.S. and MINIS, D.H. (1964). *Am. Nat.,* **98**, 261
SALISBURY, F.B. (1965). *Planta,* **66**, 1
SARGENT, M.L. and BRIGGS, W.R. (1967). *Plant Physiol.,* **42**, 1504
WARREN, D.M. (1964). *Ph.D. Thesis,* University of London
WARREN, D.M. and WILKINS, M.B. (1961). *Nature, Lond.,* **191**, 686
WILKINS, M.B. (1959). *J. Exp. Bot.,* **10**, 377
WILKINS, M.B. (1960a). *J. Exp. Bot.,* **11**, 269
WILKINS, M.B. (1960b). *Cold Spring Harbor Symp. Quant. Biol.,* **25**, 115
WILKINS, M.B. (1962a). *Plant Physiol.,* **37**, 735
WILKINS, M.B. (1962b). *Proc. R. Soc. B,* **156**, 220
WILKINS, M.B. (1967). *Planta,* **72**, 66
WILKINS, M.B. (1973). *J. Exp. Bot.,* **24**, 488

26

THE NATURE OF PHOTOPERIODIC TIME MEASUREMENT: ENERGY TRANSDUCTION AND PHYTOCHROME ACTION IN SEEDLINGS OF *CHENOPODIUM RUBRUM*

E. WAGNER

Institut für Biologie II, Universität Freiburg, D-78 Freiburg i. Br., Schänzlestrasse 9-11, West Germany

Introduction and Working Hypothesis

It is currently accepted that monophyletic evolution from prokaryotic to eukaryotic organisms was paralleled by a corresponding evolution in energy metabolism. From fermentation in the primeval anoxygenic soup, energy conservation progressed to anaerobic photosynthetic processes and then to carbon dioxide fixation with concomitant acceptance of electrons by water and evolution of oxygen. Thereafter, in a progressively oxygenic biosphere, respiration developed with oxygen as terminal electron acceptor. Eventually, light via photosynthesis became the ultimate source of energy to sustain life (Kaplan, 1972).

With these considerations in mind, we hypothesised that photoperiodism developed basically as an evolutionary adaptation of eukaryotic energy conservation and transformation, i.e. energy transduction (cf. Harold, 1972), to optimise energy harvesting by photosynthesis in respect to the yearly change in the daily light–dark cycle or cycle of energy supply from the environment (Cumming and Wagner, 1968; Wagner and Cumming, 1970). Thus, in plants, photosynthesis should be controlled through the interaction of circadian rhythms and phytochrome as the main components of photoperiodic control (e.g. Chia-Looi and Cumming, 1972; Wilkins, 1973; Hillman, 1975; Jones and Mansfield, 1975).

Circadian rhythmicity is the innate timer of eukaryotes controlling growth, differentiation and behaviour (Bünning, 1973). This endogenous rhythm of metabolic activity has a period of exactly 24 h when the organisms are subjected to the light–dark cycle of the earth. In constant conditions, however, its period is only approximately 24 h. Circadian rhythmicity seems to be a function of the metabolic control net of the whole cell, since it cannot be observed in isolated cells or organelles despite the fact that organelles as well as isolated metabolic sequences can display high-frequency (seconds to minutes) oscillations in their activities, under *in vitro* and *in vivo* conditions (Gander, 1967; Wilson and Calvin, 1955; Tornheim and Lowenstein, 1973; Yamazaki and Yokota, 1967; Fukushima and Tonomura, 1972; Gooch and Packer, 1974; Hess and Boiteux, 1971). In contrast to these high-frequency oscillations, circadian rhythms are temperature compensated and are hardly susceptible to chemical manipulation. This stability of period length is a functional prerequisite for an oscillator to act as a precise physiological clock (Pittendrigh and Caldarola, 1973). Changes in daylength as small as 5 min per week can be measured (*Table 26.1*).

Table 26.1 Latitude and daylength (After Hillman, 1969)

Latitude	Daylength (h)		Rate of change in daylength (April and August) (min/week)
	Min.	Max.	
60° N (Stockholm, Leningrad)	6	19	40
45° N (London 51°, New York 41°)	9	15.5	25
30° N (Delhi, Shanghai)	10	14	12
15° N (Manila, Dakar)	11	13	5

There is evidence that this stability of frequency is based on the regulatory effects of end-products in feedback, and rate effectors (NAD/NADH, phosphate potential or energy charge and ionic ratios; Atkinson, 1971; Bygrave, 1967; Meli and Bygrave, 1972) in allosteric control to maintain constancy or homeostasis within an organism to compensate for the static and dynamic disturbances from the environment (Watson, 1972). It is conceivable that homeostasis and rhythmicity are just two aspects of one biological principle (Wever, 1965; Hyndman, 1974). Homeostasis protects living systems against random disturbances while circadian rhythmicity provides an optimal phase relationship between oscillatory metabolic sequences and the environment through entrainment by the daily light–dark or temperature cycle.

In the search for the mechanism of photoperiodic control, we assumed that a circadian rhythm in energy transduction should be the basis for phytochrome interaction (Wagner and Cumming, 1970; Frosch, Wagner and Cumming, 1973; Wagner *et al.*, 1975). The circadian rhythm in energy metabolism is viewed as a compensatory control oscillation between glycolysis and oxidative phosphorylation, which itself is controlling photophosphorylation in photosynthetic systems (Wagner *et al.*, 1975; Deitzer *et al.*, 1974). An endogenous rhythm in energy metabolism would be likely to be reflected in a corresponding rhythm in the state of energy transducing biomembranes (Harold, 1972; Dutton and Wilson, 1974), which could integrate their binding sites for the photoreceptor(s) (Quail, Marmé and Schäfer, 1973). In this way, the interaction of photoreceptor and binding site could be controlled by energy metabolism. Photostimulation of the receptor could lead to transitory modulation (phase setting) (White and Pike, 1974; Wagner, Frosch and Kempf, 1974; Frosch, Wagner and Mohr, 1974) of

energy transduction and switching of metabolic controls.   The concept
of photoperiodic control could account for endogenous rhythmic changes
in sensitivity to phytochrome photoconversion in light (Könitz, 1958)
and darkness (Cumming, Hendricks and Borthwick, 1965; King and
Cumming, 1972; Frosch and Wagner, 1973b; Satter, Geballe and
Galston, 1974).

Although night-break experiments clearly demonstrate the interaction
of endogenous rhythms and phytochrome, they are not very likely to
represent the natural situation of phytochrome involvement in photoperiodic control.   It appears as if phytochrome control at the end of
day (Fredericq, 1964; Fredericq and De Greef, 1968; Kasperbauer,
1971; Kasperbauer, Tso and Sorokin, 1970; Lane, Cathey and Evans,
1965; Nakayama, 1958; Nakayama, Borthwick and Hendricks, 1960;
Frosch and Wagner, 1973a) is of more relevance to reality, especially
in view of the changes in the red/far-red ratio (Shropshire, 1973) and
other spectral bands (Johnson, Salisbury and Connor, 1967) at dawn
and dusk (*Figure 26.1*).   The time distance between the two signals
per day at dawn and dusk is determined by the rotation of the earth
and the seasonal changes in solar elevation.   It is independent of light
intensity and thus has the precision required for photoperiodic control
(cf. *Table 26.1*).

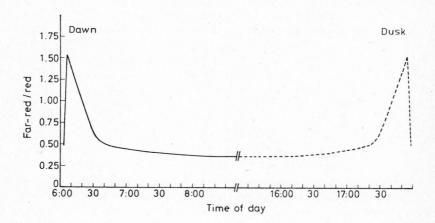

*Figure 26.1   Ratios measured for the irradiances of far-red light at 730 ± 4 nm
and red light at 660 ± 4 nm for 1-min intervals shortly after sunrise and before
sunset.   Solid line after Shropshire (1973); broken line supplemented (Shropshire,
1973; Shropshire, personal communication)*

Circadian Rhythmicity in Energy Transduction and Metabolic Activity

The existence of endogenous rhythms of energy conservation (energy uptake and storage) and transformation (i.e. energy transduction) is very likely in view of circadian rhythms in respiration and photosynthesis (Chia-Looi and Cumming, 1972; Hastings, Astrachan and Sweeney, 1960; Hillman, 1975; Jones and Mansfield, 1975; Van Den Driessche, 1966).

We proposed (Wagner and Cumming, 1970) that a circadian oscillation in energy transduction should be generated through the coupling (Pavlidis and Kauzman, 1969; Pavlidis, 1971) of high-frequency oscillations (Betz, 1968; Hess and Boiteux, 1971; Nicolis and Portnow, 1973) in different pathways of energy metabolism. The well documented high-frequency oscillations in glycolysis (Hess and Boiteux, 1971), the purine nucleotide cycle (Tornheim and Lowenstein, 1973), the peroxidase system (Yamazaki and Yokota, 1967) and in ATPase activity and phosphorylation of microsomes (Fukushima and Tonomura, 1972) will be separated and controlled by energy-transducing biomembranes which, via their shuttle mechanisms (Heber, 1974), could act as frequency filters (Hyndman, 1974), as suggested by low-frequency reciprocal activity changes between mitochondria and chloroplasts which occur in the same cell (Packer, Murakami and Mehard, 1970).

The coupling of the different sequences of energy transduction could be effected by modulation of enzymic activity (Lebherz, Sevage and Abacherli, 1973; Wieland and Portenhauser, 1974) through cofactor ratios (Atkinson, 1971), ionic balances (Bygrave, 1967; Meli and Bygrave, 1972; Bygrave, Daday and Doy, 1975; Blair, 1970), substrate availability and/or end-product feedback.

An integrated view of such regulatory mechanisms in metabolic control was advanced by Atkinson (e.g. Atkinson, 1969, 1971). Atkinson developed the 'energy charge' concept to indicate quantitatively the energy state of the cell. He defined the energy charge as

$$([ATP] + \tfrac{1}{2}[ADP])/([ATP] + [ADP] + [AMP])$$

Subsequently, it was shown that the rates of energy-expending reactions increase and the rates of energy-producing reactions decrease as the energy charge increases (Chapman, Fall and Atkinson, 1971).

In analysing the energy metabolism of synchronised and photosynthetically active seedlings of *Chenopodium rubrum* with respect to rhythmic changes in adenine and pyridine nucleotides, it was demonstrated that the pool sizes of all nucleotides changed rhythmically (Wagner, Stroebele and Frosch, 1974; Wagner and Frosch, 1974). The calculation of the energy charge and of the ratio NADPH/NADP revealed a circadian rhythm (*Figure 26.2*) in continuous darkness as well as in continuous light. Even in dark-grown seedlings which had not been subjected to cyclic environmental conditions, there is a circadian rhythm in energy charge with two sub-peaks per circadian period (*Figure 26.3*) (Wagner

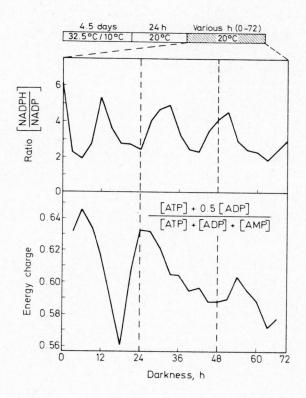

Figure 26.2 Time course of the ratio NADPH/NADP and of energy charge in Chenopodium rubrum *seedlings germinated in constant light and alternating conditions of temperature for 4.5 days. Thereafter constant at 20 °C for 24 h followed by a dark period of varied duration (3-h increments). Graphs show first 3-point moving means of one (NADPH/NADP) and of five independent experiments (energy charge). (After Wagner and Frosch, 1974; Wagner, Stroebele and Frosch, 1974)*

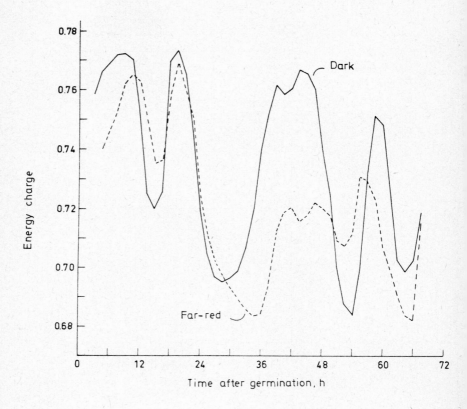

Figure 26.3   Time course of energy charge during a 3-day period in far-red light or darkness, following dark germination at constant 20 °C after a temperature step, which induced germination.   Data represent the second 3-point moving mean of three independent experiments.   (After Wagner et al., 1974)

et al., 1974).   In continuous far-red light, the pattern of oscillation is more or less identical with that in darkness, except for a change in amplitude, which is possibly due to the increasing metabolic activity of chloroplasts which are developing under these conditions of photomorphogenesis.

Circadian rhythmicity in energy charge and NAD/NADH ratio has also been observed in *Neurospora crassa* (Brody, 1973; Delmer and Brody, 1973), and this system also shows endogenous rhythms in activity of a number of enzymes (Hochberg and Sargent, 1974). Further, it had been shown that oscillations in energy charge can be generated using the molecular components of the glycolytic sequence (Goldbeter, 1974).

The demonstration of circadian rhythmicity in energy charge and redox state is particularly relevant in the search for the mechanism of circadian rhythmicity, since the adenylate system (ATP, ADP, AMP and adenylate kinase) mediates the energy metabolism of the cell by controlling glycolysis, gluconeogenesis, lipogenesis and possibly photophosphorylation (Atkinson, 1971; Chapman, Fall and Atkinson, 1971; Dynnik, Sel'kov and Semashko, 1973; Mangat, Levin and Bidwell, 1974; Coombs, Maw and Baldry, 1974; Michel and Thibault, 1973; Thibault and Michel, 1972; Miginiac-Maslow and Champigny, 1974; Heber, 1974).

Considering the interdependent network character of energy metabolism (Bennun, 1974; Reich and Sel'kov, 1974; Cardon and Iberall, 1970), it seems feasible to conceive a circadian rhythmicity in energy charge and redox potential as the result of a compensatory control oscillation in a kind of compartmental feedback (Packer, Murakami and Mehard, 1970; Heber, 1974) between glycolysis and oxidative phosphorylation.   This fundamental rhythm in energy metabolism could be conditioning the development and the capacity of the photosynthetic machinery and thus optimise energy conservation in daily light–dark cycles.   This concept is supported by the demonstration of energy charge control of photosynthetic 3-phosphoglyceric acid kinase (Pacold and Anderson, 1973, 1975) and reduction charge regulation of glucose-6-phosphate dehydrogenase (Wildner, 1975), ribulose bisphosphate carboxylase and fructose diphosphatase (Wildner, personal communication).

The endogenous rhythmic changes in nucleotide pool size levels are of great importance in relation to reciprocal activity changes in glycolytic and photosynthetic glyceraldehyde-3-phosphate dehydrogenase (GPD) in *Chenopodium rubrum* (Frosch, Wagner and Cumming, 1973). There is increasing evidence for nucleotide regulation of activity and specificity of GPDs (Pupillo and Piccari, 1973; Müller, 1970; Müller, Ziegler and Ziegler, 1969; Yang and Deal, 1969a,b; Lebherz, Sevage and Abacherli, 1973).   Nucleotide control could also account for the reciprocal activity changes in phytochrome controlled GPDs in *Sinapis alba* (Cerff, 1974; Cerff and Quail, 1974).   An analysis of GPD activities in *Chenopodium rubrum* with ammonium sulphate chromatography (Wagner, Kleinschmidt and Stühmer, in preparation) revealed three separate enzymes with different nucleotide specificity: one NAD-specific

NAD-GPD I), one NADP-specific (NADP-GPD I) and one bifunctional NAD/NADP-specific GPD activity (NAD-GPD II/NADP-GPD II). As shown in *Figures 26.4* and *26.5*, there is a change in the relation of the different GPD activities depending on the time of light and darkness at which the enzymes had been isolated from plants grown under a 12 h/12 h light–dark cycle. In the case of potential nucleotide-mediated interconversion of these enzyme activities (cf. Lebherz, Sevage and Abacherli, 1973), the changes in the pool sizes of adenine and pyridine nucleotides in leaf cells upon light–dark transition (cf. Heber, 1974) could well account for the changes in the enzyme pattern observed here.

Manipulation of energy metabolism by supplying glucose to the seedlings increases the activity of glycolytic NAD-GPD I and inhibits photosynthetic NAD/NADP-GPD II compared with the Hoagland control (*Figures 26.6* and *26.7*). Glucose inhibits photosynthesis and stimulates respiration, as shown in *Table 26.2* (Werner, personal communication).

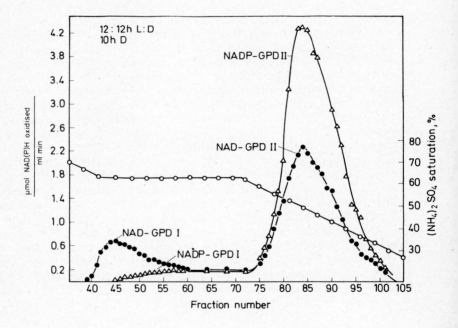

Figure 26.4 Activity profile of glyceraldehyde-3-phosphate dehydrogenases (GPD) from a leaf extract of Chenopodium rubrum *plants grown at constant 20 °C and a light–dark cycle of 12 h/12 h. Leaves were extracted at the 10th hour of darkness and subjected to ammonium sulphate chromatography. (Wagner, Kleinschmidt and Stühmer, in preparation)*

Table 26.2 Dependence of photosynthetic and respiratory values on glucose supply and the length of preceding dark period (nmol CO_2/h × seedling)

Light 6500 lx (incandescent); temperature 20 °C; humidity 75%.

Darkness (h)	Photosynthesis		G Value H Value (= 100%)	Respiration		G Value H Value (100%)	Photosynthetic quotient[2], $(LV + DV)/DV$	
	G^1	H^1		G	H		G	H
9	49.42	60.12	82	58.69	22.55	260	1.84	3.66
15	56.48	57.92	98	70.60	28.38	249	1.80	3.04
24	54.14	61.42	88	63.42	26.51	239	1.85	3.32

[1] G = 0.4 M glucose in Hoagland's solution; H = Hoagland's solution
[2] LV= light value (photosynthesis); DV = dark value (respiration)

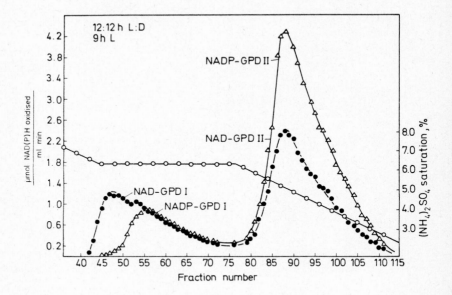

Figure 26.5 *Activity profile of glyceraldehyde-3-phosphate dehydrogenases (GPD) from a leaf extract of* Chenopodium rubrum *plants. Experimental conditions as in* Figure 26.4, *but leaves were extracted at the 9th hour of light (Wagner, Kleinschmidt and Stühmer, in preparation)*

Seedlings of *Chenopodium rubrum* were grown in continuous white light, and then subjected to 9, 15 or 24 h of darkness and either supplied or not supplied with 0.4 M glucose in Hoagland's solution. Respiration and photosynthesis were determined after the dark period. The gas exchange measurements clearly demonstrate the differential effect of the length of a dark period interrupting continuous light on the channelling of energy flow in the system in accordance with the endogenous rhythm in metabolic activity (cf. Frosch, Wagner and Cumming, 1973). The differential channelling of supplied glucose by the endogenous rhythmic system is also indicated by the fact that there is heavy accumulation of starch in plastids with 24 h of darkness but very little with 15 h of darkness. Without added glucose, there is hardly any accumulation of starch in the different treatments (Bergfeld, personal communication).

The above observations clearly demonstrate circadian rhythmicity in energy transduction. The circadian frequency is most likely generated through the interaction of different metabolic sequences of energy

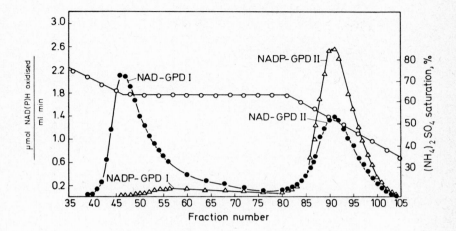

Figure 26.6   Activity profile of glyceraldehyde-3-phosphate dehydrogenases (GPD) from seedlings grown on Hoagland's medium.   (Wagner, Kleinschmidt and Stühmer, in preparation)

metabolism which oscillate with higher frequencies.   The coupling between sequences is possibly achieved through nucleotide ratios via redox shuttles (Dutton and Wilson, 1974) in the different energy-trans-ducing biomembranes which could act as frequency transformers (Hynd-man, 1974).   The coupling nucleotide ratios would be relatively inde-pendent of temperature and could thus, as rate effectors (Watson, 1972), fulfil the requirements for precise temperature-compensated time keeping.

Kinetics of Phytochrome Action Under Rhythmic and Non-rhythmic Conditions

ENDOGENOUS RHYTHMICITY AS ON/OFF CONTROL FOR PHYTOCHROME ACTION

The apparent fluctuations in the sensitivity of the photoreceptor system mediating photoperiodic control are probably closely linked to the circadian rhythm in energy transduction.   We assumed that the photo-receptor (phytochrome and/or a flavoprotein; Sargent and Briggs, 1967; Halaban, 1969) is an effector which is either membrane bound or which acts on a membrane-integrated receptor site (Wagner, Frosch and Deitzer, 1974b).   The changes in sensitivity of this photoreceptor system could

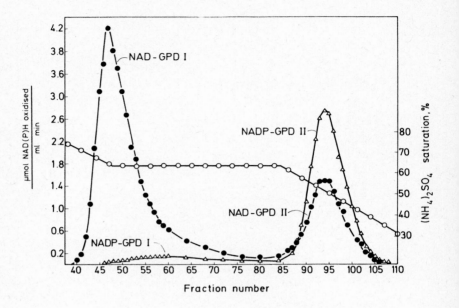

Figure 26.7   Activity profile of glyceraldehyde-3-phosphate dehydrogenases (GPD) from seedlings grown on 0.4 M glucose in Hoagland's solution. (Wagner, Kleinschmidt and Stühmer, in preparation)

be achieved through a circadian rhythm in energy state of the cellular membranes as a whole (Cumming and Wagner, 1968; Wagner and Cumming, 1970; Frosch and Wagner, 1973a; Wagner *et al.*, 1975). The photoreceptor, on the other hand, should be able to modulate vectorial metabolism (Harold, 1972) at energy-transducing biomembranes and thus directly or indirectly modulate the level or activity of enzymes involved in energy metabolism

Phytochrome modulation of NADP-dependent glyceraldehyde-3-phosphate dehydrogenase (NADP-GPD) and adenylate kinase (AK) activity is of particular relevance in view of the involvement of NADP-GPD in photosynthesis and of AK activity in fine control of energy charge.

Endogenous rhythmicity in NADP-GPD and AK activity is initiated and/or synchronised during the cyclic germination conditions (Frosch and Wagner, 1973a; Wagner and Frosch, 1971) and is free-running thereafter in constant conditions of light intensity and temperature. This is demonstrated with the experimental programme presented in *Figure 26.8.*   After 12, 24 or 36 h of constant conditions, seedlings were transferred to darkness.   At the beginning of darkness they either received 5 min of red light (*ca.* 80 per cent active phytochrome = Pfr); 5 min of far-red light (*ca.* 2.5 per cent Pfr) or 5 min of far-red light followed by 5 min of red light (*ca.* 80 per cent Pfr). The enzyme kinetics were then followed in darkness. It is obvious that the time courses of both enzyme activities are phased in accordance with an endogenous rhythm that is already operating at the beginning of each dark period. This is evident if one compares the data for the treatment with 12 and 36 h of constant light preceding darkness (*Figure 26.8a,c*) with the 24-h light treatment (*Figure 26.8b*). In the case of the 12- and 36-h light periods, the dark period begins at the same phase of a pre-existing rhythm, resulting in the same phasing of the rhythm in darkness, whereas with a 24-h light period preceding darkness, a difference of half a cycle length in relation to the 12- and 36-h treatments is observed.

Depending on the phasing of the endogenous rhythm, far-red light at the beginning of darkness decreased the amplitude of the first peak in NADP-GPD and the first sub-peak of AK activity or had no effect. The inhibition by far-red light could be completely reversed by red light.

The results demonstrate that the endogenous rhythm that controls the fluctuations in enzyme activity is an on/off switch for potential amplitude modulation by phytochrome in the first 12–15 h of darkness. Hence the results further suggest a dual control of the expression in enzyme activity: the endogenous oscillation which pre-determines the course of activity change and phytochrome which acts as a time-dependent amplitude modulator. Referring back to *Figure 26.1*, it becomes obvious that it is the phasing of the endogenous rhythm which qualitatively determines the amount of phytochrome control at the light/dark transition in photoperiodic cycles.

In relation to photoperiodic control of long- and short-day plants, we must explain how, in daily light–dark cycles, the sequence *dark period* – far-red – *light period* – far-red – *dark period* determines the phasing of the endogenous rhythm to position the triggering signals for control over growth development and behaviour. Experiments with energy charge and AK activity in different ecotypes of *Chenopodium rubrum* seem likely to help answer this question.

KINETICS OF PHYTOCHROME ACTION UNDER CONSTANT CONDITIONS OF PHOTOMORPHOGENESIS

To analyse further the action of phytochrome in control of NADP-GPD and AK activity, the kinetics of both enzyme activities were studied under so-called high-irradiance conditions of photomorphogenesis (i.e. continuous far-red light) and under inductive conditions (cf. Mohr, 1972).

If seedlings of *Chenopodium rubrum* were germinated without temperature cycles in darkness, the time course of NADP-GPD showed no rhythmic fluctuations (*Figure 26.9,* upper half); the same effect occurred with NAD-GPD (Frosch and Wagner, 1973b). After a lag phase of less than 3 h, a more or less linear increase in enzyme activity was observed for about 24 h in continuous far-red light. The activity then levelled off and began to decrease. This decline in enzyme activity is possibly due to the same mechanism controlling the endogenous rhythm in enzyme activity (Frosch and Wagner, 1973a; Frosch, Wagner and Cumming, 1973). Upon transition from far-red light to darkness, the increase in enzyme activity ceased immediately. When transferred back to far-red again, the increase in enzyme activity was immediately resumed, and reached the same level as the far-red control. Thus the photoreceptor in some activated state interacted like an on/off switch for increase in enzyme activity. AK activity (*Figure 26.9,* lower half) displayed a time course similar to that from NADP-GPD. However, the activity tended to fluctuate in darkness as well as in continuous far-red light. Phytochrome again was acting as an on/off switch and was required constantly for a steady increase in enzyme activity. The pattern of response was more complex than in NADP-GPD, possibly as a result of the interaction of compartmentalised AK isozymes (Criss, 1970). The readiness of AK to oscillate may be indicative of a very close functional relationship between the control mechanisms of the endogenous rhythm and the AK activities, especially in view of the oscillation in energy charge in non-photosynthesising seedlings of *Chenopodium rubrum* in darkness and continuous far-red light (*Figure 26.3*). From the above results, it appears that phytochrome under HIR conditions is an on/off switch which has to be continuously activated in order to maintain the increase in enzyme activity. However, phytochrome under inductive conditions also controls NADP-GPD activity, depending on the relative concentrations established by a red or far-red pulse, as shown in *Figure 26.10*.

Seedlings were germinated at constant temperature in darkness and, after 12 h of far-red light, they were transferred to darkness for 9 h. At the end of this period (0 h; *Figure 26.10*), the seedlings were either returned to continuous far-red light or remained in the dark or were given the irradiation sequences indicated in the figure. Transfer from darkness into continuous far-red light (Pfr under HIR conditions) results in an immediate increase in enzyme activity. *Figure 26.10* also indicates that the increase in enzyme activity depends on the relative levels of Pfr established by 5 min of red or 5 min of far-red light. There is complete red/far-red reversibility, as was the case at the beginning of the dark period in modulation of the amplitude in the rhythm in enzyme activity (*Figure 26.8b*).

In order to study more closely the phytochrome control of the photosynthetic machinery, we attempted to study the influence of phytochrome on isolated chloroplasts. This seemed particularly attractive in view of the possibility that the reciprocal activity changes in NAD- and NADP-GPD (Frosch, Wagner and Cumming, 1973) and phytochrome control of NADP-GPD could be mediated by phytochrome modulation of cofactor ratios, as had been suggested from oscillations in pyridine nucleotide pool-size levels (Wagner, Frosch and Kempf, 1974).

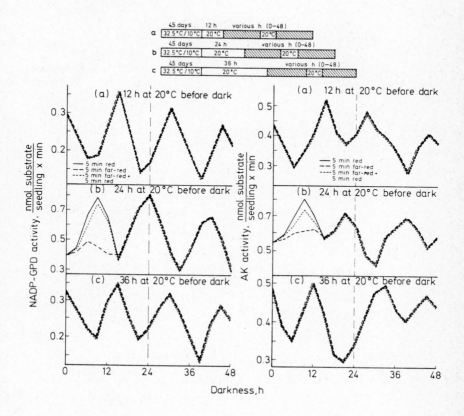

Figure 26.8   Time course of NADP-GPD and AK activity during darkness.
Chenopodium rubrum *seedlings germinated in alternating conditions of light inten-sity and temperature for 4.5 days.   Thereafter constant 20°C, 6000 lx fluorescent white light for (a) 12 h, (b) 24 h and (c) 36 h, followed by a dark period of varied duration (3-h increments).   At the beginning of the dark period, seedlings received 5 min of red, 5 min of far-red, or 5 min of far-red plus 5 min of red light.   The enzyme activities were measured after each respective dark period. (After Frosch and Wagner, 1973a)*

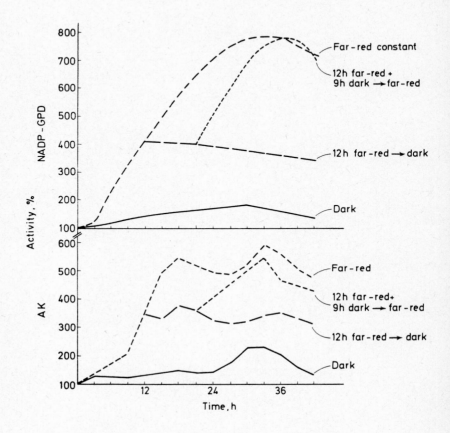

Figure 26.9 Time course of the increase in NADP-GPD and AK activity in darkness and far-red light. Acid-treated seeds of Chenopodium rubrum were germinated in darkness at 20 °C for 3 days. During the following 42 h, the seedlings remained at 20 °C and were subjected to the light—dark programmes as indicated. The enzyme activities were measured at 3-h intervals. (After Frosch and Wagner, 1973b)

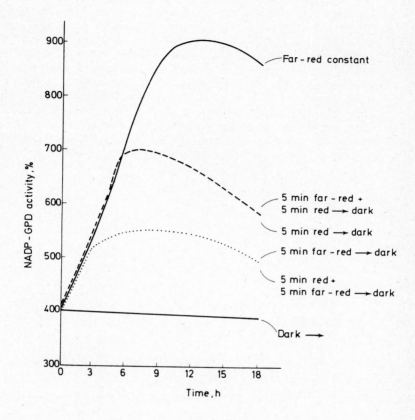

Figure 26.10 Time course of the increase in NADP-GPD activity. Acid-treated seeds of Chenopodium rubrum *were germinated for 3 days at constant 20 °C in darkness. Thereafter the seedlings received 12 h of far-red light + 9 h of darkness followed by the light–dark treatments as indicated (0 h corresponds to 21 h in Figure 26.9). The enzyme activity was determined at 3-h intervals. (After Frosch and Wagner, 1973b)*

Chloroplasts were prepared by the method of Jensen and Bassham (1966) from mature leaves of *Chenopodium rubrum* and suspended in solution 'C' at 20 °C in the dark. Thereafter they were either kept in the dark or irradiated with 5 min of red or 5 min of red + 5 min of far-red light and incubated with 10 mM ADP/P$_i$ in darkness. During the following 30 min, the time course of ATP content was followed (*Figure 26.11*) (Schwartz, personal communication). During the first 10 min of darkness, there was an increase in ATP content in all three treatments. The rates of increase in the red and red + far-red treated samples were identical but was lower in the dark controls. After this first increase in ATP content, the red + far-red treated sample and the dark control remained constant for the following 20 min. In the sample which had been induced with red light, the ATP content increased during the whole experimental period. Prior to extraction of ATP, all samples received 5 min of white light. In this context, one should

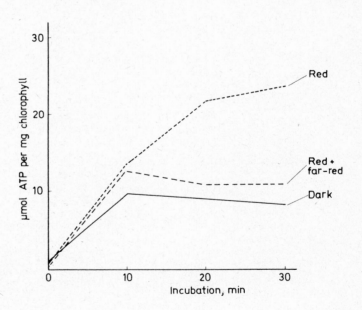

Figure 26.11 Time course of ATP content of isolated chloroplasts. Chloroplasts from Chenopodium rubrum *were isolated after the method of Jensen and Bassham (1966) and suspended in solution 'C' at 20 °C in the dark. At time zero they were irradiated with either 5 min of red, 5 min of red + 5 min of far-red, or kept in the dark and incubated with 10 mM ADP/P$_i$ in darkness. During the following 30 min, the ATP content was determined in 10-min intervals. Prior to extraction of ATP, all samples received 5 min of white light. (Schwartz, personal communication)*

recall that energy charge and reduction charge can control enzyme activities involved in photosynthesis (Pacold and Anderson, 1973, 1975; Wildner, 1975).

From the above discussion, it seems justified to envisage photoperiodic control of photosynthesis as an evolutionary adaptation to optimise energy conservation.

Membrane Oscillator as a Basis for Photoreceptor-Timer Interactions

We proposed that signal transduction in photoperiodism is achieved through interaction of phytochrome (and/or flavoprotein) with allosterically interacting binding sites organised in a matrix (Wagner, Frosch and Deitzer, 1974b). Thus we would be dealing with a membrane oscillator, capable of displaying co-operative structural changes in membrane-bound activities (e.g. redox shuttles; Dutton and Wilson, 1974) conditioned by energy transduction (Charnock, Cock and Opit, 1971) and modulated through a photochromic ligand (Bieth *et al.*, 1970). Structural changes in the receptor matrices for phytochrome could be the basis for endogenous rhythmic changes in sensitivity to phytochrome photoconversion in light (e.g. Könitz, 1958) and darkness (e.g. Cumming, 1971; Frosch and Wagner, 1973a).

Phytochrome modulation of pyridine nucleotide pools in *Chenopodium rubrum* (Wagner, Frosch and Kempf, 1974) and transitory phytochrome control of ATP levels (Junghans and Jaffe, 1972; White and Pike, 1974), although questioned by other workers (Bürcky and Kauss, 1974) seem to support the concept that phytochrome might be primarily involved in redox changes or might modulate electron flow (Fujii and Kondo, 1969; Tezuka and Yamamoto, 1972; Manabe and Furuya, 1973; Greppin, Horwitz and Horwitz, 1973; Borthwick *et al.*, 1969; Klein and Edsall, 1966; Junghans and Jaffe, 1972).

The kinetics of photomorphogenic and photoperiodic control will depend both on synthesis and on assembly of phytochrome and receptor matrices in accordance with the temporal organisation of development in a differentiating organism (etiolated seedling) as opposed to the steady-state situation in a fully developed light-grown system (Frosch, Wagner and Cumming, 1973). With the above considerations in mind, we proposed the following structural representation of the interaction between membrane oscillators and phytochrome (Wagner, Frosch and Deitzer, 1974b) (*Figure 26.12*):

(I) Membrane oscillator with receptor sites for phytochrome, oscillating between two stable configurations via a metastable transitory state in the course of energy transduction. It is conceivable that the interaction of a ligand with the receptors in a transitory state could drive the system in either direction, depending on the conformation of the ligand and thus could induce phase advances and phase delays.

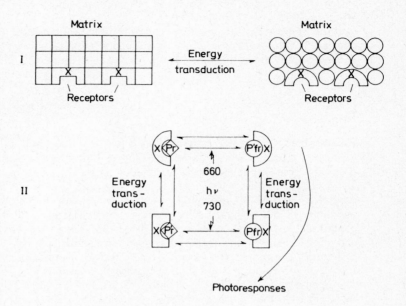

Figure 26.12 Structural representation of a membrane oscillator model of photoperiodic control. For explanation, see text. (After Wagner, Frosch and Deitzer, 1974b)

(II) Receptor site which undergoes structural change during photoconversion, resulting in signal transduction with subsequent regeneration of the receptor site. With short-term irradiations, the structural changes of the receptor sites during signal transduction are probably accompanied by transitory changes in redox and/or energy state. Long-term irradiations probably induce a more permanent change in energy flow. The on/off control by the endogenous rhythm for phytochrome action (cf. *Figure 26.8b*) could imply that there is a receptor configuration which allows no structural change during photoconversion and thus no signal transduction. Similar hypotheses of photoperiodic control had been advanced by Satter and Galston (1973) and Njus, Sulzman and Hastings (1974) and have been discussed recently in relation to our concepts (Wagner, Frosch and Deitzer, 1974b).

To evaluate this working hypothesis on the molecular basis of photoperiodic control, we proposed (Wagner, Frosch and Deitzer, 1974a) to look for circadian rhythms in phytochrome binding which possibly could be correlated with studies on 1-*N*-naphthylphthalamic acid (NPA) binding (Thomson, Hertel and Müller, 1973) and energy-dependent 1-anilinonaphthalene-8-sulphonate (ANS) fluorescence (Radda, 1971) which possibly would change according to the rhythm in energy transduction (Wagner *et al.*, 1975). Part of this programme has been verified through the demonstration of an endogenous rhythm in the amount of pelletable phytochrome (cf. Quail, Marmé and Schäfer, 1973) from seedlings of *Cucurbita pepo* by Schäfer and Jabben (personal communication) (*Figure 26.13*). There is no possibility at the moment of deciding whether this rhythm in pelletability is a function of an endogenous rhythm in the amount or the affinity of

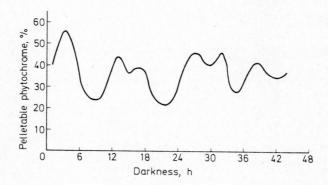

Figure 26.13 Time course of the amount of pelletable phytochrome. Seedlings of Cucurbita pepo *were germinated for 4 days at 25 °C in darkness followed by 12 h of white light plus 48 h of darkness. During the following 48 h of darkness, phytochrome was extracted at hourly intervals and pelleted after the method of Quail, Marmé and Schäfer (1973). The graph represents the first 3-point moving mean (Jabben and Schäfer, personal communication)*

binding sites for phytochrome. It is even possible that the binding is not relevant for the physiological action of phytochrome. Nevertheless, the rhythm in pelletability is indicative of an endogenous rhythm in binding structures and membrane properties which is made obvious through a photochromic marker.

Conclusion

Adenylate energy charge and reduction charge might be coupling agents in the control of growth and differentiation under photoperiodic and photomorphogenic conditions. There is a striking similarity between recent kinetic models (Deutch and Deutch, 1974; Schäfer, 1974) of the high-irradiance response (HIR) of photomorphogenesis and the kinetic and structural representation of redox shuttles integrated in energy-transducing biomembranes as components of vectorial metabolism (Dutton and Wilson, 1974). It therefore might be possible that the molecular basis for the HIR models is to be found in a membrane-organised shuttle system which could be modulated by phytochrome photoconversion.

Acknowledgements

Original work from this laboratory was supported by Deutsche Forschungsgemeinschaft: 'Biologie der Zeitmessung'. The author is indebted to Miss Christa Nolte for careful typing of the manuscript and to Christopher Beggs for critical and helpful discussion.

References

ATKINSON, D.E. (1969). *Annu. Rev. Microbiol.*, **23**, 47
ATKINSON, D.E. (1971). *Adv. Enzyme Regul.*, **9**, 207
BENNUN, A. (1974). *Ann. N.Y. Acad. Sci.*, **227**, 116
BETZ, A. (1968). In *Quantitative Biology of Metabolism*, p.205. Ed. Locker, A. Springer-Verlag, Berlin, Heidelberg, New York
BIETH, J., WASSERMAN, N., VRATSANOS, S.M. and ERLANGER, B.F. (1970). *Proc. Nat. Acad. Sci. U.S.A.*, **66**, 850
BLAIR, J. McD. (1970). *Eur. J. Biochem.*, **13**, 384
BORTHWICK, H.A., HENDRICKS, S.B., SCHNEIDER, M.J., TAYLORSON, R.B. and TOOLE, V.K. (1969). *Proc. Nat. Acad. Sci. U.S.A.*, **64**, 479
BRODY, S. (1973). *Abstr. Annu. Meet. Am. Soc. Microbiol.*, 38
BÜNNING, E. (1973). *The Physiological Clock*. English Universities Press, London
BÜRCKY, K. and KAUSS, H. (1974). *Z. Pflanzenphysiol.*, **73**, 184
BYGRAVE, F.L. (1967). *Nature, Lond.*, **214**, 667
BYGRAVE, F.L., DADAY, A.A. and DOY, F.A. (1975). *Biochem. J.*, **146**, 601
CARDON, S.Z. and IBERALL, A.S. (1970). *Curr. Mod. Biol.*, **3**, 237
CERFF, R. (1974). *Z. Pflanzenphysiol.*, **73**, 109
CERFF, R. and QUAIL, P.H. (1974). *Plant Physiol.*, **54**, 100
CHAPMAN, A.G., FALL, L. and ATKINSON, D.E. (1971). *J. Bacteriol.*, **108**, 1072
CHARNOCK, J.S., COOK, D.A. and OPIT, L.J. (1971). *Nature New Biol.*, **233**, 171
CHIA-LOOI, A.S. and CUMMING, B.G. (1972). *Can. J. Bot.*, **50**, 2219
COOMBS, J., MAW, S.L. and BALDRY, G.W. (1974). *Planta*, **117**, 279
CRISS, W.E. (1970). *J. Biol. Chem.*, **245**, 6352
CUMMING, B.G. (1971). *Proc. Int. Symp. Circadian Rhythmicity (Wageningen)*, 33

CUMMING, B.G. and WAGNER, E. (1968). *Annu. Rev. Plant Physiol.*, **19**, 381

CUMMING, B.G., HENDRICKS, S.B. and BORTHWICK, H.A. (1965). *Can. J. Bot.*, **43**, 825

DEITZER, G.F., KEMPF, O., FISCHER, S. and WAGNER, E. (1974). *Planta*, **117**, 29

DELMER, D.P. and BRODY, S. (1973). *Abstr. Annu. Meet. Am. Soc. Microbiol.*, 38

DEUTCH, B. and DEUTCH, B.I. (1974). *Physiol. Plant.*, **32**, 273

DUTTON, P.L. and WILSON, D.F. (1974). *Biochim. Biophys. Acta*, **346**, 165

DYNNIK, V.V., SEL'KOV, E.E., and SEMASHKO, L.R. (1973). *Stud. Biophys.*, **41**, 193

FREDERICQ, H. (1964). *Plant Physiol.*, **39**, 812

FREDERICQ, H. and DE GREEF, J.A. (1968). *Physiol. Plant.*, **21**, 346

FROSCH, S. and WAGNER, E. (1973a). *Can. J. Bot.*, **51**, 1521

FROSCH, S. and WAGNER, E. (1973b). *Can. J. Bot.*, **51**, 1529

FROSCH, S., WAGNER, E. and CUMMING, B.G. (1973). *Can. J. Bot.*, **51**, 1355

FROSCH, S., WAGNER, E. and MOHR, H. (1974). *Z. Naturforsch.*, **29c**, 392

FUJII, T. and KONDO, N. (1969). *Dev. Growth Differ.*, **11**, 40

FUKUSHIMA, Y. and TONOMURA, Y. (1972). *J. Biochem.*, **72**, 623

GANDER, J.E. (1967). *Ann. N.Y. Acad. Sci.*, **138**, 730

GOLDBETER, A. (1974). *Febs Lett.*, **43**, 327

GOOCH, Van D. and PACKER, L. (1974). *Arch. Biochem. Biophys.*, **163**, 759

GREPPIN, H., HORWITZ, B.A. and HORWITZ, L.P. (1973). *Z. Pflanzen-physiol.*, **68**, 336

HALABAN, R. (1969). *Plant Physiol.*, **44**, 973

HAROLD, F.M. (1972). *Bacteriol. Rev.*, **36**, 172

HASTINGS, J.W., ASTRACHAN, L. and SWEENEY, B.M. (1960). *J. Gen. Physiol.*, **45**, 69

HEBER, U. (1974). *Annu. Rev. Plant Physiol.*, **25**, 393

HESS, B. and BOITEUX, A. (1971). *Annu. Rev. Biochem.*, **40**, 237

HILLMAN, W.S. (1969). In *Physiology of Plant Growth and Development*, p.559. Ed. Wilkins, M.B. McGraw-Hill, London

HILLMAN, W.S. (1975). *Photochem. Photobiol.*, **21**, 39

HOCHBERG, M.L. and SARGENT, M.L. (1974). *J. Bacteriol.*, **120**, 1164

HYNDMAN, B.W. (1974). *Kybernetik*, **15**, 227

JENSEN, R.G. and BASSHAM, J.A. (1966). *Proc. Nat. Acad. Sci. U.S.A.*, **56**, 1095

JOHNSON, T.B., SALISBURY, F.B. and CONNOR, G.I. (1967). *Science, N.Y.*, **155**, 1663

JONES, M.B. and MANSFIELD, T.A. (1975). *Sci. Prog., Oxf.*, **62**, 103

JUNGHANS, H. and JAFFE, M.J. (1972). *Plant Physiol.*, **49**, 1

KAPLAN, R.W. (1972). *Der Ursprung des Lebens.* Georg Thieme Verlag, Stuttgart

KASPERBAUER, M.J. (1971). *Plant Physiol.*, **47**, 775

KASPERBAUER, M.J., TSO, T.C. and SOROKIN, T.P. (1970). *Phytochemistry*, **9**, 2091

KING, R.W. and CUMMING, B.G. (1972). *Planta*, **108**, 39

KLEIN, R.M. and EDSALL, P.C. (1966). *Plant Physiol.*, **41**, 979

KÖNITZ, W. (1958). *Planta*, **51**, 1

LANE, H.C., CATHEY, H.M. and EVANS, L.T. (1965). *Am. J. Bot.*, **52**, 1006

LEBHERZ, H.G., SEVAGE, B. and ABACHERLI, E. (1973). *Nature New Biol.*, **245**, 269

MANABE, K. and FURUYA, M. (1973). *Plant Physiol.*, **51**, 982

MANGAT, B.S., LEVIN, W.B. and BIDWELL, R.G.S. (1974). *Can. J. Bot.*, **52**, 673

MELI, J. and BYGRAVE, F.L. (1972). *Biochem. J.*, **128**, 415

MICHEL, J.P. and THIBAULT, P. (1973). *Biochim. Biophys. Acta*, **305**, 390

MIGINIAC-MASLOW, M. and CHAMPIGNY, M.-L. (1974). *Plant Physiol.*, **53**, 856

MOHR, H. (1972). *Lectures on Photomorphogenesis*. Springer, Berlin, Heidelberg, New York

MÜLLER, B. (1970). *Biochim. Biophys. Acta*, **205**, 102

MÜLLER, B., ZIEGLER, I. and ZIEGLER, H. (1969). *Eur. J. Biochem.*, **9**, 101

NAKAYAMA, S. (1958). *Ecol. Rev.*, **14**, 325

NAKAYAMA, S., BORTHWICK, H.A. and HENDRICKS, S.B. (1960). *Bot. Gaz.*, **121**, 237

NICOLIS, G. and PORTNOW, J. (1973). *Chem. Rev.*, **73**, 365

NJUS, D., SULZMAN, F.M. and HASTINGS, J.W. (1974). *Nature, Lond.*, **248**, 116

PACKER, L., MURAKAMI, S. and MEHARD, C.W. (1970). *Annu. Rev. Plant Physiol.*, **21**, 271

PACOLD, I. and ANDERSON, L.E. (1973). *Biochem. Biophys. Res. Commun.*, **51**, 139

PACOLD, I. and ANDERSON, L.E. (1975). *Plant Physiol.*, **55**, 168

PAVLIDIS, T. (1971). *J. Theor. Biol.*, **33**, 319

PAVLIDIS, T. and KAUZMAN, W. (1969). *Arch. Biochem. Biophys.*, **132**, 33

PITTENDRIGH, C.S. and CALDAROLA, P.C. (1973). *Proc. Nat. Acad. Sci. U.S.A.*, **70**, 2697

PUPILLO, P. and PICCARI, G.J. (1973). *Arch. Biochem. Biophys.*, **154**, 324

QUAIL, P.H., MARMÉ, D. and SCHÄFER, E. (1973). *Nature New Biol.*, **245**, 189

RADDA, G.K. (1971). *Biochem. J.*, **122**, 385

REICH, J.G. and SEL'KOV, E.E. (1974). *Febs Lett.*, **40**, 119

SARGENT, M.L. and BRIGGS, W.R. (1967). *Plant Physiol.*, **42**, 1504

SATTER, R.L. and GALSTON, A.W. (1973). *BioScience*, **23**, 407

SATTER, R.L., GEBALLE, G.T. and GALSTON, A.W. (1974). *J. Gen. Physiol.*, **64**, 431

SCHÄFER, E. (1974). *J. Math. Biol.*, in the press

SHROPSHIRE, W. Jr., (1973). *Sol. Energy*, **15**, 99

TEZUKA, T. and YAMAMOTO, Y. (1972). *Plant Physiol.*, **50**, 458

THIBAULT, P. and MICHEL, J.P. (1972). In *Proc. II Int. Congr. Photosynth., Stresa*, p.599. Dr W. Junk N.V. Publ., The Hague

THOMSON, K.-S., HERTEL, R. and MÜLLER, S. (1973). *Planta*, **109**, 337

TORNHEIM, K. and LOWENSTEIN, J.M. (1973). *J. Biol. Chem.*, **248**, 2670

VAN DEN DRIESSCHE, T. (1966). *Exp. Cell Res.*, **42**, 18

WAGNER, E. and CUMMING, B.G. (1970). *Can. J. Bot.*, **48**, 1

WAGNER, E. and FROSCH, S. (1971). *Can. J. Bot.*, **49**, 1981

WAGNER, E. and FROSCH, S. (1974). *J. Interdiscip. Cycle Res.*, **5**, 231

WAGNER, E., FROSCH, S. and DEITZER, G.F. (1974a). In *Proc. Annu. Eur. Symp. Plant Photomorphogenesis, Antwerp, Abstracts*, p.15. Ed. De Greef, J.A.

WAGNER, E., FROSCH, S. and DEITZER, G.F. (1974b). *J. Interdiscip. Cycle Res.*, **5**, 240

WAGNER, E., FROSCH, S. and KEMPF, O. (1974). *Plant Sci. Lett.*, **3**, 43

WAGNER, E., STROEBELE, L. and FROSCH, S. (1974). *J. Interdiscip. Cycle Res.*, **5**, 77

WAGNER, E., DEITZER, G.F., FISCHER, S., FROSCH, S., KEMPF, C. and STROEBELE, L. (1975). *BioSystems*, **7**, 68

WAGNER, E., TETZNER, J., HAERTLÉ, U. and DEITZER, G.F. (1974). *Ber. Dtsch. Bot. Ges.*, **87**, 291

WATSON, M.R. (1972). *J. Theor. Biol.*, **36**, 195

WEVER, R. (1965). In *Circadian Clocks*, p.74. Ed. Aschoff, J. North-Holland, Amsterdam

WHITE, J.M. and PIKE, C.S. (1974). *Plant Physiol.*, **53**, 76

WIELAND, O.H. and PORTENHAUSER, R. (1974). *Eur. J. Biochem.*, **45**, 577

WILDNER, G.F. (1975). *Z. Naturforsch.*, **30c**, 756

WILKINS, M.B. (1973). *J. Exp. Bot.*, **24**, 488

WILSON, A.T. and CALVIN, M. (1955). *J. Am. Chem. Soc.*, **77**, 5948

YAMAZAKI, I. and YOKOTA, K. (1967). *Biochim. Biophys. Acta*, **132**, 310

YANG, S.T. and DEAL, W.C. Jr. (1969a). *Biochemistry*, **8**, 2806

YANG, S.T. and DEAL, W.C. Jr. (1969b). *Biochemistry*, **8**, 2814

VI

ECOLOGICAL ASPECTS OF PHOTOMORPHOGENESIS

27

SPECTRAL DISTRIBUTION OF LIGHT IN LEAVES AND FOLIAGE

J.L. MONTEITH
Department of Physiology and Environmental Studies, University of Nottingham, School of Agriculture, Sutton Bonington, Loughborough, Leics. LE12 5RD, U.K.

Introduction

For the past 25 years, the study of light distribution in plant canopies has been one of the main growing points in crop ecology. When this subject was reviewed a few years ago (Monteith, 1969), I pointed out that theoretical studies of canopy geometry in relation to the penetration of light seemed to be almost as numerous as reports of measurements in the field. The contents of several more recent reviews tend to support this contention (Monsi, Uchijima and Oikawa, 1973; Lemeur and Blad, 1974; Norman, 1975; Ross, 1975). Apart from the imbalance between measurements and models, the literature on light in canopies deals almost exclusively with the relation between leaf irradiance and photosynthesis rate. The importance of light microclimate in relation to photomorphogenic processes has rarely been discussed and very few relevant measurements have been reported. In particular, most measurements of spectral composition in and below canopies are limited to simple divisions of the solar spectrum between photosynthetically active radiation or PAR (400–700 nm) and near-infrared radiation or NIR (700–3 000 nm) (Szeicz, 1974a). A few workers, however, have published spectra of radiation within crop stands, distinguishing between shade and sunfleck areas (e.g. Yocum, Allen and Lemon, 1964; Federer and Tanner, 1966), and McRee (1968) has pointed out the usefulness of false colour photography in this context. The chapter by Holmes and McCartney (Chapter 29) is apparently the first attempt to relate spectral composition quantitatively to a specific photomorphogenic process or state.

Criticism may also be directed at the marked reluctance of plant physiologists to study the behaviour of phytochrome in green plants growing in a natural sunlit environment. It is standard practice to grow seedlings in the dark and to measure responses mediated by phytochrome when they are illuminated by light in a discrete waveband centred at 660 or 730 nm.   Simple standard conditions may be essential to study and to analyse the diverse effects of a very complex pigment, but how does the phytochrome system behave when the leaves of plants grown in sunlight are exposed to a much more complex light regime in which the solar spectrum is modified by scattering and absorption within a canopy? Until this question is answered, agriculture and ecology are unlikely to benefit much from the great wealth of information about phytochrome accumulated in laboratory studies and reviewed in this volume.

Scattering and absorption act at two levels: within the canopy when light strikes a leaf, a stem or some other organ; and within a leaf when light passes through a membrane or impinges on a chloroplast. This chapter is an attempt to estimate the relative quantum content of red and far-red light both in the canopy of a field crop and in a leaf mesophyll, which may be a small step towards estimating the photostationary state of phytochrome in the leaves of field crops.

Symbols and Nomenclature

Many different symbols are used by crop ecologists for solar irradiance (the amount of radiant energy received per unit area and per unit time).   In studies of radiation climate, it is convenient to use S with appropriate subscripts.   The spectral irradiance at a wavelength λ can then be written as $(dS/d\lambda)_\lambda$.

The quantum content of radiation per unit wavelength is given by

$$Q = (dS/d\lambda)_\lambda (\lambda/hc) \qquad (27.1)$$

where h is Planck's constant and c is the velocity of light.   This chapter is concerned with the ratio of quantum contents at two specific wavelengths, λ_1 = 660 nm and λ_2 = 730 nm.   To avoid repetition of these numbers, it is convenient to write the corresponding quantum contents simply as Q_1 and Q_2 and their ratio as

$$\zeta = Q_1/Q_2$$

a symbol chosen because it has no other standard connotation in microclimatology.   It follows that ζ is related to spectral irradiance by

$$\zeta = (\lambda_1/\lambda_2)[(dS/d\lambda)_1/(dS/d\lambda)_2] \qquad (27.2)$$

where λ_1/λ_2 = 660/730 = 0.90.

To relate the quantum flux density within a canopy to the foliage area, it is convenient to use the leaf area index, which is the area of leaves (one side only) above unit area of ground.   If the quantum flux density below a leaf area index of L is written as $Q_1(L)$, the flux density of radiation at the top of the canopy will be $Q_1(0)$. Corresponding values of the relative quantum flux are $\zeta(L)$ and $\zeta(0)$.

A spectral transmission coefficient τ_f can be defined by

$$\tau_f = \frac{\zeta(L)}{\zeta(0)} = [Q_1(L)/Q_1(0)]/[Q_2(L)/Q_2(0)] \qquad (27.3)$$

Similarly, when radiation whose spectral quality is specified by $\zeta(L)$ strikes the epidermis of a leaf, the relative quantum flux in the mesophyll at a specified distance from the epidermis can be expressed as

$$\zeta_m = \tau_m \, \zeta(L) \qquad (27.4)$$

where τ_m is a mesophyll transmission coefficient. It follows that the spectral quality of radiation incident on a phytochrome molecule is given by

$$\zeta_m = \tau_m \, \tau_f \zeta(0) \qquad (27.5)$$

The Quality of Radiation

ABOVE A CANOPY

Measurements of the spectral quality of solar radiation have been reported by a number of workers (Henderson, 1970) and the shape of the spectrum can be calculated from a knowledge of atmospheric composition, in particular the amount of water vapour and dust. *Figure 27.1* is an example from a large number of spectral distributions calculated by Avaste, Moldau and Shifrin (1962). In general, the spectral distribution is conservative between 660 and 730 nm so that the value of $\zeta(0)$ changes little with time of day, season of the year, cloudiness or latitude. From measurements with a Gamma spectrophotometer at Sutton Bonington (Holmes and McCartney, Chapter 29; McCartney, 1975), the average value of $\zeta(0)$ is about 1.1. For radiation in the direct solar beam, $\zeta(0)$ ranges from about 0.9 to 1.1 and tends to increase with solar elevation. For radiation from blue sky, which is relatively rich in visible light, $\zeta(0)$ is about 1.2. The fact that the ratio of PAR to total radiation is more or less independent of site and season (Szeicz, 1974b) implies that these values of ζ may be taken as representative of the light to which most vegetation is exposed during its growth. Moreover, we shall see that the variability of ζ within a canopy and within a single leaf is very much greater than that of the ratio for unmodified sunlight.

WITHIN A CANOPY

Complex analysis is needed to describe, in detail, the changes in spectral composition which occur when sunlight filters through foliage (Ross, 1975), but for field studies of phytochrome in their current primitive stage of development, the advantages of physical precision are entirely offset by biological uncertainties. We shall therefore consider one of the simplest ways possible of describing how light quality depends on canopy geometry and leaf optics.

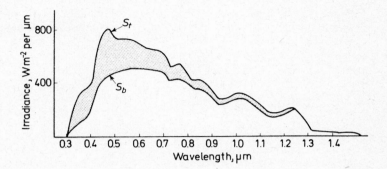

Figure 27.1 Spectral distribution of solar radiation under a cloudless sky calcu-lated by Avaste, Moldau and Shifrin (1962) for a solar elevation of 30°, atmos-pheric water content of 21 mm and average turbidity. The lower curve is the direct component of radiation S_b, the upper curve is total radiation S_t and the hatched area between the curves represents blue skylight

The treatment to be followed is valid for a 'uniform' crop stand, i.e. a stand in which all plants are of the same height and size. It is assumed, however, that leaf laminae are randomly distributed in space (Nilson, 1971, treated the more realistic case of leaves which were clumped or arranged in a regular pattern, corresponding to under-dispersion or overdispersion in the statistical sense). For random foliage with a leaf area index of L, the probability that any individual ray of light should penetrate to the ground without intercepting a leaf is $\exp(-KL)$ where K is a non-dimensional extinction coefficient deter-mined by the direction of the incident ray in relation to the architec-ture of the canopy. When the sky is overcast, or when the sun is unobscured and more than 30° above the horizon, K ranges from approximately unity for species with predominantly horizontal leaves to about 0.3 for species with a vertical leaf habit (Monteith, 1969).

Any ray of light which is intercepted by a leaf will be scattered both backwards and forwards (i.e. it will be reflected and transmitted) as well as being absorbed. We shall assume, however, that transmitted light continues in the same direction so that the ray may penetrate 1, 2, 3 or more leaves before it finally reaches the ground. The pro-bability of intercepting any integral number of leaves is given by the terms of a Poisson distribution, i.e. by the expression

$$e^{-KL} \cdot e^{+KL} \quad = \quad 1$$

or

$$e^{-KL} \cdot \left(1 + KL + \frac{K^2 L^2}{2!} + \dots \frac{K^n L^n}{n!}\right) = 1$$

The probability of intercepting *no* leaves is the first term $\exp(-KL)$, as stated already; the probability of *one* interception is $KL \exp(-KL)$; and the probability of n interceptions is $K^n L^n \exp(-KL)/n!$. The sum of all these probabilities is unity, as required.

If we now consider a beam of sunlight falling on a stand, we can interpret each of these terms in a slightly different way. The quantity $\exp(-KL)$, corresponding to the probability of zero interception, will be the fractional area of sunflecks below a leaf area of L, $KL \exp(-KL)$ will be the fractional area of shade cast by one leaf, $K^2 L^2 \exp(-KL)/2!$ the fractional area of shade cast by two leaves, and so on.

The concept of sunlit and shaded areas of leaf must be defined more exactly at this stage. In the remainder of the chapter, the term 'sunlit' will refer to sections of leaf illuminated directly from the sun *or* by diffuse light from the blue sky or from clouds. A sunfleck is an area of sunlit leaf in the same sense. A 'shaded' leaf receives light which has been transmitted or reflected by at least one other leaf, i.e. 'complementary' radiation in the nomenclature of Ross (1975). At any given depth in the canopy, the horizontal distribution of scattered light is regarded as uniform so that sunflecks will also receive a small contribution from the complementary flux. The irradiance in sunflecks near the top of a canopy is sometimes greater than the irradiance above the canopy for this reason, and the effect is most prominent in the near-infrared spectrum.

If $Q_1(0)$ is the flux density of quanta above the canopy, the quantity $e^{-KL} Q_1(0)$ will be the contribution made by sunflecks to the mean quantum flux per unit area of a horizontal surface at a leaf-area index L below the top of the canopy. To find the quantum flux density in *shaded* areas beneath the canopy, we assume that the fraction of incident quanta transmitted by a leaf at wavelength λ_1 is τ_1. Then the fraction of the quantum flux penetrating n leaves is $\tau_1^n K^n L^n \exp(-KL)/n!$. The total quantum flux density below a leaf area index of L is given by

$$e^{-KL} \cdot \left(1 + \tau_1 KL + \frac{\tau_1^2 K^2 L^2}{2!} + \dots \frac{\tau_1^n K^n L^n}{n!}\right) Q_1(0)$$

$$= e^{-KL} \cdot e^{\tau_1 KL} \cdot Q_1(0)$$

$$= e^{-(1-\tau_1)KL} \cdot Q_1(0) \tag{27.6}$$

This expression allows for the fact that leaves receive radiation by transmission through higher leaves, but it does not take account of radiation received by reflection from lower leaves. To a first approximation, the combined effects of transmission and reflection can be allowed for by replacing τ_1 in equation 27.6 by $(\tau_1 + \rho_1)$. The term $(1 - \tau_1)$ then becomes $(1 - \tau_1 - \rho_1) = a_1$, which is the fraction of incident quanta absorbed by the leaf. The mean quantum flux density, $\overline{Q}_1(L)$, below a leaf-area index of L then assumes the form used by Kuroiwa (1968) and others, i.e.

$$\overline{Q}_1(L) = Q_1(0)e^{-a_1 KL} \qquad (27.7)$$

To find the contribution which transmitted and reflected light make to this quantum flux, the sunfleck component must be subtracted to give

$$Q_1(L, \text{ shade}) = Q_1(0)(e^{-a_1 KL} - e^{-KL}) \qquad (27.8)$$

The quantum flux density in sunflecks is the sum of two components: a contribution from direct sunlight or $Q_1(0)$ and a contribution from transmitted and reflected light given by equation 27.8. The sum can be expressed as

$$Q_1(L, \text{ sun}) = Q_1(0)(1 + e^{-a_1 KL} - e^{-KL}) \qquad (27.9)$$

We can now use this set of equations to find the *relative* quantum flux at any level in the canopy specified by L.

For the *average* relative flux at any level, equations 27.3 and 27.7 give

$$\overline{\tau_f} = e^{(a_2 - a_1)KL} \qquad (27.10)$$

For the relative quantum flux in the *shade,* i.e. in transmitted and reflected light, equations 27.3 and 27.8 give

$$\tau_f(\text{shade}) = \frac{e^{-a_1 KL} - e^{-KL}}{e^{-a_2 KL} - e^{-KL}} \qquad (27.11)$$

For the relative flux in *sunflecks,* equations 27.3 and 27.9 give

$$\tau_f(\text{sun}) = \frac{Q_1(L, \text{ sun})}{Q_2(L, \text{ sun})} = \frac{1 + e^{-a_1 KL} - e^{-KL}}{1 + e^{-a_2 KL} - e^{-KL}} \qquad (27.12)$$

The implications of equations 27.10–27.12 are illustrated in *Figure 27.2*, where the three spectral transmission coefficients were calculated for a leaf with $a_1 = 0.88$ and $a_2 = 0.075$, derived as mean values from the spectra of 20 agricultural crops recorded by Gausman *et al.* (1973) (*see Figure 27.3*). The value of K was taken as 0.7, in the middle of the range for crop species, and the maximum value of $L = 6$ is rarely exceeded in farm crops. Irrespective of the value of L, the value of τ_f in sunflecks is an order of magnitude greater than the value in shade because leaves absorb radiation very strongly at 660 nm but are almost completely translucent at 730 nm. The value of τ_f corresponding to the mean quantum flux density approaches the value for sunflecks near the top of the canopy, where there is little

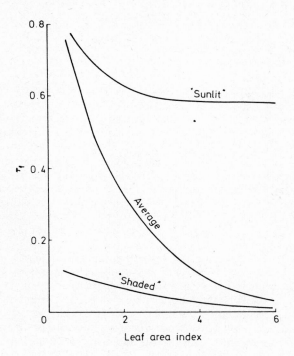

Figure 27.2   Relation between spectral transmission coefficient, τ_f, and leaf-area index for a canopy with an extinction coefficient K = 0.7

shade light, but decreases to the shade value of τ_f as L increases because the relative area of sunflecks decreases with depth in the canopy.

The relative quantum flux density at different levels in the canopy is $\tau_f \zeta(0)$. Taking $\zeta(0)$ as 0.9, the relative quantum flux will be 10 per cent less than the value of τ_f in *Figure 27.2*. The value of τ_f calculated for values of K ranging from 0.3 to 1.0 is of the same order of magnitude as for $K = 0.7$. It follows that for many types of canopy architecture displayed by plant communities, the value of ζ for sunlit leaves will be close to the value $\zeta(0)$ above the canopy, but for leaves in shade, below a leaf area index of about 3, ζ will be an order of magnitude less than $\zeta(0)$.

One of the main assumptions in deriving values of ζ is the equality of the extinction coefficient, K, for direct sunlight, diffuse skylight and complementary radiation. Significant differences in K for direct and diffuse radiation have been demonstrated both theoretically and experimentally.   Error from this assumption is minimised by considering relative quantum fluxes rather than absolute fluxes, and it is encouraging to find that the values of ζ from *Figure 27.2* are consistent with those quoted by Ross (1975), which were derived much more rigorously.

WITHIN LEAVES

The response of plants to light is usually discussed in terms of the external light climate, in particular the flux densities of radiant energy incident on leaves and the spectral distribution of that energy.   For some purposes, such as the calculation of a photosynthetic rate, it is not essential to account for the distribution of light energy within the mesophyll.   In phytochrome studies, however, the changes in spectral composition which must occur, in principle, when light diffuses through the mesophyll are just as important as the measurable changes which occur when light filters through a canopy.   As a logical extension of the calculations for canopy light, we shall now consider how the quality of light within a leaf is likely to be modified by absorption and scattering.

The literature on the optical properties of leaves published before 1971 has been thoroughly reviewed by Kumar (1972), but the treatment which follows is based almost exclusively on a series of papers from a USDA Agricultural Research Service group in Texas.   This group has recently published a series of very detailed measurements of the spectra of leaves from 20 agricultural crops, including figures for transmission and reflection at intervals of 50 nm from 500 to 1 500 nm (Gausman *et al.*, 1973).   *Figure 27.3*, showing spectra for wheat exposed to diffuse radiation, has a number of features common to all 20 species.

(1) In the visible spectrum, the reflectance is about 10 per cent and the transmittance is less.
(2) Transmittance and reflectance both increase sharply as the wavelength increases beyond 660 nm and reach maximum values at about 730 nm.
(3) From 730 to about 1 250 nm, absorption is only a few per cent and almost equal amounts of radiation are transmitted and reflected.
(4) Between 1 250 and 2 500 nm, there is marked absorption by liquid water (e.g. at 1 400 and 1 900 nm) but the similarity of transmittance and reflectance is still evident.

It appears that chlorophyll and other pigments responsible for the absorption of light between 400 and 700 nm act as a filter to produce maximum discrimination between the two wavelengths of 660 and 730 nm which have been identified with the action spectra of phytochrome in its red and far-red states.   How does the corresponding ratio of the quantum flux change across the mesophyll?

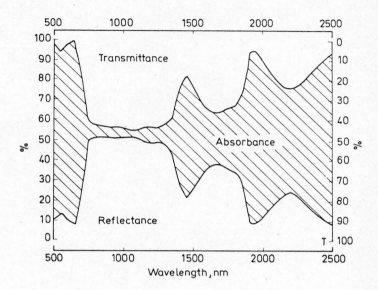

Figure 27.3 Transmission, reflection and absorption spectra for wheat leaves. (From Gausman et al., 1973)

To answer this question, it is necessary to use a model for the transmission of light in a leaf. After testing a number of models, Allen *et al.* (1969) found that the transmission of diffuse light by a compact leaf with few intercellular spaces could be simulated by treating the leaf as a translucent plate with rough parallel surfaces. [In later papers (e.g. Gausman *et al.*, 1970), the model was extended to a non-compact leaf but the additional algebra is not essential to the present discussion.] *Figure 27.4* illustrates the main features of the compact leaf model: light incident on a leaf may be reflected from the upper epidermis with a reflectance $R_{1,2}$ or may be transmitted with a transmittance $T_{1,2}$. Part of the transmitted light is absorbed in the mesophyll and the remainder is internally reflected ($R_{2,1}$) or transmitted ($T_{2,1}$) by the lower epidermis. The reflected component then suffers a series of internal reflections. In practice, any ray of light passing through a leaf is scattered by pigment molecules, cell walls, etc. (Sinclair, Schreiber and Hoffer, 1973) so that the radiation field is diffuse and not confined to a specific direction as in *Figure 27.4*. However, Stern (1964) extended classical electromagnetic theory to calculate the average reflectance and transmittance for all directions of incident light and of polarisation. His paper contains a table of

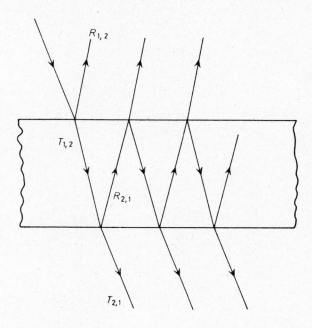

Figure 27.4 Model for multiple reflection of light within a leaf treated as an absorbing medium bounded by two rough parallel surfaces. $R_{1,2}$ and $T_{1,2}$ are reflection and transmission coefficients for a ray moving from medium 1 (air) to medium 2 (leaf tissue)

transmittances, $T_{1,2}$, at an interface between air and a medium with a refractive index n. According to Allen, Gausman and Richardson (1970), the refractive index of cuticular wax is likely to be about 1.5 (at 489 nm) and the corresponding value of $T_{1,2}$ is 0.91. It follows that $R_{1,2}$ = 0.09, i.e. about 9 per cent of diffuse radiation incident on a leaf will be reflected from the epidermis. This figure is consistent with minimum values reported by Gausman *et al.* (1973) for wavelengths at which light is strongly absorbed in the mesophyll. The value of $T_{2,1}$ is given by $T_{1,2}/n^2$ = 0.40 and is less than $T_{1,2}$ because rays which strike the lower epidermis at an angle of incidence greater than $\sin^{-1}(1/n)$ = 42° will be totally internally reflected. It follows that $R_{2,1}$ = 0.60.

It remains to estimate the fractional absorption of radiation within the leaf using, in Beer's law, a transmission coefficient obtained experimentally, and a pathlength which is twice the leaf thickness to allow for the diffuse nature of the radiation. From the absorption spectra given by Gausman *et al.* (1973), and for a representative leaf with a thickness of 0.15 mm, it was estimated that the fractional absorption of radiation passing once through a leaf would be approximately 90 per cent at 660 nm and 2 per cent at 730 nm.

Figure 27.5 summarises the estimates based on these figures relative to 100 units of incident radiation. If it is assumed that phytochrome molecules absorb a very small fraction of the light passing through a leaf, the average irradiance of a molecule can be taken as the mean of the upward and downward fluxes at the appropriate depth in the mesophyll. *Table 27.1* shows the corresponding values of τ_m calculated on this basis. The figures imply that in cells close to an illuminated epidermis, the quantum flux densities at. 660 and 730 nm will be similar to the ratio in the incident light. Close to the opposite epidermis, however, the ratio is smaller by an order of magnitude. By implication, the value of ζ in a green leaf must decrease by an order of magnitude across the mesophyll. In a leaf which lacks chlorophyll,

Table 27.1 Relative quantum flux densities in leaf mesophyll based on a multiple internal reflection model

Relative quantum flux	Wavelength (nm)		τ_m
	660	730	
Incident on upper epidermis	100	100	
Immediately below upper epidermis:			
Downward	91	91	
Upward	0.5	32	
Mean	46	62	0.74
Immediately above lower epidermis:			
Downward	9	89	
Upward	5	34	
Mean	7	62	0.11

however, there is little absorption of light at 660 nm and so ζ cannot change much across the mesophyll. It follows that the mean red to far-red ratio in the mesophyll of a green leaf will be much smaller than the ratio in unpigmented tissue exposed to the same light environment. The physiological implications of this major difference in the light crypto-climate of phytochrome in etiolated and in green leaves are considered below.

Conclusions

The figures derived in the above sections can now be combined to find the values of ζ_m within leaves at different points in a leaf canopy by using equation 27.5. *Figure 27.6* illustrates this synthesis, distinguishing between leaves which receive direct sunlight and those which are shaded by upper leaves. Although it has been necessary to make a number of assumptions and approximations in order to derive values

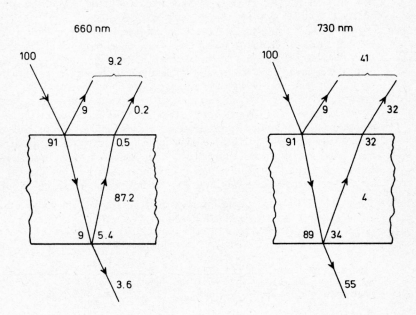

Figure 27.5   Summary of quantum flux densities for diffuse light at 660 and 730 nm.   Figures are relative to an incident flux of 100 units striking the upper epidermis of a leaf; absorption figures circled.   For details, see text

for τ_f and τ_m, the ratios in *Figure 27.6* are accurate enough to make two important generalisations about the quantum ratio of red to far-red light in uniform canopies.

(1) If the ratio above the canopy is $\zeta(0)$, the ratio within sunlit leaves will be the same order of magnitude as $\zeta(0)$ close to the sunlit epidermis but will be an order of magnitude smaller than $\zeta(0)$ close to the opposite epidermis;

(2) For shaded leaves, assuming the same diffuse illumination on both surfaces, the value of ζ_m for cells close to either epidermis will be one to two orders of magnitude less than $\zeta(0)$.   By inference, it will be three orders of magnitude less than $\zeta(0)$ for cells midway between the two epidermises.

Until the physiological implications of these figures have been evaluated, there seems little point in trying to refine them by intro-ducing greater physical realism or complexity.   For example, detailed information is available about the orientation of leaves in crop stands

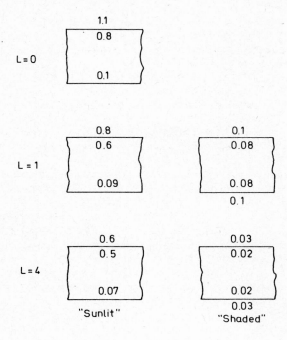

Figure 27.6   Values of ζ, the relative quantum flux density for 'sunlit' and 'shaded' leaves in a canopy based on calculations shown in Figures 27.2 and 27.5. L *is the leaf area index above the leaf*

(Lang, 1973), about the diurnal movements of leaves (Shell, Lang and Sale, 1974), about the fluctuating nature of light in canopies as a result of leaf fluttering (Desjardins, Sinclair and Lemon, 1973) and about the significance of the finite size of the sun's disc which is responsible for penumbral effects (Miller and Norman, 1971). The relevant measurements and models will be available for ecologists to use when they are needed. In the meantime, the calculations summarised in this chapter coupled with the measurements reported by Holmes and McCartney (Chapter 29) provide a basis for ecologically relevant studies of phytochrome function, either in natural field conditions or in the laboratory using light sources with an appropriate spectral composition.

References

ALLEN, W.A., GAUSMAN, H.W. and RICHARDSON, A.J. (1970). *J. Opt. Soc. Am.*, **60**, 542

ALLEN, W.A., GAUSMAN, H.W., RICHARDSON, A.J. and THOMAS, J.R. (1969). *J. Opt. Soc. Am.*, **59**, 1376

AVASTE, O., MOLDAU, H. and SHIFRIN, K.S. (1962). *Akad. Nauk Est. SSR Inst. Phys. Astron. Rep.*, No.3

DESJARDINS, R.L., SINCLAIR, T. and LEMON, E.R. (1973). *Agron. J.*, **65**, 904

FEDERER, C.A. and TANNER, C.B. (1966). *Ecology*, **47**, 555

GAUSMAN, H.W., ALLEN, W.A., CARDENAS, R. and RICHARDSON, A.J. (1970). *Appl. Opt.*, **9**, 545

GAUSMAN, H.W., ALLEN, W.A., WIEGAND, C.L., ESCOBAR, D.E., RODRIGUEZ, R.R. and RICHARDSON, A.J. (1973). *Tex. Agric. Exp. Stn Tech. Bull.*, No.1465

HENDERSON, S.T. (1970). *Daylight and its Spectrum.* Hilger, London

KUMAR, R. (1972). *Radiation from Plants.* Report, Purdue University School of Aeronautics, Astronautics and Engineering Sciences

LANG, A.R.G. (1973). *Agric. Meteorol.*, **11**, 37

LEMEUR, R. and BLAD, B.L. (1974). *Agric. Meteorol.*, **14**, 255

McCARTNEY, H.A. (1975). *PhD Thesis,* University of Nottingham

McCREE, K.J. (1968). *Agric. Meteorol.*, **5**, 203

MILLER, E.E. and NORMAN, J.M. (1971). *Agron. J.*, **63**, 739

MONSI, M., UCHIJIMA, Z. and OIKAWA, T. (1973). *Annu. Rev. Ecol. Syst.*, **4**, 301

MONTEITH, J.L. (1969). In *Physiological Aspects of Crop Yield,* pp.89-109. Ed. Eastin, J.D., Haskins, F.A., Sullivan, C.Y. and van Bavel, C.H.M. American Society of Agronomy and Crop Science Society of America, Madison, Wisc.

NILSON, T. (1971). *Agric. Meteorol.*, **8**, 25

NORMAN, J. (1975). In *Heat and Mass Transfer in the Biosphere,* p.187. Ed. de Vries, D.A. and Afgan, N.H., Scripta Book Co., Washington, D.C.

ROSS, J. (1975). In *Vegetation and the Atmosphere,* Vol.1, pp.13-55. Ed. Monteith, J.L. Academic Press, London

SHELL, G.S.G., LANG, A.R.G. and SALE, P.J.M. (1974). *Agric. Meteorol.*, **13**, 25

SINCLAIR, T.M., SCHREIBER, M.M. and HOFFER, R.M. (1973). *Agron. J.*, **65**, 276

STERN, F. (1964). *Appl. Opt.*, **3**, 111

SZEICZ, G. (1974a). *J. Appl. Ecol.*, **11**, 1117

SZEICZ, G. (1974b). *J. Appl. Ecol.*, **11**, 617

YOCUM, C.S., ALLEN, L.H. and LEMON, E.R. (1964). *Agron. J.*, **56**, 249

28

PRACTICAL APPLICATIONS OF ACTION SPECTRA

K.J. McCREE
Soil and Crop Sciences Department, Texas Agricultural Experiment Station, Texas A & M University, College Station, Texas 77843, U.S.A.

Introduction

In photobiology, the classical use of action spectra is to identify the photochemical compound which causes the observed biological response. In most cases, only the shape of the action spectrum is used, although for a truly quantitative analysis the absolute magnitude of the response must be known (Hartmann and Unser, 1972).

In plant ecology, action spectra can be used for the practical purpose of predicting the biological response to be expected in a given situation. If both the action spectrum for the response and the spectral irradiance of the incoming radiation are known, the expected response is the integral of the product of the two, over all wavelengths at which the product is greater than zero.

Absolute numbers for the action spectrum and the spectral irradiance must be used in these calculations. Also, since most biological responses are non-linear, the action spectrum must be determined over the full range of natural irradiances, and not only at the low end of the range. Another complication is that the biological effects at different wavelengths may not be strictly additive, because more than one mechanism is involved. Finally, since the irradiance at the surface of the organism is used in the calculations, the action spectrum must be related to the irradiance at that surface, and not at some internal point. The spectral irradiance at the photoreceptor itself will usually be different from that at the surface. For example, an action spectrum of a chloroplast suspension cannot be used to calculate leaf responses.

This view of action spectra is obviously different from that of the analytical spectroscopist, and often different data and techniques are needed. Most of the action spectra in the literature on photomorphogenesis, for example, are plotted in relative units, or were obtained only at very low irradiances where linearity is assumed to hold, or were obtained with unnaturally small amounts of screening pigments in the tissue. Therefore, to illustrate the present case for a different type of action spectra for use in solving the practical problems of crop ecology, the author's action spectra for photosynthesis in the leaves of crop plants (McCree, 1972a,b,c) will be used.

Action Spectra for Photosynthesis

At the time of this investigation, most of the action spectra had been obtained with algae, which had been chosen because they contained many different pigments, combined in many interesting ways in the photosynthetic apparatus. The chloroplasts of crop plants, on the other hand, contain a fairly constant mixture of pigments. Therefore, it was not surprising to find that the action spectra were all similar in shape (McCree, 1972a) (*Figure 28.1*).

A more surprising feature, perhaps, was the fact that they did not resemble the sharply peaked absorption spectrum of chlorophyll. This result must be due to the efficient energy transfer from the carotenoids. The curves in *Figure 28.1* indicate that the quantum yield (carbon dioxide uptake per photon absorbed) is virtually constant between 400 and 700 nm. This has the practical result that the photon flux in this waveband, which is a purely physical entity, is a very good measure of 'photosynthetically active radiation', from the point of view of the 'average' crop plant (McCree, 1972b). It is not necessary to base the measurement on the actual action spectrum for photosynthesis.

At the shortwave end of the waveband, the quantum yield of photons incident on the surface of the leaf decreases because of absorption in the epidermis, not because of any limitation at the chloroplast level (McCree and Keener, 1974). At the longwave end, it decreases because absorption by the chloroplast pigments decreases to zero. Eventually, the yield per quantum absorbed also decreases. Although several experiments have shown that it is possible to extend this limit by applying radiation of more than one wavelength at a time (enhancement effect), in practice the effects of these and other wavelength interactions are negligible. The measured photosynthetic response to polychromatic radiation was shown to be equal to that calculated from the monochromatic action spectrum and the spectral distribution of the radiation (McCree, 1972c). The particular combination of monochromatic radiations which was used in the enhancement experiments does not occur in any natural or artificial light source.

As in many photosynthesis experiments, the relationship between photosynthetic rate and photon flux density was found to be hyperbolic, so that the yield decreased continuously with increasing flux. However, by taking the data at constant photosynthetic rate, rather than at constant photon flux, it was possible to eliminate the effects on non-linearity from the action spectrum measurements. In effect, the photosynthetic apparatus was made to operate at the same rate at each wavelength, so that the same portion of the hyperbolic curve was used. In both the original measurements and the subsequent applications, it was assumed that only one of the constants of the response curve,

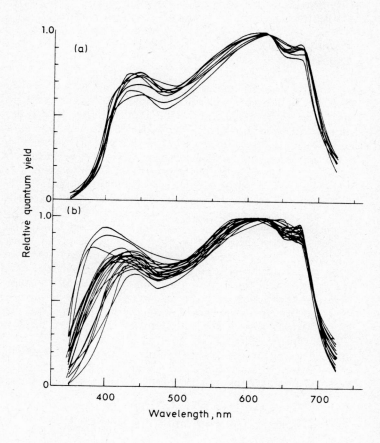

Figure 28.1 Relative spectral quantum yield for photosynthesis in the leaves of eight species of crop plant grown in the field (a), and 20 species of crop plant grown in a growth chamber (b). (From McCree, 1972a)

the slope at the origin (quantum yield), was a function of wavelength. The other constant, the value of the response at infinite photon flux density, was assumed to be independent of wavelength.

Applications

Once all of the parameters of a practical action spectrum have been determined, it becomes possible to calculate the response of the organism to any known radiation field. For example, it was shown that large increases in atmospheric aerosols, due to dust and pollution, are unlikely to decrease the photosynthetic rates of leaves by more than a few per cent (McCree and Keener, 1974; Unsworth and McCartney, 1973). The decrease in the photosynthetic photon flux from the sun's beam was almost balanced by an increase in the flux from the sky.

The photosynthetic efficiencies of different types of electric lamp were compared in another set of calculations (McCree, 1972b). The calculations showed that the high-pressure sodium lamp produced the greatest photosynthetic photon flux per unit of power consumed. It is relatively easy to produce solar levels of photon fluxes in controlled-environment chambers, using these lamps in commercial fixtures.

Commercial instruments for measuring photosynthetic photon fluxes have been developed, and the readings from these instruments are being used increasingly in applied photosynthesis studies. The practical advantage of having such instruments available is that the plant ecologist need not concern himself with spectral measurements, which are difficult to make, even under the best laboratory conditions, and almost impossible under changing field conditions. If calculated spectral data are available, the photosynthetic photon flux can easily be calculated from them.

Action Spectra for Photomorphogenesis

At present, the prospects do not seem good for finding a single, all-embracing action spectrum for photomorphogenesis, which can be used in the same way as the photosynthesis curve. There is a great diversity of action spectra (Evans, 1973; Mohr, 1972). Even in those cases where phytochrome is thought to be the sole receptor, it seems likely that the action spectrum will not resemble the absorption spectrum of this compound, because of strong screening effects. The etiolated tissues which are used in studies on response mechanisms cannot be expected to yield action spectra which can be applied to chlorophyll-bearing tissues. Problems of non-linearity and additivity of the effects of different wavelengths are still largely unexplored.

The most obvious practical problems occur when plants are grown under artificial lights. This is a long-standing problem, since new lamps with different spectral outputs are continually being developed. In the absence of a recognised, well established action spectrum on which to base calculations, we must rely on observations and measurements of the growth habits of plants grown under these lamps (Bickford and Dunn, 1972).

A universal problem with these practical tests is that lamps differ not only in their relative spectral distributions but also in the absolute magnitudes of their outputs. Since it is well known that the way in which a plant distributes its biomass depends not only photomorphogenic effects but also on how much biomass is available to distribute (Evans, 1973), it would be logical to adjust the lighting system so that equal photosynthetic photon fluxes are maintained at plant level at all times. This should take care of gross differences in growth rate due to photosynthesis. It is possible that some of the differences which have been observed in the past would disappear if the tests were made in this way. With present-day discharge lamps, it is possible to maintain solar levels of photosynthetically active radiation (about 2 mE s^{-1} m^{-2}), while the maximum possible level with fluorescent lamps is about 25 per cent of this value. Perhaps fewer problems will occur with the new lamps, simply because they operate at these more realistic levels of useful radiation.

References

BICKFORD, E.D. and DUNN, S. (1972). *Lighting for Plant Growth,* 221pp. Kent State University Press, Kent, Ohio

EVANS, L.T. (1973). In *Plant Response to Climatic Factors (Proc. Uppsala Symposium),* p.21. Ed. Slatyer, R.O. UNESCO, Paris

HARTMANN, K.M. and UNSER, I.C. (1972). *Ber. Dtsch. Bot. Ges.,* **85,** 481

McCREE, K.J. (1972a). *Agric. Meteorol.,* **9,** 191

McCREE, K.J. (1972b). *Agric. Meteorol.,* **10,** 443

McCREE, K.J. (1972c). *Plant Physiol.,* **49,** 704

McCREE, K.J. and KEENER, M.E. (1974). *Agric. Meteorol.,* **13,** 349

MOHR, H. (1972). *Lectures on Photomorphogenesis.* Springer, Berlin, Heidelberg, New York

UNSWORTH, M.H. and McCARTNEY, H.A. (1973). *Atmos. Environ.,* **7,** 1173

29

SPECTRAL ENERGY DISTRIBUTION IN THE NATURAL ENVIRONMENT AND ITS IMPLICATIONS FOR PHYTOCHROME FUNCTION

M.G. HOLMES
Department of Physiology and Environmental Studies, University of Nottingham, School of Agriculture, Sutton Bonington, Loughborough, Leics. LE12 5RD, U.K.

H.A. McCARTNEY
Department of Mechanical and Production Engineering, Brighton Polytechnic, Mouslecoomb, Brighton, Sussex, U.K.

Most experimental work on phytochrome has concentrated on the physiological action of phytochrome in dark-grown plants. While the findings have been extremely successful in pointing to the site and mode of action of phytochrome, the experiments have usually required the use of etiolated plants and unnatural durations of narrow wavebands of light. Under natural irradiation conditions, plants are subject to a spectrum which covers the entire waveband over which phytochrome absorbs. It can be expected that any changes in the distribution of this radiant energy will result in an altered photoequilibrium of the two forms of phytochrome. Investigations of the spectral energy distribution (SED) within forests (Stoutjesdijk, 1964; Federer and Tanner, 1966) have led to suggestions that phytochrome may be involved in the detection of shading by other plants (Cumming, 1963; Evans, 1966; Taylorson and Borthwick, 1969; Kasperbauer, 1971; Smith, 1973; Holmes and Smith, 1975). The suggestions are based on the high absorption of the blue and red wavebands and the low absorption of green and far-red wavebands by leaf canopies. The greater depletion of red light in comparison with far-red light when radiation is attenuated by leaves may be expected to decrease the equilbrium proportion of phytochrome existing as Pfr.

In many agricultural situations, a high degree of adaptability to large variations in the quality and amount of light (Sinclair and Lemon, 1973) is shown by underlying weed species. This investigation was designed to determine the nature of these variations in SED and of the resultant effects on weed development. The study was based on measurements of the SED of daylight above and below two crop canopies, of corresponding phytochrome photoequilibria and of the growth patterns of weeds grown under artificial irradiation conditions. Wheat (*Triticum aestivum* L.) and sugar beet (*Beta vulgaris* L.) were used as representative crop species owing to their contrasting canopy architecture. Wheat provided a closely spaced, narrow-leaved and erect canopy, while sugar beet provided a broad leaved and widely spaced system. *Chenopodium album* L. (fat hen) and *Tripleurospermum maritimum* ssp. *inodorum* L. (scentless mayweed) were used as representative weed species owing to their ubiquity in cultivated land and suitability for experimental assay.

Measurement of SED between 400 and 800 nm of both natural and artificial light regimes were made with a scanning spectroradiometer system (Gamma, San Diego). To reduce disturbance of the crop canopy to a minimum, the instrument was fitted with a flexible fibre-optic light guide connected to a miniature cosine-corrected receptor head. This system yields data on spectral irradiance after calibration against a standard light source. Spectral irradiance values ($W\ m^{-2}\ nm^{-1}$) were multiplied by $8.359 \times 10^{-3} \lambda$ (where λ is the wavelength in nanometres) to obtain units of spectral quantum flux ($\mu E\ m^{-2}\ s^{-1}\ nm^{-1}$). For describing the SED in the red and far-red wavebands, we have expressed the ratio of the quantum flux in 10-nm wide wavebands in the red and far-red, respectively (ζ). This symbol is equivalent to and supersedes the $E_{660}:E_{730}$ definition used previously (Holmes and Smith, 1975) and has been chosen in order to simplify and standardise the definition of this ratio (Monteith, Chapter 27).

On clear days, the SED of global radiation above the crops changed little throughout the daylight hours although the magnitude of irradiance changed considerably. On overcast days, there was a relatively larger fraction of blue light compared with clear days but the SED in the red and far-red wavelengths was almost unchanged. Values of ζ above crops on overcast and cloudless days were therefore very similar except at sunrise and sunset, when significant changes were observed.

Sunrise was preceded by a general increase at the blue end of the spectrum, followed by an increase in the red and far-red regions in which the rate of increase in far-red was higher. This resulted in a decrease in ζ. As the solar zenith angle decreased, the red contribution increased more rapidly than the far-red, thus increasing ζ to the equilibrium level maintained throughout the day. At sunset, a reversal of this procedure occurred, resulting in a decrease in ζ prior to darkness (*Figure 29.1*). The rate, magnitude and duration of these red/far-red shifts depended on atmospheric conditions. In general, the smallest and most gradual shifts (*ca.* 15 per cent from mid-day values) were observed under overcast skies while the largest and most erratic shifts (*ca.* 45 per cent) occurred with clear or partially clear skies. Shropshire (1973) has reported larger shifts in collimated direct solar radiation at twilight. However, the smaller shifts in total global radiation reported here are a closer approximation to the radiation reaching the plant.

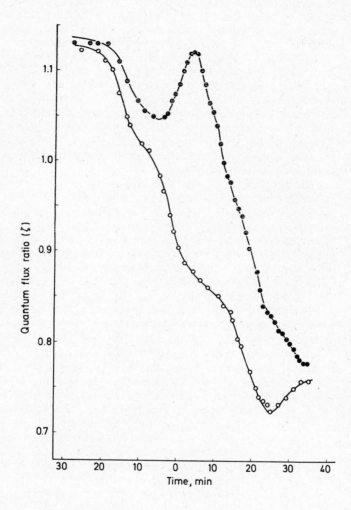

*Figure 29.1   Changes in ζ during sunset (denoted by time zero) at latitude
52° 45′, longitude 1° 15′ on partially (○) and totally (●) overcast days.   Solar
elevation decreased at 8° 44′ per hour under the partially overcast sky (13.2.74)
and at 7° 38′ per hour under the totally overcast sky (6.3.74)*

The largest natural variations in ζ were observed within the two crop canopies. The horizontal distribution of both intensity and SED was non-uniform in both types of canopy. Although mean values of SED could be calculated in the wheat canopy by integration of spectra in shade and in sunflecks, this method was not practical for the sugar beet owing to the wide spacing used in its cultivation. Consequently, no attempt was made to measure the relative distribution of sunfleck and shade in either time or space. Typical SEDs at various levels in the wheat canopy and at the ground within the sugar beet canopy are illustrated in *Figures 29.2* and *29.3*. In wheat, ζ decreases with increasing canopy depth. Before the onset of senescence, ζ shows a marked dependence on leaf area index (LAI) and position in the canopy (*Figure 29.4*). In the senescing crop, however, the difference between the red and far-red radiation penetrating the crop decreased with time at constant LAI.

The decrease in ζ with increasing depth in the actively growing wheat canopy corresponded to a gradual decrease in the transmission of photosynthetically active radiation (PAR). This attenuation of PAR within crop canopies is clearly another important factor in determining plant growth where the PAR flux becomes limiting for photosynthesis. *Figure 29.5* shows ζ as a function of PAR in the wheat canopy for two clear days. As the crop senesced, ζ increased at a greater rate than PAR transmission, indicating a change in the optical properties of the canopy during senescence.

Whether the observed SED shifts have any effect on phytochrome-controlled development of plants growing beneath these crops depends primarily on the ability of phytochrome to respond to these light distributions. Owing to the high levels of chlorophyll present, it is not possible at present to determine the photoequilibrium of phytochrome in green plants. Photoequilibria were therefore measured in dark-grown plant material which had been exposed for short periods to the relevant irradiation environments within the crops. The values of Pfr as a percentage of total phytochrome (Φ) obtained by this method were used as an indication of the photoequilibrium in the green plant. These values should be regarded only as an approximation. The intracellular irradiation environments of dark-grown and light-grown plant material will not be the same owing to the differences in pigment content resulting in differences in the proportions of red and far-red radiation within the tissues. Similarly, differences may exist in the molecular environment of phytochrome; such environments cannot be assumed to be identical in light-grown and etiolated plants (Mumford and Jenner, 1971). Finally, no provision has been made for the accumulation of phototransformation intermediates under high irradiances (Kendrick and Spruit, 1972). It seems likely that the irradiance levels found in natural conditions are high enough to cause significant accumulation of such intermediates.

Bearing these qualifications in mind, photoequilibria were determined by exposing etiolated bean (*Phaseolus vulgaris*) hypocotyl hook sections

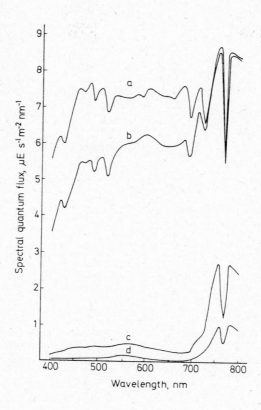

Figure 29.2 SED of daylight above and within the wheat canopy. Measurements were made under clear skies between 11.47 and 12.00 GMT on 22.6.73: (a) above the canopy (ς = 1.06); (b) sunfleck 60 cm above ground (ς = 0.86); (c) shade 60 cm above ground (ς = 0.27); (d) shade at ground level (ς = 0.14). Canopy height 85 cm

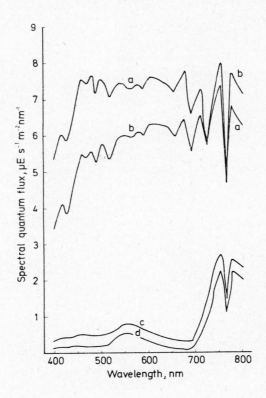

Figure 29.3 SED of daylight above and at ground level below sugar beet canopy. Measurements were made under clear skies at about 11.00 GMT on 1.7.73: (a) above the canopy (ξ = 1.18); (b) sunfleck (ξ = 0.96); (c) shade between rows (ξ = 0.22); (d) shade between plants (ξ = 0.12)

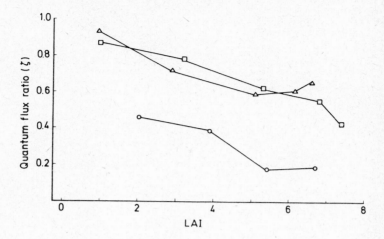

Figure 29.4 Relationship between ζ and LAI in the wheat canopy under clear skies on 7.6.73 (○), 22.6.73 (□) and 5.7.73 (△)

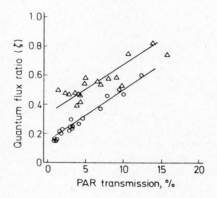

Figure 29.5 Relationship between ζ and PAR transmitted by the wheat canopy under clear skies on 7.6.73 (○) and 5.7.73 (△)

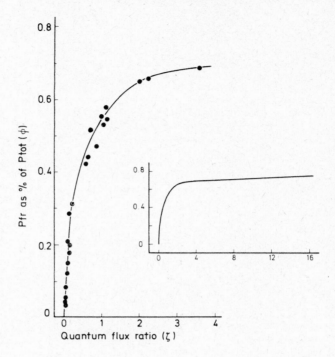

Figure 29.6   Phytochrome photoequilibria (Φ) measured in dark-grown Phaseolus vulgaris *L. hypocotyl hook sections as a function of* ζ

in a plastic cuvette at 0 °C to the various natural radiation distributions which had been measured within and above the two crop canopies. A series of dose response measurements was made for each light situation to ensure equilibrium had been achieved. After a saturating dosage, ·Φ values were determined with a modified Perkin-Elmer 156 dual-wavelength spectrophotometer using standard procedures (Hillman, 1965; Kendrick and Frankland, 1968). It is evident from *Figure 29.6* that relatively small changes in ζ within the range found in natural conditions (*ca.* 0.05–1.2) elicit large shifts in Φ from *ca.* 0.04 to 0.55. For higher values of ζ (which are found only under artificial light sources), much larger changes in ζ are required in order to effect shifts in phytochrome photoequilibria of similar magnitude.

The effect of the photoequilibrium shifts which appear to occur in the natural environment were tested by growing *T. maritimum* and *C. album* under two forms of artificial lighting which produce different

Table 29.1 Developmental changes in *T. maritimum* and *C. album* over 15 days at 25 ± 1 °C under fluorescent (low far-red) and incandescent (high far-red) light sources

The light sources were adjusted to produce equal quantum fluxes in the 400–700 nm waveband. *T. maritimum* seedlings were grown in a glasshouse until 9 weeks old, then transferred to fluorescent lighting in a growth chamber for 1 week before starting the treatments which were carried out in 8-h days. *C. album* were collected from an open situation in the field, potted and grown under fluorescent lighting for 10 days. The plants were then divided equally between the fluorescent and incandescent sources; 16-h photoperiods were used.

Property	T. maritimum		C. album	
	Fluorescent	*Incandescent*	*Fluorescent*	*Incandescent*
Height (cm)	29.7 ± 1.2	59.4 ± 1.4	15.0 ± 0.9	28.4 ± 1.5
Internode length (cm)	0.84 ± 0.09	3.53 ± 0.18	–	–
Leaf dry weight (g)	0.483 ± 0.029	0.464 ± 0.024	0.338 ± 0.016	0.310 ± 0.016
Stem dry weight (g)	0.329 ± 0.017	0.583 ± 0.032	0.104 ± 0.006	0.197 ± 0.009
Leaf:steam dry weight ratio	1.47	0.80	3.25	1.57
Leaf area (dm^2)	–	–	1.071 ± 0.012	0.785 ± 0.009

phytochrome photoequilibria. The quantum flux densities between 400 and 700 nm were equalised under both light sources in an attempt to equalise the photosynthetic rates in each treatment (McCree, 1973). After 15 days of treatment, the pattern of growth was detectably different in the two treatments (*Table 29.1*). The lower Φ level resulted in a re-direction of growth towards increased stem elongation and decreased leaf size. It should be noted that the fluorescent source produces a higher Φ level than occurs in natural conditions; however, the results do indicate a change in ontogeny which cannot be due solely to photosynthetic effects. Such evidence appears to support the hypothesis that the function of phytochrome in the natural environment is the detection of shading caused by plant material and the initiation of appropriate changes in the pattern of development. Such a mechanism would offer considerable competitive advantage in dense plant communities.

Acknowledgements

This work was supported by grants from the Agricultural Research Council and from the Royal Society. We gratefully acknowledge the advice of Professor J.L. Monteith, Professor H. Smith and Dr M.H. Unsworth throughout the course of this work.

References

CUMMING, B.G. (1963). *Can. J. Bot.,* **41**, 1211
EVANS, G.C. (1966). In *Light as an Ecological Factor,* pp.53–76. Ed. Bainbridge, R., Evans, G.C. and Rackham, O. Blackwell, Oxford
FEDERER, C.A. and TANNER, C.B. (1966). *Ecology,* **47**, 555
HILLMAN, W.S. (1965). *Physiol. Plant.,* **18**, 346
HOLMES, M.G. and SMITH, H. (1975). *Nature, Lond.,* **254**, 512
KASPERBAUER, M.J. (1971). *Plant Physiol.,* **47**, 775
KENDRICK, R.E. and FRANKLAND, B. (1968). *Planta,* **82**, 317
KENDRICK, R.E. and SPRUIT, C.J.P. (1972). *Nature New Biol.,* **237**, 281
McCREE, K.J. (1973). *Agric. Meteorol.,* **10**, 443
MUMFORD, F.E. and JENNER, E.L. (1971). *Biochemistry,* **10**, 98
SHROPSHIRE, W. (1973). *Solar Energy,* **15**, 99
SINCLAIR, T.R. and LEMON, E.R. (1973). *Solar Energy,* **15**, 89
SMITH, H. (1973). In *Seed Ecology,* pp.219–232. Ed. Heydecker, W. Butterworths, London
STOUTJESDIJK, P.H. (1964). *Versl. Meded. Kon. Nad. Bot. Ver. 1963,* 42
TAYLORSON, R.B. and BORTHWICK, H.A. (1969). *Weed Sci.,* **17**, 48

30

PHYTOCHROME CONTROL OF SEED GERMINATION IN RELATION TO THE LIGHT ENVIRONMENT

B. FRANKLAND

Department of Plant Biology and Microbiology, Queen Mary College, University of London, Mile End Road, London E1 4NS, U.K.

Introduction

Seed germination may be promoted or inhibited by white light. Positive responses to both short and long periods of light show an action peak in the red region of the spectrum, corresponding to the absorption peak of the Pr form of phytochrome. Reversal of the red effect by far-red light confirms that phytochrome is the photoreceptor. This was first demonstrated in seeds of *Lactuca sativa* (Borthwick *et al.,* 1952). Seeds of some species, such as *Limonium vulgare* (Bowes, 1967), are stimulated only by a long period of red light, suggesting that there is a requirement for Pfr action over a prolonged period.

In species such as *Amaranthus caudatus,* where seed germination is inhibited by prolonged exposure to light, far-red and blue are the most effective parts of the spectrum *(Figure 30.1).* The inhibitory effect of 720-nm light can be nullified by mixing it with either 655- or 755-nm light, suggesting that here phytochrome is acting in its high-irradiance mode to counteract the otherwise promoting effects of Pfr. Seeds exposed to prolonged white light (high Pfr/P) germinate on transfer to darkness, whereas seeds exposed to prolonged far-red light (low Pfr/P) do not germinate on transfer to darkness unless given an irradiation with red light. Many inhibitory effects of low-intensity far-red light can be interpreted in terms of the maintenance of a low level of Pfr. This is so for *Amaranthus* (Kendrick and Frankland, 1969), *Lactuca* (Mancinelli and Borthwick, 1964) and other species. Such effects indicate that Pfr is present in dark-imbibed seeds and can appear in that form over a period in the dark.

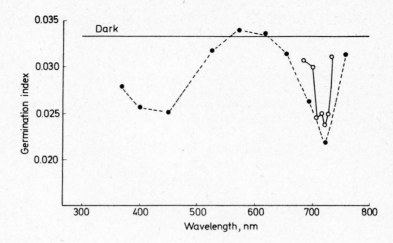

Figure 30.1   Photoinhibition of germination in Amaranthus caudatus *by light of various wavelengths.   Seeds transferred to light after 20 h of dark imbibition. Results using interference filters of bandwidth 20 nm and irradiance 5-7 W m^{-2} (•) or interference filters of bandwidth 10 nm and irradiance 1 W m^{-2} (○). Germination index in h^{-1} plotted against wavelength.   Germination index $\bar{g} = (1/N)\Sigma\frac{1}{t}$, where N = total number of seeds and t = time at which each seed germinates.   It is related to final germination percentage (P) and time to half final germination percentage (t'). Thus $\bar{g} = \frac{1}{t'} \cdot \frac{P}{100}$, although it is calculated empirically thus:*

$$\bar{g} = \frac{1}{N} \left[\tfrac{1}{2}(\tfrac{1}{t_2} + \tfrac{1}{t_1})(n_2 - n_1) + \tfrac{1}{2}(\tfrac{1}{t_3} + \tfrac{1}{t_2})(n_3 - n_2) + \dots \right]$$

where n_1, n_2, n_3, etc., are the number of seeds germinated after times t_1, t_2, t_3, etc.

Phytochrome Control of Germination in *Sinapis Arvensis*

A detailed study has been made of phytochrome control of germination in *Sinapis arvensis* (charlock or wild mustard).   Like many weed species such as *Plantago major* (plantain), *Rumex obtusifolius* (dock) and *Chenopodium album* (goosefoot), *Sinapis* produces seeds which have a light requirement for germination.   In a given batch of seeds, a proportion will be dormant, this effect being relative dormancy, the proportion failing to germinate in the dark being higher at 25 °C than at 15 °C. All seeds germinate following removal of the seed coats or following treatment with 3 × 10^{-3} M gibberellic acid (Edwards, 1968).   Freshly harvested seeds are very dormant and there is a reduction in dormancy during the first 6 months of dry storage; experiments are usually carried out with seeds which have been stored for at least 9 months. Seeds were collected from plants grown in glasshouses at the Queen

Mary College Field Station near Brentwood, Essex, the original seed stocks being obtained from the May and Baker Ongar Research Station. The seeds tended to be less dormant than those collected from natural populations of plants.

Different batches of seed differ in the proportion of dormant seeds and in their light sensitivity, although they usually show a positive response to a brief irradiation with red light. This variation is found even amongst seed samples collected from individual plants grown from a single batch of seeds (*Table 30.1*). There is evidence from selection

Table 30.1 Variation in germination behaviour amongst seeds from 20 individual plants of *Sinapis arvensis*

200 seeds per treatment. Seeds maintained in the dark or irradiated with 10 min of red light after 6 h of imbibition. Temperature 25 °C. Data given are germination percentages after 7 days, although the germination response is virtually complete after 2 days.

Plant no.	Dark	10 min red light
1	0	1
2	0	15
3	0	19
4	0	29
5	2	38
6	4	57
7	4	79
8	7	42
9	7	70
10	10	80
11	12	39
12	12	55
13	12	61
14	16	73
15	20	81
16	20	87
17	20	92
18	23	83
19	29	83
20	52	91

experiments that at least some of this variation is genetic. The effect of red light can be wholly or partially reversed by a subsequent brief irradiation with far-red light and, as shown in *Table 30.2*, there is the repeatable red/far-red reversibility characteristic of a phytochrome-controlled response (Ben Hameda, 1970). Partial reversion is due to far-red light itself having a promoting effect on germination, indicating that a proportion of seeds are capable of responding to the low levels of Pfr (Pfr/P = 0.05) established by the far-red source. This promoting effect of far-red light is particularly noticeable at 15 °C (*Table 30.3*) and has been observed in some other species, such as *Eragrostis curvula* (Toole and Borthwick, 1968). As might be expected, blue light has a promoting effect on germination which is greater than that of far-red but less than that of red light.

Table 30.2   Red/far-red reversibility in the photocontrol of germination in three batches of *Sinapis arvensis* seed
Data given are germination percentages after 7 days

Light treatment	Batch A	Batch B	Batch C
D	8	20	46
R	20	56	84
FR	10	20	50
FR-R	20	60	84
R-FR	12	24	52
R-FR-R	24	58	86
FR-R-FR	12	22	52
FR-R-FR-R	22	56	86
R-FR-R-FR	10	24	54

Table 30.3   Effect of temperature on germination
Data are germination percentages after 7 days

Batch	25 °C			15 °C		
	D	R	FR	D	R	FR
A	8	20	9	22	72	54
B	20	57	22	32	87	64
C	45	87	52	64	91	81

Experiments in which seeds were exposed to different red-light irradiances for different periods showed that the germination response is related to the logarithm of the incident light energy. Dose response curves for three different batches are shown in *Figure 30.2*. The three batches were obtained by pooling seed samples, taken from individual plants, with high (A), medium (B) and low (C) proportions of dormant seeds. It can be seen that the higher is the proportion of dormant seeds, the higher are the doses of red light required to stimulate germination. Each batch, of course, consists of individual seeds which vary in the amount of light required to trigger germination. This variation conforms to a normal distribution (*Figure 30.3*), which explains why the dose response curves are sigmoid. It follows that there is a linear relationship between the probit of the germination percentage and the logarithm of light energy (*Figure 30.4*). The curves in *Figure 30.2* were drawn on the assumption that in all three batches the probit of germination increases by 0.35 for each 10-fold increase in red-light energy dose. The use of the probit transformation simplifies mathematical operations on germination data and allows high and low germination percentages in different batches to be combined in a meaningful way. In *Figure 30.4*, data from 100 seed samples were combined.

The variation in light sensitivity probably reflects the variation in the requirement of individual seeds for a particular minimum amount of Pfr for germination. It is difficult to detect phytochrome

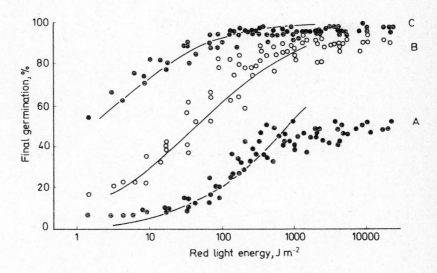

Figure 30.2 Germination of three batches of Sinapis arvensis *seed in relation to red-light dose. Seeds exposed to red light for 5 s to 3 h at intensities ranging from 0.3 to 3 J m⁻² s⁻¹. Red-light treatment begun after 6 h of imbibition. Temperature 25 °C*

spectrophotometrically in seeds, but Pfr/P ratios can be measured using dark-grown seedlings of *Sinapis arvensis*. There is a linear relationship between the probit of the proportion of phytochrome transformed by red light and the logarithm of red light energy (*Figure 30.5*). It follows that there must be a linear relationship between the probit of the germination percentage and the probit of the percentage of phytochrome transformed (*Figure 30.6*). On the dubious assumption that phytochrome in seedling tissue behaves in the same way as phytochrome in seed tissue, it is possible by extrapolation to estimate the Pfr/P ratio in dark-imbibed seeds as about 0.005. Seeds with no light requirement presumably germinate in response to this low level of Pfr, while seeds which fail to germinate in the light are those which require higher Pfr levels than can be established by a saturating dose of red light. Such seeds are capable of responding to Pfr, as can be seen by their ability to germinate in response to red light, both at lower temperatures and in the presence of sub-optimal gibberellin concentrations. Such treatments increase the sensitivity to Pfr of all seeds in a batch.

Reversion of the effect of a red-light irradiation by a subsequent far-red irradiation does not take place if there is a sufficiently long

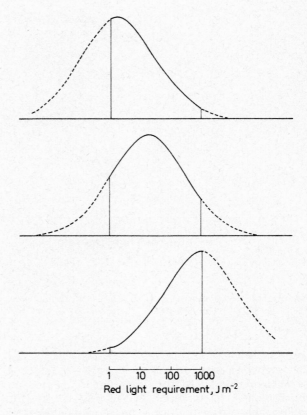

Red light requirement, J m^{-2}

Dark-germinating   Light-requiring   Non-germinating
seeds   seeds   seeds

Figure 30.3   Diagrammatic representation of variation in light requirement amongst individual seeds in three batches

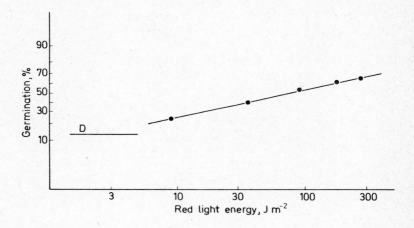

Figure 30.4 *Germination percentage on a probit scale plotted against red-light dose on a logarithmic scale. Pooled data from 100 different seed samples*

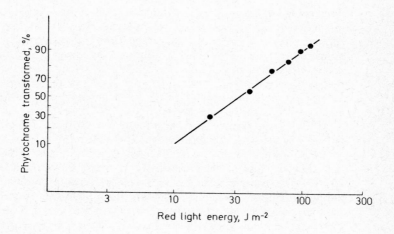

Figure 30.5 *Phototransformation of phytochrome by red light in dark-grown seedlings of* Sinapis arvensis. *Pfr as a percentage of amount formed by saturating red light plotted on a probit scale against red-light dose plotted on a logarithmic scale*

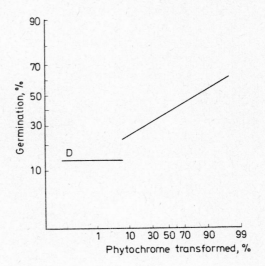

Figure 30.6   Germination percentage on a probit scale plotted against percentage of phytochrome transformed by red light also plotted on a probit scale

intervening dark period. This 'escape' from far-red reversion is of interest since it can be taken to represent the time course of Pfr action. *Figure 30.7* illustrates escape from far-red reversion at two temperatures. At 25 °C, light-requiring seeds need a period of Pfr action ranging from 30 min to 3 h; at 15 °C, the range is from a few minutes to 6 h. Presumably, seeds with a low Pfr requirement for germination can be stimulated to germinate by high Pfr levels acting over a fairly short period.

As shown in *Figure 30.8,* maximum sensitivity to red light is found with 4–8 h of dark imbibition at 25 °C and with 4–30 h of imbibition at 15 °C. These temperature differences are probably related to the fact that the loss of light sensitivity is much slower at 15 °C than at 25 °C. The process whereby the seeds become responsive to red light during imbibition probably represents the progressive rehydration of phytochrome. Seeds are fully imbibed after 5 h and during this time the water content increases from about 10 per cent to about 50 per cent.

There are interesting differences in response to far-red light applied after different periods of dark imbibition. Although far-red light promotes germination when applied after 6 h of imbibition, it tends to inhibit germination when applied before 4 h of imbibition (*Figure 30.9*).

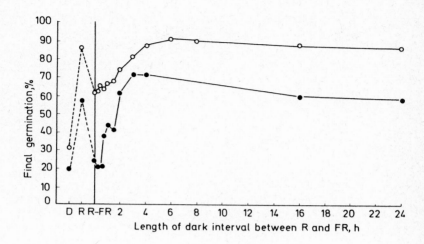

Figure 30.7 Effect of length of intervening dark period on escape from far-red reversion of a previous red irradiation. Data for 15 °C (○) and 25 °C (●)

These early inhibitory effects of far-red light have been confirmed with several different batches of seed. Clearly, far-red light decreases the Pfr in the early stages of imbibition but increases the Pfr when given at later stages of imbibition. It is possible that the Pfr/P ratio is relatively high during the early stages of dark imbibition, although a better explanation of these results is provided by recent observations on the formation of phytochrome intermediates in dehydrated tissue (Kendrick, 1974; Kendrick and Spruit, Chapter 4). A simplified scheme showing the intermediates involved in the Pr → Pfr and Pfr → Pr transformations is presented in *Figure 30.10*. The first step in the transformations involves a photoinduced change in the chromophore and can take place in dehydrated phytochrome. Later steps involve protein conformation changes and so cannot take place in the dehydrated state. In partially hydrated phytochrome, such as would exist in seeds during early stages of imbibition, far-red light can remove Pfr by conversion to P650 but cannot readily form Pfr by conversion from Pr. An inhibitory effect of far-red light in a situation where there is no promotive effect of red light may arise from the fact that the dark reversion of P650 to Pfr is much slower than the dark reversion of P698 to Pr.

    It is clear that phytochrome can be manipulated in dry seeds and it has been recently demonstrated for *Lactuca sativa* that germination responses can arise from such manipulations (Kendrick and Russell, 1975).

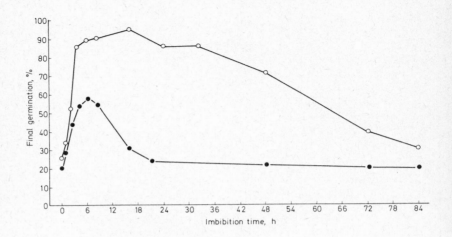

Figure 30.8   Changes in sensitivity to red light during imbibition.   Data for 15 °C (○) and 25 °C (●)

Data for *Sinapis arvensis* are presented in *Table 30.4*.   Far-red light has only a small effect on normal dry seeds, even when the irradiation is applied immediately before sowing.   In this particular experiment, the effects were not statistically significant.   However, an effect of far-red light on dry seeds can be shown more clearly in seeds with an artificially high Pfr level produced by red-light irradiation of imbibed seeds followed by re-drying.   The effect of far-red light can be partially reversed by red light, probably owing to conversion of P650 back to Pfr.   In contrast to the situation in *Lactuca, Sinapis* and many other species with light-sensitive seeds, germination in *Pinus banksiana* can be stimulated by red-light irradiation of dry seeds   (Orlandini and Bulard, 1972).   This effect of red light is far-red reversible.   Either the phytochrome in dry *Pinus* seeds is in a sufficiently hydrated state to allow the full interconversions of Pr and Pfr or the red-light effect is due to absorption by P650.

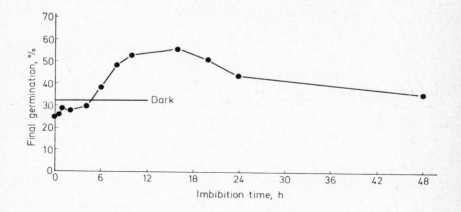

Figure 30.9 Changes in sensitivity to far-red light during imbibition at 25 °C

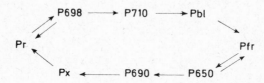

Figure 30.10 Intermediates involved in the interconversions of Pr and Pfr. Simplified version of scheme proposed by Kendrick and Spruit (1973). The conversions of Pr to P698 and Pfr to P650 are photoinduced

Table 30.4 Inhibitory effects of far-red light on dry seeds of *Sinapis arvensis*

Data given are germination percentages after 2 days at 25 °C. 200 seeds per treatment. Differences of 12 per cent or more are statistically significant at the 5 per cent level of risk

Light treatment	Normal, dry seeds	Seeds imbibed for 6 h, irradiated with red light and allowed to dry for 18 h in dark
D	32	73
R	31	73
FR	26	46
FR-R	29	64
R-FR	25	52

Ecological Implications

Most weed species are ruderals, i.e. annual plants adapted to habitats where the soil is regularly disturbed, such as river banks or cliffs. Cultivation of the land by man produced new habitats suitable for such plants. A light requirement for germination is a common feature of weed seeds and it can be argued that this is one reason why such plants are weeds. Their seeds tend not to germinate while buried in the soil but do so when exposed at the surface by cultivation.

Wesson and Wareing (1969a) have studied the effects of soil disturbance on the emergence of weed seedlings under field conditions and have shown that light is very important in determining germination. *Sinapis arvensis* was one of the weed species they studied and *Table 30.5* shows the results of a simple experiment with *Sinapis* seeds sown

Table 30.5 Effect of depth in soil on seed germination in dark and light

Experiments carried out in pots under glasshouse conditions at 20 ± 5 °C. Dark: pots wrapped in metal foil. Daylight: pots wrapped in transparent polythene. Data are germination percentages after 7 days

Depth in soil (cm)	Dark	Daylight
0	37	93
1	30	45
3	0	0
6	0	0

at different depths of soil in pots. Germination at the surface shows the stimulatory effect of light. Very little light penetrates to a depth of 1 cm and seeds germinating at this level are mainly those which do not require light. At 3- and 6-cm depths, germination is greatly inhibited by factors other than the absence of light, possibly factors linked to aeration. After 2 weeks, the soil containing seeds at a depth of 6 cm was mixed (*Table 30.6*). A few seeds now germinated, particularly in pots in the light. The number is small but this is to be expected as only a small proportion will be finally located at or near the surface of the soil. Two weeks later, the soil was spread out thinly in trays. In the dark no seeds germinated, but there was a marked response to light. It is interesting that burial in soil induced a deeper state of dormancy while preserving the ability of the seeds to respond to light. Seeds imbibed on moist filter paper lose their light sensitivity within a few days. The induction of dormancy can be simulated by exposing seeds to a high carbon dioxide concentration (30 per cent carbon dioxide, 20 per cent oxygen, 50 per cent nitrogen) for 3 days. However, under these conditions, the ability of seeds to respond to light is lost. Wesson and Wareing (1969b) noted that a light requirement can actually be induced in the seeds of some species buried in soil for 1 year. A possible explanation of such effects is that Pfr in the seeds reverts to Pr under the conditions of burial in the soil.

Table 30.6 Effect of soil disturbance on seed germination in dark and light

	Dark	*Daylight*
Seeds buried 6 cm deep in soil: germination after 3 wk (%)	0	0
Soil disturbed: germination after 2 wk (%)	1	3
Soil spread to a depth of 0.5 cm: germination after 2 wk (%)	1	22

Seeds at or near the surface of the soil may spend a significant time in a dry or incompletely imbibed condition. An understanding of the transformations of phytochrome in various states of hydration may be important in explaining the behaviour of seeds under these conditions. The situation is complex as the seeds are in a very heterogeneous environment. Germination will depend on the length of time a seed is exposed at the surface of the soil, the degree of hydration both at the time of exposure and subsequently, and many other factors. The fact that freshly harvested seeds are very dormant increases the probability of their becoming incorporated into the soil before they germinate in response to light. Variation in germination behaviour confers a selective advantage in that it ensures that only a proportion of seeds germinate in a given environmental situation and at a given time.

There is clear evidence that Pfr is present in dark-imbibed seeds. During the maturation, and consequent dehydration, of seeds on the parent plant, it might be expected that a large proportion of phytochrome would be trapped as Pfr or as intermediates which can form Pfr on rehydration. It has been demonstrated, for instance, that high levels of Pfr can persist in dry *Lactuca* seeds for at least 1 year (Vidaver and Hsiao, 1972). This suggests that light conditions during seed maturation might affect subsequent germination by affecting the amount of Pfr trapped in the dry seed. In *Arabidopsis thaliana* (Shropshire, 1973) and *Cucumis* species (Gutterman and Porath, 1975), differences in light quality during seed maturation certainly do lead to differences in subsequent germination behaviour. However, effects arising from daylength treatments of the fruiting parent plants in a number of species do not appear to involve phytochrome differences amongst the seeds (Karssen, 1970; Gutterman, 1974). Also, in *Sinapis arvensis* there is no evidence that variation in germination behaviour is due to variation in Pfr levels within the seeds and the problem becomes that of explaining the very low levels of Pfr which are found. Hsiao and Vidaver (1973) have shown that in *Lactuca* seeds, Pfr can revert in the dark to Pr under conditions of partial hydration which do not allow germination to proceed. This process may take place during seed maturation and account for the low levels of Pfr found in many seeds.

Germination of some seeds at the soil surface within dense vegetation may be determined by light quality rather than by the simple presence and absence of light. Seeds in such a situation will be exposed to light which has passed through green leaves and so is depleted much more in red light than in far-red light (*Figure 30.11*). Cumming (1963)

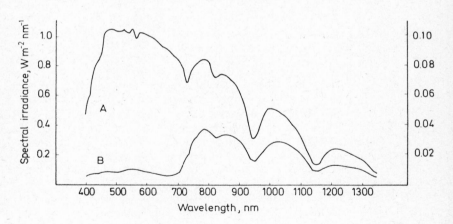

Figure 30.11   Spectral distribution of sunlight (A) and light within a woodland (B).   Measurements made with an ISCO spectroradiometer during July 1974 at the Queen Mary College Field Station at Dytchleys and in Mores Wood, near Brentwood, Essex.   Left-hand scale refers to curve A and right-hand scale to curve B

studied effects of light of various red/far-red ratios on germination of *Chenopodium* species and related those effects to germination of seeds in natural conditions under a leaf canopy. The possibility of effects due to differences in spectral quality in the natural light environment has also been pointed out in review papers by Black (1969) and Smith (1972).

Acknowledgements

The data in *Tables 30.1* and *30.2* and in *Figures 30.2, 30.4, 30.5, 30.6, 30.7* and *30.8* were obtained by Mr D.R. Waddoups during his tenure of a Natural Environment Research Council studentship at Queen Mary College, 1971-74. The data in *Figure 30.1* were obtained at Freiburg University in collaboration with Dr K.M. Hartmann. The data in *Figure 30.11* were obtained by Mr R.J. Letendre. The author is also indebted to Mr M. Tasnif and Mr H.A.M. Mostafa for their assistance with certain aspects of the work. Thanks are also due to Dr B.M. Savory of May and Baker Ltd. for providing the original batches of seed.

References

BEN HAMEDA, A. (1970). *Ph.D Thesis,* University of London

BLACK, M. (1969). *Symp. Soc. Exp. Biol.,* **23**, 193

BORTHWICK, H.A., HENDRICKS, S.B., PARKER, M., TOOLE, E.H. and TOOLE, V.K. (1952). *Proc. Nat. Acad. Sci. U.S.A.,* **38**, 662

BOWES, G. (1967). *Ph.D Thesis,* University of London

CUMMING, B.G. (1963). *Can. J. Bot.,* **41**, 1211

EDWARDS, M.M. (1968). *J. Exp. Bot.,* **19**, 583

GUTTERMAN, Y. (1974). *Oecologia,* **17**, 27

GUTTERMAN, Y. and PORATH, D. (1975). *Oecologia,* **18**, 37

HSIAO, A.I. and VIDAVER, W. (1973). *Plant Physiol.,* **51**, 459

KARSSEN, C.M. (1970). *Acta Bot. Neerl.,* **19**, 81

KENDRICK, R.E. (1974). *Nature, Lond.,* **250**, 159

KENDRICK, R.E. and FRANKLAND, B. (1969). *Planta,* **85**, 326

KENDRICK, R.E. and SPRUIT, C.J.P. (1973). *Photochem. Photobiol.,* **18**, 153

KENDRICK, R.E. and RUSSELL, J.P. (1975). *Plant Physiol.,* **56**, 332

MANCINELLI, A.L. and BORTHWICK, H.A. (1964). *Ann. Bot., Rome,* **28**, 9

ORLANDINI, M. and BULARD, C. (1972). *Biol. Plant.,* **14**, 260

SHROPSHIRE, W. Jr. (1973). *Solar Energy,* **15**, 99

SMITH, H. (1972). In *Seed Ecology, Proceedings of University of Nottingham 19th Easter School in Agricultural Science,* p.219. Ed. Heydecker, W. Butterworths, London

TOOLE, V.K. and BORTHWICK, H.A. (1968). *Plant Cell Physiol.,* **9**, 125

VIDAVER, W. and HSIAO, A.I. (1972). *Can. J. Bot.,* **50**, 687

WESSON, G. and WAREING, P.F. (1969a). *J. Exp. Bot.,* **20**, 402

WESSON, G. and WAREING, P.F. (1969b). *J. Exp. Bot.,* **20**, 414

31

THE MECHANISM OF ACTION AND THE FUNCTION OF PHYTOCHROME

HARRY SMITH

Department of Physiology and Environmental Studies, University of Nottingham, School of Agriculture, Sutton Bonington, Loughborough, Leics. LE12 5RD, U.K.

Introduction

My original intention in presenting this paper was to attempt to sum up the whole Easter School. Having listened to the uniformly excellent papers, however, I now see that to try to distil a potable liqueur from such heady wines would be an impossible task. I intend, therefore, to take advantage of my privileged position as organiser of the conference to present some personal views and speculations on what are probably the three most important areas of future phytochrome research, namely the molecular mechanism of phytochrome action, the processes through which phytochrome regulates development and the function of phytochrome in plants growing in the natural environment.

Mechanism of Phytochrome Action

The 4 years since the last international conference devoted to phytochrome, i.e. the NATO Advanced Study Institute held at Eretria, Greece, in 1971 (Mitrakos and Shropshire, 1972), have been particularly fertile. At that conference, it was generally agreed that phytochrome must be associated with membranes or at least must act upon them to control metabolism and development. This conclusion was based almost solely on indirect physiological evidence, such as the *Mougeotia* work by Haupt and colleagues, and the many observations of very rapid responses to phytochrome photoconversion. We now know, thanks to the work of Marmé, Quail, Briggs, Schäfer and Boisard, that under certain circumstances a substantial proportion of the cellular phytochrome can be recovered associated with, or bound to, a membrane fraction, even though it is still a matter of doubt whether this reflects the *in vivo* situation (cf. Marmé, Bianco and Gross, Chapter 8, and Quail and Gressel, Chapter 9).

These findings have not yet led to the formulation of a specific molecular mechanism of phytochrome action. Admittedly, a succession of formalistic models have been proposed, the latest and most comprehensive of which is presented by Schäfer in Chapter 5. Such models undoubtedly serve a useful purpose if they help to concentrate attention on ideas which can be tested experimentally. One cannot help wondering, however, whether mathematical models based on possibly spurious phytochrome/membrane interactions (*see* Quail and Gressel, Chapter 9) might not themselves turn out to be spurious. Further, although the mathematical approach can be very powerful, formalistic models really satisfy only the physicist and mathematician; ordinary mortals such as biochemists and physiologists need to be able to visualise actual processes!

Any proposed mechanism of phytochrome action must naturally be able to withstand rigorous mathematical analysis and thus a very full description will ultimately be required. Probably the most promising approach to achieve such a level of understanding is an intensive study of phytochrome-controlled phenomena in cell-free systems. Two such *in vitro* systems have been described at this conference. Miss Evans of Sutton Bonington has presented the first evidence that etioplasts contain phytochrome, and that the phytochrome controls, *in vitro,* the efflux of gibberellin from the etioplasts (Chapter 10). In the second case (Chapter 11), Furuya and Manabe elaborated upon earlier reports of the presence of phytochrome in isolated mitochondria, where it is thought to control pyridine nucleotide coenzyme transport. Thus, there are already two cell-free experimental systems available in which phytochrome controls the transport of a specific metabolite across an organelle membrane. Others may soon be found.

A question of great importance here is whether phytochrome, *in vivo,* is a normal component of membranes, as suggested by the *Mougeotia* experiments (Haupt and Weisenseel, Chapter 6) or migrates to and from the membranes depending on whether it is present as Pfr or Pr (Pratt, Coleman and Mackenzie, Chapter 7; Marmé, Bianco and Gross, Chapter 8). There is as yet no evidence for photoreversible binding of phytochrome to the etioplasts studied by Evans (Chapter 10), but Furuya and Manabe (Chapter 11) report that pre-treatment of pea segments with red light results in enhancement of the phytochrome content of subsequently isolated mitochondria. With the exception of the *in vivo* immunological observations of Pratt, Coleman and Mackenzie (Chapter 7), the evidence for photoreversible binding of free phytochrome to membranes is also consistent with phytochrome being a normal component of the membrane but being more readily dissociated from the membrane as Pr than as Pfr.

Whatever the correct interpretation of the phytochrome-binding experiments, it seems indisputable that phytochrome acts by altering membrane properties in some way and the evidence suggests that this leads to changes in transmembrane transport. Apart from the etioplast and mitochondria work already referred to, Galston and Satter (Chapter

12) have provided direct evidence for the *in vivo* phytochrome-controlled transport of K^+ ions across the plasmalemmae of the motor cells of *Albizzia* pulvinules.

It is possible to envisage several different ways in which phytochrome could regulate transmembrane transport, but these can be broadly divided into two categories — general and specific. It would be most simple, and therefore most satisfying, to believe that phytochrome controls the transport of only one, highly important, substance, but the fact that the movement of at least three different substances is already known to be under phytochrome control (i.e. K^+, NAD^+ and gibberellin) persuades otherwise. On the other hand, if phytochrome were to control a specific electrogenic ion pump, as suggested by Satter and Galston (1973), the transmembrane potential established could be used to drive the transport of other substances. Equally, the transport of a specific membrane-active substance across a membrane could result in changes in the properties of the membrane such that other substances could subsequently be transported. Thus, it is not inconceivable for a specific primary action to lead to a more general effect on the transport of a range of substances.

General effects could also be caused by changes in energy metabolism at the membrane, or by structural changes in the membrane itself. Haupt and Weisenseel, in their verbal presentation, speculated on a model in which the phytochrome molecules were considered to be only peripherally associated with the membrane in the Pr form but to extend across the membrane in the Pfr form. They also proposed (Chapter 6) that Pfr functions within the membrane as a Ca^{2+} carrier, although they admitted that no direct evidence for this idea is yet available. In fact, it may not be necessary to propose a specific ion carrier mechanism. If phytochrome is an elongated non-globular protein, as suggested by the elegant work of Briggs and colleagues (reviewed by Briggs and Rice, 1972), then a transition from a peripheral to a transmembrane position could produce aqueous pores in the membrane. Such pores might be expected to be extensive, in view of the evidence from *Mougeotia* of dichroic orientation of the pigment molecules. This concept of phytochrome action (*Figure 31.1*) is consistent with the fluid mosaic model of membrane structure proposed by Singer and Nicholson (1972), in which certain proteins are restricted to the membrane surface, some are inserted to varying extents into the membrane matrix and others, having a dumbell configuration, protrude right across the width of the membrane. Further, the change in dichroic orientation from Pr to Pfr could be accounted for, and the proposed loose peripheral location of Pr could explain the smaller amounts of phytochrome found associated with membrane fractions prepared without prior treatment with red light. Since migration of the phytochrome molecules into and out of the membrane matrix would depend on the physical state of the membrane, one would expect to see sharp changes in the rates of processes intimately linked to phytochrome action, at or near the temperature of the transition of the membrane lipids from a liquid crystalline to the gelled condition. Although Schäfer and Schmidt (1974) have found such transitions in phytochrome reversibility at relatively high temperatures (*ca.* 20 °C) and Galston and Satter (Chapter 12) have found effects of temperature on K^+ flux in pulvinules, there has as yet been no extensive investigation of the effects of temperature on phytochrome-controlled transport processes in etioplasts and mitochondria.

The creation of aqueous pores in membranes would result in the transport of hydrophilic substances across the membrane in accordance with concentration gradients and there would be no obvious requirement for metabolic energy. Further, the specificity of the transport would not be expected to be very high and different substances could be moved across the membranes of different cellular compartments. Both of these characteristics could be tested using the cell-free etioplast and mitochondrial phytochrome-controlled transport phenomena.

The other major possibility for a relatively non-specific transport mechanism is the modulation of energy metabolism at the membrane. Sandmeier and Ivart (1972) were the first to show very rapid changes in ATP and ADP levels after treatment with red light. In their tissue, *Avena* mesocotyls, they observed an increase in ATP within 1 min of the onset of red light, coupled with a stoichiometric decrease in ADP. This effect has since been confirmed in other tissues; in particular, very rapid increases in ATP levels after treatment of bean epicotyl hooks with red light have been reported at this conference by De Greef *et al.* (Chapter 19). Here again it should be reasonably straightforward to determine whether or not phytochrome-mediated transport in isolated etioplasts and mitochondria is associated with changes in energy charge. Indeed, Wagner (Chapter 26) has described red/far-red reversible changes in ATP content of isolated chloroplasts, although the phytochrome-mediated changes were rather slow (*ca.* 10–20 min).

It seems, therefore, that the evidence at present available does not directly support a highly specific effect of phytochrome on the transport of a single critical metabolite as implied in the specific permease hypothesis I proposed some years ago (Smith, 1970). The transport phenomena studied so far, however, have shown detectable lags between the onset of irradiation and the first detectable signs of transport, and thus may not be immediately connected to the primary event. Little real information has yet been gathered, and the speculations indulged in here rest on a very flimsy foundation; there are hopeful signs, however, that by the time the next international meeting on phytochrome is held, a solid basis of fact will have been established upon which to form a reasoned judgement.

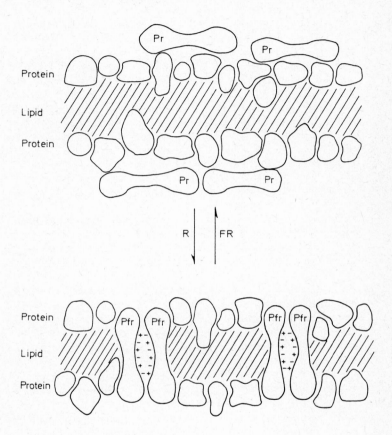

Figure 31.1   One possible mechanism by which phytochrome could regulate trans-membrane transport with relatively low specificity.   Phytochrome in the Pr form is proposed to be loosely and peripherally associated with the membrane; upon the formation of Pfr, arrays of Pfr molecules may be formed stretching across the membrane forming aqueous pores

Phytochrome Control of Development

Perhaps the biggest gap in our knowledge of phytochrome is the link between the primary events at the membrane level and the ultimate processes which constitute the developmental changes. The magnitude of the problem is best appreciated by a consideration of the enormously wide range of metabolic and physiological changes which have been reported to occur in various tissues as a result of treatment with red light.

It is now known that a considerable number of phenomena occur within a few minutes of the onset of red light, and a further large number which have time lags of 1–1½ h or more are known. It may be useful, for purposes of discussion only, to distinguish arbitrarily three different categories of responses. Those which occur rapidly (i.e. within 0–15 min) may be considered as 'membrane-associated events'. Responses which have time lags of *ca.* 1–3 h probably involve changes in amounts or activities of various enzymes, and may be thought of as 'gene expression events'. Those responses, normally even slower, which are measurable as overt changes in growth, may be considered as the final 'developmental responses'. The second and third categories of responses are clearly closely related and may stem from common causative processes. A major question exists, however, as to whether the 'membrane-associated events' are in any way causally related to the later responses. Virtually no information is available on what happens in the 'lost hour' between the rapid membrane-associated events and those events which may be related to the regulation of gene expression and ultimately cell development.

The great plurality and diversity of phytochrome-controlled biochemical changes pose major difficulties for the construction of a unified hypothesis of the phytochrome control of development. There is now evidence for the phytochrome control of polyribosome formation (Smith, 1976), ribosome function (Travis, Key and Ross, 1974), enzyme synthesis (Attridge, 1974; Acton, Drumm and Mohr, 1974), and enzyme activation/inactivation (Attridge, Johnson and Smith, 1974), although there is some difference of opinion on the last point. Johnson (1976) has recently reported phytochrome control of nitrate reductase activity in mustard cotyledons which is so rapid as to be almost certainly due to enzyme activation. As yet there is no direct evidence for the regulation of mRNA synthesis by phytochrome. Clearly, then, it is no longer possible to hold to the view that phytochrome controls development by regulating gene expression at one locus only.

There are basically two general hypotheses to account for the manifold and rapid effects of phytochrome on cell metabolism and development. The first proposes that one or more 'second messengers' act as intermediaries between the single primary action of phytochrome at the membrane, and its ultimate effects on enzyme synthesis and activity (Smith, 1970, 1976). The alternative view is that more than one primary reaction of phytochrome is possible, depending perhaps on its location within the cell. Mohr's group has recently advanced evidence that at least two different types of phytochrome responses are distinguishable, i.e. 'graded' responses, such as the control of anthocyanin accumulation, and 'threshold' responses, such as the control of lipoxygenase activity (Drumm and Mohr, 1974). Mohr and Oelze-Karow (Chapter 17) report evidence which is claimed to prove that the threshold behaviour is associated with the action of phytochrome at the membrane, and is not a function of some later step leading to changes in lipoxygenase synthesis or activity. Even if this is the case, and the argument is by no means clear-cut, it is still not logically necessary to invoke separate primary actions of phytochrome for threshold and graded responses. It is certainly not inconceivable for a single primary action involving transport of one or more substances across a membrane, whether it be of a threshold or a graded nature itself, to bring about *both* threshold and graded secondary responses — it all depends on the relationship between the secondary response mechanism and the transported substances. For example, a highly cooperative allosteric activation (or inactivation) of an enzyme, with the transported substance(s) as the effector ligand(s), would appear as a threshold response. On the other hand, the induction of enzyme synthesis by the transported substance(s) would appear as a graded response.

The actual mechanism of the lipoxygenase response probably involves inactivation of the enzyme as it is synthesised. The fact that activity 'catches up' rapidly after the Pfr-mediated prevention of 'apparent synthesis' has been relieved suggests that enzyme synthesis continues above the threshold even though increased activity is not seen. Thus, it seems possible that the threshold nature of this response is due to a highly cooperative allosteric inactivation of the enzyme — a possibility which could presumably be readily investigated.

The phytochrome control of lipoxygenase activity in mustard cotyledons is indeed a unique phenomenon — the changes in rate of activity increase are instantaneous and total, the response is only sensitive to light for a precise period beginning exactly 33.25 h after sowing, and the changes correlate, not with phytochrome in the cotyledon, but in the hook. Until other similarly unusual, but fascinating, phenomena are found, it may be unwise to use the lipoxygenase response as the main basis of an argument for the rejection of the conceptually simple hypothesis of a single primary action of phytochrome.

The 'second messenger' idea clearly stems directly from the discovery of the multivalent role of cyclic AMP in animal cells, but it is not necessary to restrict our thoughts to a single substance. Indeed, we may need to look no further than the familiar range of plant hormones whose levels are now known to be regulated in subtle ways via phytochrome. The papers presented at this conference by Evans (Chapter 10), Wareing and Thompson (Chapter 18), De Greef *et al.* (Chapter 19) and Black and Shuttleworth (Chapter 20) have shown that both identifiable hormones, and as yet unidentified transportable morphogens, are subject to very rapid control by phytochrome. It is true that there is no really acceptable evidence that hormones can substitute for light in photomorphogenesis, but it may be impossible to reproduce, by exogenous hormone applications, the subtle and complex changes in the overall pattern of hormones induced by light. Everything we know about plant hormones indicates that they do not act singly to initiate specific differentiational processes, but that they interact extensively, and it is the overall pattern of hormones which determines the pathway of development to be taken by each individual cell.

If phytochrome action at the membrane is the broadly specific effect on transport which seems most likely at the moment, it is easy to see how, by the movement of different hormones across the membranes of different compartments, or even across cell membranes, profound changes in the balance of hormones within and between cells could be established. These could, secondarily, bring about the observed wide range of metabolic changes (RNA synthesis, polysome formation, enzyme synthesis, enzyme activation, etc.) which appear to us as isolated events suitable for investigation, but in reality may be integrated partial processes of the overall developmental change initiated by light.

The Function of Phytochrome

Virtually all of the papers presented at this conference on the physiology and biochemistry of phytochrome action [with the notable exceptions of those by Galston and Satter (Chapter 12), Black and Shuttleworth (Chapter 20) and some of the work described by Wareing and Thompson (Chapter 18)] have been devoted to experiments using etiolated seedlings. The concentration of phytochrome research on etiolated seedlings is understandable, since such seedlings contain vast amounts of phytochrome (relatively speaking) and show marked responses to brief irradiation treatments. One cannot help feeling, however, that the adaptive value of phytochrome must be more than merely its capacity to detect light and thus to initiate de-etiolation upon seedling emergence. Indeed, it is well known that blue light is a more potent de-etiolating factor than red or far-red light. It seems much more likely that the major evolutionary advantage endowed by phytochrome is associated with some aspect of the adaptation of the plant to the natural radiation environment (*see also* Vince-Prue, 1973).

Information on the nature of this advantage may be sought by trying to match the properties of phytochrome with those of the natural radiation environment (Smith, 1972). The most striking property of phytochrome is its photoreversibility, which confers upon it the capacity to detect changes in the quality of light in the 600–800-nm region through the establishment of different photoequilibria between Pr and Pfr. It is clearly very difficult to obtain data which would prove definitively that phytochrome functions in the natural environment to detect shading from other plants and to redirect development accordingly.

However, the results presented by Holmes and McCartney (Chapter 29) and Monteith's theoretical analysis (Chapter 27) are surely consistent with this view. Really critical evidence will not be obtainable until a technical breakthrough is made such that phytochrome photoequilibria can be determined in green tissues; unfortunately, this seems a long way off at the present time.

The idea that phytochrome detects mutual shading is not new, of course, but this conference has seen the presentation of the first data directly relating photoequilibrium values with the spectral energy distributions of natural radiation. If the overall function of phytochrome is to detect mutual shading, it is certainly within the bounds of possibility that the metabolic and developmental changes caused by red and far-red light treatment in etiolated seedlings are merely spurious manifestations of this central function. If this is so, we may not achieve a satisfactory understanding of phytochrome until all of the careful, critical work carried out on etiolated tissues is repeated with plants grown in the light. In any case, it is surely time for phytochrome studies to be liberated from the restrictions of the etiolated seedling.

Conclusions

The picture that emerges from this selective and personal analysis is possibly prejudiced, certainly idiosyncratic. It places functional phytochrome in the membranes of certain, possibly all, cellular compartments where it controls, by virtue of its photoequilibrium, the transport of one, or probably more, important metabolites, perhaps including hormones. This action is seen as a manifestation of the function of phytochrome in plants in the natural environment, where, during continuous irradiation, it continuously monitors spectral quality, thereby controlling the flux and distribution of the hormones or other critical metabolites mentioned above. The patterns of metabolism and development in any cell or tissue would thus be determined by the intra- and inter-cellular distribution of the metabolites and/or hormones as controlled by phytochrome, but only over a relatively long time course, thus allowing for averaging of the signal. A corollary of this view of the role of phytochrome, which is presented here as a conceptual framework upon which to design experiments rather than a fully fledged theory, is that phytochrome is a modulating agent controlling rates and fluxes rather than an initiator, or 'on-off switch', evoking completely new processes. This is in direct contrast to the concept of phytochrome as a regulator of gene expression, switching on some genes and switching off others, but I believe it is more consistent with the facts of photomorphogenesis.

References

ACTON, G.J., DRUMM, H. and MOHR, H. (1974). *Planta,* **121,** 39
ATTRIDGE, T.H. (1974). *Biochim. Biophys. Acta,* **362,** 258
ATTRIDGE, T,H., JOHNSON, C.B. and SMITH, H. (1974). *Biochim. Biophys. Acta,* **343,** 440
BRIGGS, W.R. and RICE, H.V. (1972). *Annu. Rev. Plant Physiol.,* **23,** 293
DRUMM, H. and MOHR, H. (1974). *Photochem. Photobiol.,* **20,** 151
JOHNSON, C.B. (1976). *Planta,* **128,** 127
MITRAKOS, K. and SHROPSHIRE, W. Jr. (1972). *Phytochrome.* Academic Press, London
SANDMEIER, M. and IVART, J. (1972). *Photochem. Photobiol.,* **16,** 51
SATTER, R.L. and GALSTON, A.W. (1973). *BioScience,* **23,** 407
SCHÄFER, E. and SCHMIDT, W. (1974). *Planta,* **116,** 257
SINGER, S.J. and NICHOLSON, G.L. (1972). *Science, N.Y.,* **175,** 720
SMITH, H. (1970). *Nature, Lond.,* **227,** 655
SMITH, H. (1972). In *Seed Ecology,* pp.219–230. Ed. Heydecker, W. Butterworths, London
SMITH, H. (1975). *Phytochrome and Photomorphogenesis,* pp.117-119. McGraw-Hill (UK) Ltd., Maidenhead
SMITH, H. (1976). *Eur. J. Biochem.,* in the press
TRAVIS, R.L., KEY, J.L. and ROSS, C.W. (1974). *Plant Physiol.,* **53,** 28
VINCE-PRUE, D. (1973). *An. Acad. Bras. Cienc.,* **45,** Suppl., 93

LIST OF PARTICIPANTS

Acton, Dr J.

Department of Botany, University of Glasgow, Glasgow

Ancliffe, Mr C.B.

McGraw-Hill Book Co. (U.K.) Ltd., Shoppenhangers Road, Maidenhead, Berkshire

Andres Canadell, Dr J.

Fisiologia Vegetal, Facultad de Ciencias, Universidad Autonoma Barcelona, Bellaterra, Spain

Attridge, Dr T.H.

Department of Biological Sciences, North East London Polytechnic, Romford Road, Stratford, London, E15

Banbury, Dr G.H.

Department of Botany, University of Durham, Durham

Bastin, Prof. R.J.V.

Université Catholique de Louvain, Place Croix de Sud 4, 1348 Louvain La Neuve, Belgium

Beggs, Mr C.J.

Institut für Biologie II, Universität Freiburg, 78 Freiburg i. Br., Schänzle-strasse 9-11, West Germany

Bienfait, Dr H.F.

Lindengracht 171, Amsterdam, The Netherlands

Bjorn, Prof. L.O.

Department of Plant Physiology, University of Lund, S-220 07 Lund, Sweden

Bjorn, Mrs G.S.

Department of Plant Physiology, University of Lund, S-220 07 Lund, Sweden

Black, Dr M.

Department of Biology, University of London, Queen Elizabeth College, Campden Hill Road, London, W8

Blair, Mr I.M.	Department of Applied Physics, University of Strathclyde, 107 Rottenrow, Glasgow G4 0NG
Boisard, Dr J.	Laboratoire de Photobiologie, 76 Mont Saint Aignan, France
Bolton, Mr G.W.	Department of Botany, University of Minnesota, 220 Biological Sciences Center, St. Paul, Minnesota 55108, U.S.A.
Bosselaers, Mr J.P.H.	Limburgs Universitair, Centrum Universitaire Campus, 3610 Diepenbeek, Belgium
Bradbeer, Prof. J.W.	Department of Plant Sciences, University of London, King's College, 68 Half Moon Lane, London SE24 9JF
Bretz, Mr N.	Botanisches Institut, Universität Erlangen-Nürnberg, D-852 Erlangen, Schlossgarten 4, West Germany
Briggs, Dr W.R.	Department of Plant Biology, Carnegie Institution of Washington, Stanford, California 94305, U.S.A.
Brown, Dr N.A.C.	Department of Botany, University of Natal, Pietermaritzburg, South Africa
Bulard, Prof. C.	Laboratoire de Physiologie Végétale, Université de Nice, 28 Avenue Valirose, 06 Nice, France
Caubergs, Dr R.	R.U.C.A., University of Antwerp, Groenenborgerlaan 171, B-2020 Antwerp, Belgium
Cocking, Prof. E.	Department of Botany, University of Nottingham, University Park, Nottingham
Cockshull, Dr K.E.	Glasshouse Crops Research Institute, Worthing Road, Littlehampton, Sussex
Cooke, Dr R.J.	Department of Plant Biology, The University, Newcastle upon Tyne NE1 7RU
Coult, Mr D.A.	Department of Physiology and Environmental Studies, University of Nottingham, School of Agriculture, Sutton Bonington, Loughborough, Leics. LE12 5RD

De Greef, Prof. J.A.   R.U.C.A., University of Antwerp, Groenenborgerlaan 171, B-2020 Antwerp, Belgium

De Proft, Mr M.P.   Department of Biology, U.I.A., University of Antwerp, Groenenborgerlaan 171, B-2020 Antwerp, Belgium

Deutch, Prof. B.   Department of Physics, University of Aarhus, Aarhus, Denmark

Deutch, Mrs. B.   Institute of Plant Physiology, University of Aarhus, Aarhus, Denmark

Digby, Dr J.   Biology Department, University of York, Heslington, York

Dring, Dr M.J.   Department of Botany, Queen's University Belfast, Northern Ireland

Duell, Miss N.   Institut für Biologie II, Universität Freiburg, 78 Freiburg i. Br., Schänzlestrasse 9-11, West Germany

Dullforce, Dr W.M.   Department of Agriculture and Horticulture, University of Nottingham, School of Agriculture, Sutton Bonington, Loughborough, Leics. LE12 5RD

Ebephardt, Miss U.   Long Ashton Research Station, Long Ashton, Bristol

Edwards, Dr M.M.   Department of Physiology and Environmental Studies, University of Nottingham, School of Agriculture, Sutton Bonington, Loughborough, Leics. LE12 5RD

Engelsma, Dr G.   Philips Research Laboratories, Eindhoven, The Netherlands

Evans, Dr A.   Department of Physiology and Environmental Studies, University of Nottingham, School of Agriculture, Sutton Bonington, Loughborough, Leics. LE12 5RD

Firn, Dr R.D.   Department of Biology, University of York, Heslington, York

Fischer, Mr P.

Department of Plant Physiology and Anatomy, Royal Veterinary and Agricultural University, Thorvaldsensvej 40, DK-1871 Copenhagen V, Denmark

Frankland, Dr B.

Department of Plant Biology and Microbiology, University of London, Queen Mary College, Mile End Road, London E1 4N

Fredericq, Prof. H.

Laboratorium voor Plantenfysiologie, Ledeganckstraat 35, B-9000 Ghent, Belgium

Furuya, Prof. M.

Department of Botany, University of Tokyo, Hongo, Tokyo 113, Japan

Galston, Prof. A.W.

Department of Biology, 904 Kline Biology Tower, Yale University, New Haven, Connecticut 06520, U.S.A.

Grierson, Dr D.

Department of Physiology and Environmental Studies, University of Nottingham, School of Agriculture, Sutton Bonington, Loughborough, Leics. LE12 5RD

Grill, Dr R.E.M.

Department of Biology 2, University of Ulm, Oberer Eselsberg, West Germany

Gross, Mr J.

Institut für Biologie III, Universität Freiburg, 78 Freiburg i. Br., Schänzlestrasse 9-11, West Germany

Gutterman, Dr Y.

Hebrew University of Jerusalem, Jerusalem, Israel

Guttridge, Dr C.G.

Long Ashton Research Station, Long Ashton, Bristol

Hansen, Mr J.

Department of Plant Physiology and Anatomy, Royal Veterinary and Agricultural University, Thorvaldsensvej 40, DK-1871 Copenhagen V, Denmark

Harris, Mr P.J.C.

Department of Botany, University of Glasgow, Glasgow

Hartmann, Prof. E.	Institüt für Allg. Botanik der Universität, 6500 Mainz, Saarstrasse 21, West Germany
Harvey, Dr B.M.R.	Field Botany Research Division, Department of Agriculture, Newforge Lane, Belfast, Northern Ireland
Haupt, Prof. W.	Botanisches Institut, Universität Erlangen-Nürnberg, D-852 Erlangen, Schlossgarten 4, West Germany
Hebblethwaite, Mr P.	Department of Agriculture and Horticulture, University of Nottingham, School of Agriculture, Sutton Bonington, Loughborough, Leics. LE12 5RD
Heydecker, Dr W.	Department of Agriculture and Horticulture, University of Nottingham, School of Agriculture, Sutton Bonington, Loughborough, Leics. LE12 5RD
Hillman, Dr W.S.	Biology Department, Brookhaven National Laboratory, Upton, New York 11973, U.S.A.
Hoad, Dr G.V.	Long Ashton Research Station, Long Ashton, Bristol
Holmes, Dr M.G.	Department of Physiology and Environmental Studies, University of Nottingham, School of Agriculture, Sutton Bonington, Loughborough, Leics. LE12 5RD
Hong, Mr Y.-N.	Botanisches Institut II, Universität Freiburg, 78 Freiburg i. Br., Schänzlestrasse 9-11, West Germany
Hopkins, Dr D.	Institut für Biologie III, Universität Freiburg, 78 Freiburg i. Br., Schänzlestrasse 9-11, West Germany
Hughes, Dr J.G.	Department of Biologial Sciences, University of Aston in Birmingham, Birmingham
Jabben, Dr M.	Institut für Biologie II, Universität Freiburg, 78 Freiburg i. Br., Schänzlestrasse 9-11, West Germany

Jaffe, Dr M.J.

Botany Department, Ohio University, Athens, Ohio 45701, U.S.A.

Janssens, Miss D.

Laboratorium voor Plantenfysiologie, Ledeganckstraat 35, B-9000 Gent, Belgium

Jones, Mr M.G.

Department of Botany, University College of Wales, Aberystwyth, Wales

Jose, Dr A.M.

Department of Agricultural Botany, Plant Science Laboratory, University of Reading, Whiteknights, Reading

Kadman-Zahavi, Dr A.

The Volcani Center, A.R.O., Bet-Dagan, Israel

Kendrick, Dr R.E.

Department of Plant Biology, University of Newcastle upon Tyne, Newcastle upon Tyne NE1 7RU

Kigel, Dr J.

Faculty of Agriculture, Hebrew University of Jerusalem, Jerusalem, Israel

Kinet, Dr J.M.

Centre de Physiologie Végétale Appliquée, Department de Botanique, Université de Liege, Sart Tilman, B-400 Liege, Belgium

Kordan, Dr H.A.

Department of Botany, University of Birmingham, P.O. Box 363, Birmingham B15 2TT

Kost, Dr H.P.

Botanisches Institut, Universität München, D-8 München 19, Menzingerstrasse 67, West Germany

Kost, Mrs E.

Botanisches Institut, Universität München, D-8 München 19, Menzingerstrasse 67, West Germany

Kraml, Dr M.

Botanisches Institut, Universität Erlangen-Nürnberg, D-852 Erlangen, Schlossgarten 4, West Germany

Lehmann, Mr U.

Institut für Biologie II, Universität Freiburg, 78 Freiburg i. Br., Schänzle-strasse 9-11, West Germany

Lisansky, Dr S.G.	Tate & Lyle Ltd., Group Research and Development, Philip Lyle Memorial Research Laboratory, P.O. Box 68, Reading, Berkshire RG6 2BX
Lucas, Dr J.A.	Department of Plant Biology, University of Newcastle upon Tyne, Newcastle upon Tyne NE1 7RU
McCartney, Mr H.A.	Department of Mechanical and Production Engineering, Brighton Polytechnic, Moulsecoomb, Brighton, Sussex
McCree, Dr K.J.	Department of Soil and Crop Sciences, Texas A & M University, College Station, Texas 77843, U.S.A.
McDougall, Mr J.	Department of Botany, University of Glasgow, Glasgow
Marmé, Dr D.	Institut für Biologie III, Universität Freiburg, 78 Freiburg i. Br., Schänzlestrasse 9-11, West Germany
Marston, Dr M.E.	Department of Agriculture and Horticulture, University of Nottingham, School of Agriculture, Sutton Bonington, Loughborough Leics. LE12 5RD
Mayson, Mr J.	Butterworth & Co. (Publishers) Ltd., Borough Green, Sevenoaks, Kent TN15 8PH
Mohr, Prof. H.	Institut für Biologie II, Universität Freiburg, 78 Freiburg i. Br., Schänzlestrasse 9-11, West Germany
Monteith, Prof. J.L.	Department of Physiology and Environmental Studies, University of Nottingham, School of Agriculture, Sutton Bonington, Loughborough, Leics. LE12 5RD
Montes, Mr G.	Department of Plant Sciences, University of London, King's College, 68 Half Moon Lane, London SE24 9JF
Orlandini, Mrs M.	Laboratoire de Physiologie Végétale, Université de Nice, 28 Avenue Valrose, 06 Nice, France

Pecket, Dr R.C.	Department of Botany, The University, Manchester
Porath, Dr D.	George S. Wise Center for Life Sciences, The Institute for Nature Conservation Research, Tel Aviv University, Ramat Aviv, Israel
Pratt, Prof. L.H.	Vanderbilt University, P.O. Box 1550, Station B, Nashville, Tennessee 37235, U.S.A.
Purse, Mr J.G.	Shell Research Ltd., Milstead Laboratory, Sittingbourne, Kent
Quail, Dr P.H.	Research School of Biological Science, Australian National University, Canberra A.C.T. 2601, Australia
Queiroz, Dr O.	C.N.R.S. Phytotron, 91190 Gif-sur-Yvette, France
Rethy, Dr R.F.E.	Laboratorium voor Plantenfysiologie, Ledeganckstraat 35, B-9000 Gent, Belgium
Rieger, Mr D.S.	Butterworth & Co. (Publishers) Ltd., Borough Green, Sevenoaks, Kent TN15 8PH
Robertson, Mr J.	Department of Botany, University of Glasgow, Glasgow
Rombach, Dr J.	Laboratory for Plant Physiological Research, Agricultural University, Generaal Foulkesweg 72, Wageningen, The Netherlands
Ryle, Dr G.J.	Grassland Research Institute, Hurley, Maidenhead, Berkshire
Satter, Dr R.L.	912 Kline Biology Tower, Yale University, New Haven, Connecticut 06520, U.S.A.
Schäfer, Dr E.	Institut für Biologie II, Universität Freiburg, 78 Freiburg i. Br., Schänzlestrasse 9-11, West Germany
Schonbohm, Prof. E.	Department of Biology, University of Marburg, 355 Marburg, Lahnberge, West Germany

Schopfer, Dr P.	Institut für Biologie II, Universität Freiburg, 78 Freiburg i. Br., Schänzlestrasse 9-11, West Germany
Schuber, Dr F.	Institut de Botanique, 28 rue Goethe, 67000 Strasbourg, France
Schwabe, Prof. W.W.	Department of Horticulture, University of London, Wye College, Nr. Ashford, Kent TN25 5AH
Scott, Dr R.K.	Department of Agriculture and Horticulture, University of Nottingham, School of Agriculture, Sutton Bonington, Loughborough, Leics. LE12' 5RD
Shuttleworth, Miss J.E.	Department of Biology, Queen Elizabeth College, Campden Hill, London W8 7AH
Skrivanova, Miss R.	Institute for General Botany, Kunstlergasse 16, 8006 Zurich, Switzerland
Smith, Prof. H.	Department of Physiology and Environmental Studies, University of Nottingham, School of Agriculture, Sutton Bonington, Loughborough, Leics. LE12 5RD
Steinitz, Mr B.	Institut für Biologie II, Universität Freiburg, 78 Freiburg i. Br., Schänzlestrasse 9-11, West Germany
Tan, Dr K.K.	Department of Botany, University College, Dublin 4, Eire
Thien, Mr W.	Botanisches Institut II, Universität Freiburg, 78 Freiburg i. Br., Schänzlestrasse 9-11, West Germany
Thompson, Mr A.G.	Department of Botany, University College of Wales, Aberystwyth, Wales
Tregunna, Prof. E.B.	Botany Department, University of British Columbia, Vancouver, British Columbia, Canada
Tychsen, Mr K.	Department of Plant Physiology and Anatomy, Royal Veterinary and Agricultural University, Thorvaldsensvej 40, DK-1871 Copenhagen V, Denmark

Van Hoof, Dr R.

R.U.C.A., University of Antwerp,
Groenenborgerlaan 171, B-2020 Antwerp,
Belgium

Van Onckelen, Dr H.A.

R.U.C.A., Laboratory Plantkunde,
University of Antwerp, Groenenborgerlaan
171, B-2020 Antwerp, Belgium

Veierskon, Mr B.

Department of Plant Physiology and
Anatomy, Royal Veterinary and Agricul-
tural University, Thorvaldsensvej 40,
DK-1871 Copenhagen V, Denmark

Verbelen, Mr J.P.

Department of Biology, University of
Antwerp, Universiteitspleain 1, B-2610
Wilrijk, Belgium

Verbelen, Mrs E.

Department of Biology, University of
Antwerp, Groenenborgerlaan 171, B-2020
Antwerp, Belgium

Vince-Prue, Dr D.

Department of Botany, Plant Science
Laboratories, University of Reading,
Whiteknights, Reading, Berkshire

Wagner, Dr E.

Institut für Biologie II, Universität
Freiburg, 78 Freiburg i. Br., Schänzlestrasse 9-11, West Germany

Wareing, Prof. P.F.

Department of Botany and Microbiology,
University College of Wales, Penglais,
Aberystwyth, Wales

Weir, Mr J.A.C.

Electricity Council, 30 Millbank,
London, SW1

Widell, Miss E.

Department of Plant Physiology,
University of Lund, S-220 07 Lund,
Sweden

Wilkins, Prof. M.B.

Department of Botany, University of
Glasgow, Glasgow

Yung, Dr K.H.

Department of Biochemistry, Cambridge
University, Tennis Court Road, Cambridge

Zimmermann, Dr H.E.

Botanisches Institut, Universität München,
D-8 München 19, Menzingerstrasse 67,
West Germany

INDEX